T0310205

FRAGILE NETWORKS

FRAGILE NETWORKS

Identifying Vulnerabilities and Synergies in an Uncertain World

Anna Nagurney

Qiang Qiang

University of Massachusetts
Amherst, Massachusetts

A JOHN WILEY & SONS, INC., PUBLICATION

Published by John Wiley & Sons, Inc., Hoboken, New Jersey.
Published simultaneously in Canada.

Library of Congress Cataloging-in-Publication Data:

Nagurney, Anna.
Fragile networks : identifying vulnerabilities and synergies in an uncertain world / Anna Nagurney, Qiang Qiang.
p. cm.
Includes bibliographical references and index.
ISBN 978-0-470-44496-2 (cloth)
1. Network analysis (Planning) 2. System analysis. I. Qiang, Qiang, 1974- II. Title.
T57.85.N34 2009
003'.72--dc22 2009004198

Printed in the United States of America.

10 9 8 7 6 5 4 3 2 1

The authors dedicate this book to their families.

Contents in Brief

CONTENTS

List of Figures

List of Tables

PART I

NETWORK FUNDAMENTALS, EFFICIENCY MEASUREMENT, AND VULNERABILITY ANALYSIS

CHAPTER 1

INTRODUCTION AND OVERVIEW

Networks provide the foundations for transportation and logistics, for communication, energy provision, social interactions, and financing. The study of networks spans many disciplines, due to their wide application and importance [see Beckmann, McGuire, and Winsten (1956), Ford and Fulkerson (1962), Sheffi (1985), Ahuja, Magnanti, and Orlin (1993), Nagurney and Siokos (1997), Nagurney (1999, 2006a), Ran and Boyce (1996), Nagurney and Dong (2002a), Yerra and Levinson (2005), Roughgarden (2005), and Newman, Barabási, and Watts (2006)]. Today, the subject has garnered renewed interest since a spectrum of catastrophic events such as 9/11, the North American electric power blackout in 2003, Hurricane Katrina in 2005, the Minneapolis bridge collapse in 2007, the Mediterranean cable disruption in 2008, Cyclone Nargis in 2008, and the Sichuan earthquake in China in 2008 have drawn great attention to the study of network vulnerability and fragility. Moreover, because most critical infrastructure networks are large-scale and complex in nature, they are liable to be faced with disruptions. As argued by Haimes and Longstaff (2002), "[infrastructure systems] are invariably large-scale dynamic 'systems of systems' with numerous components, they are nonlinear, and they are spatially distributed with multiple agents/decisionmakers ... and they are dominated by multiple conflicting and competing objectives."

Fragile Networks. By Anna Nagurney and Qiang Qiang

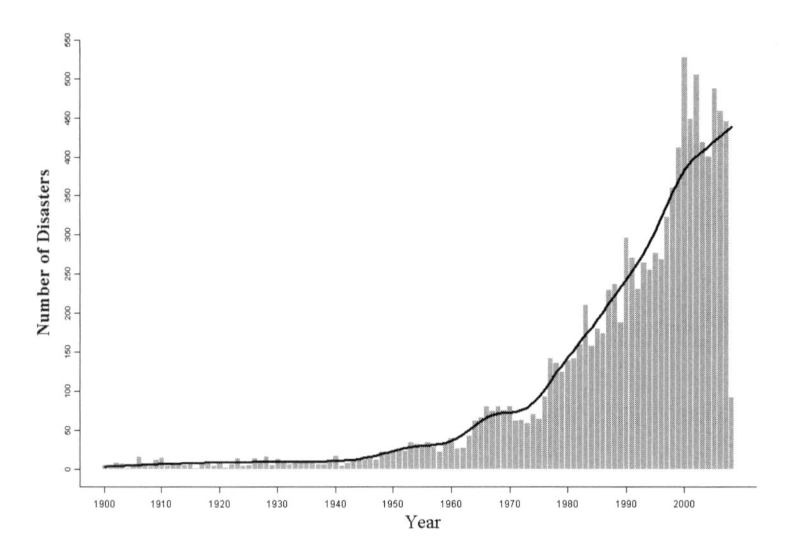

Figure 1.1: Frequency of disasters [Source: Emergency Events Database (2008)]

Indeed, disasters have brought an unprecedented impact on human lives in the 21st century. According to a recent study [Braine (2006)], from January to October 2005, an estimated 97,490 people were killed in disasters globally; 88,117 of them lost their lives because of natural disasters. Figure 1.1 clearly shows the increasing number of disasters over the past century.

The Emergency Events Database (2008) defines a *disaster* as an event that fits at least one of the following criteria: 1). 10 or more people killed; 2). 100 or more people affected; 3). declaration of a state of emergency; 4). call for international assistance. According to the Federal Emergency Management Agency (FEMA) (1992), a catastrophe disaster is "An event that results in large numbers of deaths and injuries; causes extensive damage or destruction of facilities that provide and sustain human needs; produces an overwhelming demand on state and local response resources and mechanisms; causes a severe long-term effect on general economic activity; and severely affects state, local, and private-sector capabilities to begin and sustain response activities." From these definitions, we see that although disasters may have different meanings, depending on the specific domain, they have one thing in common: they have a catastrophic effect on human lives and a region's or even a nation's resources.

Due to the enormous impact of disasters, disaster management has become a topic that is drawing attention from researchers in various disciplines. For example, Altay and Green (2006) conducted a bibliographical analysis in which they summarized operations research and management science studies in the disaster management literature. The authors suggested that there is a lack of scientific approach in disaster management research, especially because disaster responses differ drastically from daily decision-making and, therefore, the conventional wisdom may not be applicable to disaster relief efforts. For a comprehensive review of the disaster management literature, with a focus on analytical models and solution approaches, refer to Altay and Green (2006) and the references therein. The report by the Committee on Disaster Research in the Social Sciences (2006), in turn, emphasizes the need for integrated research on disasters that incorporates the human dimension from a social science perspective.

Lewis (2006) discussed methodologies that can be used to analyze various infrastructure systems from the perspective of protecting infrastructure from damage due to disasters and from terrorist attacks. Richardson, Gordon, and Moore (2006) collected reports describing the economic effects of terrorist attacks and related government mitigation policies. The topics range from transportation system disruptions, to power system failures in the United States, to housing market price analyses after terrorist attacks. Murray and Grubesic (2007) collected studies on infrastructure reliability and vulnerability with a wide-ranging transportation, regional science, and geographic scope.

In this book, we focus on networks, viewed as complex systems, and ranging from transportation and logistical networks, in the form of supply chains, to financial networks and the Internet. Our goals include the conceptualization, definition, and construction of mathematically rigorous, computer-based tools for the assessment of network performance and efficiency, along with robustness and vulnerability analysis. In addition, we identify synergies that can be expected to occur through network integration among firms or organizations in the setting of mergers and acquisitions. Hence, in terms of disaster management, the focus of this book is not only on disaster mitigation through risk reduction and disaster and emergency preparedness but also on disaster recovery through the identification of possible synergies. The relevance of our network framework spans the spectrum of physical to economic disasters.

The recent theories of scale-free and small-world networks in complex network research have significantly enhanced our understanding of the behavior as well as the vulnerability of many real-world networks [see Amaral et al. (2000), Chassin and Posse (2005), and Holmgren (2007)]. However, the majority of network vulnerability studies have focused on the topological characteristics of the networks, such as the connectivity or the shortest path length of the network (see, e.g., Callaway et al. (2000) and the references therein). Although the topological structure of a network provides critical information regarding network vulnerability, the flow on a network is also an important indicator, as are the economic aspects, such as the flow-induced costs and the behavior of users both before and after any disruptions. As pointed out by Barabási (2003), "To achieve that [understanding of complexity] we must move beyond structure and topology and start focusing on the dynamics that take place along the links. Networks are only skeletons of complexity, the highways for various processes that make our world hum."

Latora and Marchiori (2001, 2002, 2003, 2004) proposed a network efficiency measure that was shown to have advantages over several existing network measures. The authors then applied their measure to study the Boston subway network (MBTA) and the Internet. Nevertheless, their measure considers only geodesic information and, therefore, does not capture such important factors as network flows, the associated costs, and users' behavior, be it according to centralized or decentralized decision-making principles.

In this book, we lay out, beginning with Part I, the theoretical and practical foundations for a new network performance/efficiency measure that extends the Latora-Marchiori measure to incorporate such crucial network characteristics as decision-making induced flows and costs to assess the importance of network components in a plethora of network systems. We demonstrate that the new network measure has significant advantages over several existing measures, and it captures the reality of real-world networks in that it explicitly considers congestion. Furthermore, the measure can handle both fixed and elastic demand network problems plus time-dependent, dynamic networks; the latter is of particular relevance to the Internet and to electric power generation and distribution networks. Moreover, the new measure allows for the ranking of network components, that is, the nodes and links, or combinations thereof, in terms of their importance from an economic performance standpoint. This has implications for not only planning and maintenance purposes but also for national security.

In addition, instead of quantifying and assessing the impact of the complete disruption/removal of a network component, we also formalize the concept of network robustness, another important aspect associated with network vulnerability. For example, network robustness is concerned with the reduction in network resources, such as link capacity. As defined by the Institute of Electrical and Electronics Engineers in 1990, the robustness of a system is "the degree to which a system or component can function correctly in the presence of invalid inputs or stressful conditions." This topic is especially relevant and timely because it has been consistently reported that the once world-envied U.S. infrastructure is now experiencing tremendous aging and deterioration, which exposes the networks and populations to additional vulnerability. In particular, over one quarter of the nation's 590,750 bridges have been rated as structurally deficient or functionally obsolete. The degradation of transportation networks due to poor maintenance, natural disasters, deterioration over time, and unforeseen attacks now leads to estimates of $94 billion in the United States in terms of needed repairs for roads alone. Poor road conditions in the United States cost motorists $54 billion in repairs and operating costs annually. On the other hand, the number of motorists in the country has risen by 157 million (or 212.16%) since 1960, while the population of licensed drivers has grown by 109 million (or 125.28%) [U.S. Department of Transportation Federal Highway Administration (2004)], which adds more stress to the already fragile U.S. transportation networks.

Due to the breakdowns of U.S. transportation networks and the increasing number of vehicles, American commuters now spend 3.5 billion hours stuck in traffic, which translates to a cost of $63.2 billion a year to the economy [American Society of Civil Engineers (2005)]. At the same time, a recent report from the U.S. Department of Transportation Federal Highway Administration (2006a) states that the country is experiencing a freight capacity crisis that threatens the strength and productivity of the U.S. economy. According to the Road & Transportation Builders Association [see Jeannert (2006)], nearly 70% of U.S. freight is carried on highways, and bottlenecks are causing truckers 243 million hours of delay annually, with an estimated associated cost of $8 billion. It is noted that the U.S. government is facing a $1.6 trillion deficit over 5 years in terms of infrastructure repair and reconstruction according to a recent estimate [Environment News Service (2008)].

Hence the construction of suitable network robustness measures is of theoretical and practical importance. Moreover, this book proposes network robustness measures under different user behaviors with direct relevance to transportation networks. This is especially relevant due to the Braess (1968) paradox in which the addition of a new road, under user-optimizing behavior, can increase travel time/cost for all! In addition, the book provides environmental impact assessment indicators to quantify the effects of network degradation on the environment. The book also codifies measures as to the impacts of network infrastructure investments in terms of capacity enhancements.

Our economies and societies depend on such critical infrastructures as transportation and logistical networks, on communication networks, including the Internet, on electric power generation and distribution networks, and on financial networks. Hence rigorous tools that enable the identification of which nodes and links really matter in such network systems and should, thus, be better maintained and/or enhanced provide essential information to decision-makers from governmental employees and policymakers to planners, engineers, and scientists, to corporate and organizational leaders.

Part I of this book consists of three chapters, beginning with this introduction and overview chapter. Chapter 2 lays down the fundamental methodologies, the network models that capture distinct decision-making behaviors, the tools for qualitative analysis, and the algorithms for the computation of solutions. Throughout the book, realizations of appropriate algorithms are presented for the solution of specific models and applications. Chapter 3 introduces the unified network efficiency measure, assuming decentralized, user-optimizing behavior of the network users and relates the measure to several existing ones. This chapter also explores network robustness using the unified measure as a basis and then provides alternative measures based on total network cost and either user-optimizing or system-optimizing decision-making behavior. Both deteriorating network capacity scenarios and investment/capacity enhancement scenarios are explored.

Part II exploits the recently established connections between transportation networks and different critical networks [cf. Nagurney and Dong (2002a), Nagurney (2006b), Liu and Nagurney (2007), Nagurney, Parkes, and Daniele (2007), Wu et al. (2006)] and demonstrates how the new network measures and robustness indices can be applied to different network systems, such as transportation networks, supply chain networks in the case of disruptions and random demands, multitiered financial networks, and, in the dynamic context, to the Internet. For example, in the case of supply chain networks alone, several recent major disruptions and associated effects on the business world have vividly demonstrated the need to address supply-side risk. A case in point is a fire in the Phillips Semiconductor plant in Albuquerque, New Mexico, which caused its major customer, Ericsson, to lose $400 million in potential revenues. On the other hand, another major customer, Nokia, managed to arrange alternative supplies and, therefore, mitigated the impact of the disruption [cf. Latour (2001)]. Another example concerns the effect of Hurricane Katrina, with the consequence that 10 - 15% of the total U.S. gasoline production was halted, which raised the oil price not only in the United States, but also overseas [see, e. g., Canadian Competition Bureau (2006)]. As emphasized by Sheffi (2005), one of the main characteristics of disruptions in supply networks is "the seemingly unrelated consequences and vulnerabilities stemming from global connectivity." Indeed, supply chain disruptions may have impacts that propagate not only locally but globally and hence a holistic, system-wide approach to supply chain network modeling and analysis is essential to be able to capture the complex interactions among decision-makers. Such an approach/perspective is demonstrated in Part II of this book.

In addition, in Part II, we study financial networks from a vulnerability perspective. The advances in information technology and globalization have shaped today's financial world into a complex network, which is characterized by distinct sectors, the proliferation of new financial instruments, and increasing international diversification of portfolios. Recently, financial networks have been studied using network models with multiple tiers of decision-makers, including intermediaries [cf. Nagurney (2003)]. As we are seeing today, our financial networks are highly interconnected and interdependent; consequently, any disruption that occurs in one part of the network may produce consequences in other parts of the network, which may not only be in the same region but be miles away in other countries.

The world is now reeling from the effects of the financial credit crisis, with leading financial services and banks closing, including the investment bank Lehman Brothers; others merging; and the financial landscape changing for forever. The domino effects of the U.S. economic troubles have rippled through overseas markets and have pushed countries such as Iceland to the verge of bankruptcy. The root of the financial problems was considered to have stemmed from the U.S. housing market and the huge number of bad loans that could not be repaid. Many of the subprime loans had been bundled and then sold to investment banks, some of whom had, in turn, borrowed money for these transactions, thus further complexifying the number of decision-makers or agents and the financial linkages. Ultimately, businesses could not obtain financial resources because so many banks and financial institutions were failing due the large number of unpaid loans, which caused millions of people to lose their homes. The number of home foreclosures in the United States is, as of 2009, still increasing. It is, therefore, crucial for the decision-makers in financial systems, including managers, executives, and regulators, to be able to identify a financial network's vulnerable components to protect the functionality of the network.

The Internet, in turn, as a global telecommunications network, with linkages to other network systems, including transportation and financial ones, as well as energy systems, is fundamental to the functioning of our modern societies and economies. Hence an appropriate efficiency measure and a means of identifying its critical nodes and links are essential not only to the management of such a dynamic network but also to its very security. Clearly, those nodes and links that are deemed most important are those that also merit greatest protection. Coffman and Odlyzko (2002) note that Internet traffic is approximately doubling each year, which is extremely fast growth.

At the same time, the number of security attacks against the Internet [cf. Bagchi and Tang (2005)] is also growing rapidly, and the number of global security incidents rose to 73,359 in 2002, far surpassing the 52,658 events of the previous year. Internet disruptions need not be due to terrorist attacks but may occur as a result of accidents or natural disasters. For example, the severance of two cables in the Mediterranean, on January 30, 2008, apparently by ship anchors, caused major Internet service problems for days in the Middle East and in parts of Asia [see Reed (2008)]. In terms of the financial impact of a major Internet disruption, a report by the Business Roundtable [see Insurance Journal (2007)] estimated a cost to the global economy of $250 billion.

In particular, Part II of this book is organized as follows. Chapter 4 explores, in depth, the methodological tools and network measures unveiled in Chapters 2 and 3. In Chapter 4, large-scale, real-world transportation networks are investigated and the importance of their nodes and links identified and the network components ranked. In addition, because global climate change is affecting our network infrastructure and our communities, we provide environmental impact assessment indices for transportation networks. With such indices, one can then answer questions as to how the deterioration of link capacities affects the environmental emissions. Interestingly, we discover that, in certain real-world transportation networks, if travelers select their routes of travel in a decentralized, user-optimized manner, lower environmental emissions may result than those that occur under centralized behavior! We also relate our proposed robustness indices based on total network cost and distinct user behaviors to the price of anarchy [Roughgarden (2005)].

Chapter 5 turns to supply chain networks under disruptions. Here we quantify disruption risks and consider random demands for products. We introduce a weighted supply chain performance measure that provides sufficient flexibility that it can be used for products as varied as toys to healthcare/medical products. In this chapter, as in all the application-based ones, solutions to specific numerical examples are computed to demonstrate the models and the accompanying methodological tools.

Chapter 6 focuses on complex financial networks with intermediation and demonstrates how an extension of the unified network performance measure can be applied to determine the importance and ranking of financial nodes (from financial source agents to intermediaries) and financial links (either physical or electronic-based). This chapter also discusses such concepts as financial contagion. Chapter 7 contains more advanced material on dynamic networks with the Internet serving as the par excellence example. This chapter presents extensions of the unified network performance measure to the dynamic network domain. In this chapter we also discuss, at a high level, electric power generation and distribution networks and how the tools in this book can be used to identify and rank nodes and links in such critical infrastructure networks.

The final part of the book turns to the exploration of synergy associated with network systems and focuses on supply chain networks in the case of mergers (and acquisitions). As noted by Langabeer (2003) the use of mergers and acquisitions (M&As) continues to grow exponentially; more than 6,000 M&A transactions were conducted world-wide in 2001, with a value of over a $1 trillion. Nevertheless, many scholars argue whether mergers achieve their objectives. For example, Marks and Mirvis (2001) found that fewer than 25% of all mergers achieve their stated objectives. Langabeer and Seifert (2003) determined a direct correlation between how effectively supply chains of merged firms are integrated and how successful the merger is. Furthermore, they state, based on the empirical findings in Langabeer (2003), which analyzed hundreds of mergers over the past decade, that improving supply chain integration between merging companies is the key to improving the likelihood of post-merger success. With the present financial and economic crisis, companies as well as organizations need to be able to quantifiably determine what gains, if any, can be expected if they do merge. By representing firms as networks [see also Nagurney (1999)] of their economic activities with the associated costs, one can graphically depict the network structure of a given merger.

In Part III, we thus envision each firm as a network of economic activities consisting of manufacturing, which is conducted at the firm's plants or manufacturing facilities; distribution, which occurs between the manufacturing plants and the distribution centers, which also store the product produced by each firm; and the ultimate distribution of the product to the retailers. Associated with each such economic activity is a link in the network with a total associated cost that depends on the flow of the product on the link. The links, be they manufacturing, shipment, or storage, have capacities on the flows. We assume, as given, the demand for the product at each retailer. We provide a *system-optimization* perspective for supply chains through the prism of system-optimization as arising in network systems. We quantify the synergy associated with mergers and also explore environmental synergies. In the case of multiproduct supply chains, we discuss similarities and differences between corporate and humanitarian supply chains, as they apply to disaster situations. We then turn, in the last chapter, to the modeling, analysis and solution of competitive firms in the form of network oligopolies. We investigate the formation of coalitions and the merger paradox.

Specifically, in Chapter 8, we provide a basic framework for capturing the economic activities of a firm and associated costs. The benefits of proposed mergers or acquisitions can then be quantified by modeling the integration of the firms and their network activities. The formalism, which is presented from a supply chain network perspective, is also applicable to mergers and acquisitions in different industries from transportation to telecommunications to services, such as financial services. Chapter 9 then considers not only cost synergy associated with a proposed merger or acquisition but also the environmental synergy. In this chapter we use a multicriteria decision-making approach for modeling the firms both before and after the network integration. We evaluate numerically the effects of proposed hostile and friendly M&As. Chapter 9 hence adds to our analytical tools for environmental impact assessment associated with networks, a topic that was also explored in Chapter 4 for transportation networks but in the presence of capacity degradation (or investments but not explicit mergers).

Chapter 10 formulates network integration and associated synergy in the case of M&As of multiproduct firms. It explores not only corporate applications but also demonstrates how cooperation and partnerships between organizations can yield cost synergies. It provides quantifiable means for assessing the possible cost synergies associated with supply chain network integration for humanitarian logistical operations. With the number of disasters increasing over the last century, the timely delivery of essential goods to needy and vulnerable populations after disasters is essential.

Chapter 11 models coalition formation among any number of competing firms (or organizations) within a network framework and quantifies the associated possible gains in terms of cost reduction, profit increases, etc. It aims to illuminate why, despite the existing economic theory and the merger paradox, mergers between as few as two firms in an industry may be profitable. Hence this chapter addresses the question of specific gains from cooperation as opposed to pure competition.

In summary, this book provides a rigorous, unified treatment of network systems and identifies how to quantify performance, vulnerabilities, and synergy. It incorporates numerous network examples and applications and is accompanied by an appendix, which provides the necessary technical foundations of optimization on which the book builds. As governmental decision-makers, policymakers, regulators; and network designers, engineers, and scientists; and managerial executives and leaders turn to revitalizing and investing in network infrastructure, enhanced decision-making is now possible, because of the concepts and tools developed in this book.

The problems that we face today are uniquely challenging. We hope that this book, with its interdisciplinary, network-based approach, provides some solutions. We envision, and expect, that researchers, practitioners, and students for generations to come will add to the knowledge base of this book.

The research on which this book is based was supported, in part, and over the years, by the National Science Foundation (NSF) through NSF Grant No.: IIS-0002647 with Anna Nagurney as the Principal Investigator. Additional support for the relevant research was provided by the Radcliffe Institute for Advanced Study at Harvard University in Cambridge, Massachusetts, while Anna Nagurney was a 2005-2006 Science Fellow at the Institute and on sabbatical leave from the University of Massachusetts Amherst. Special thanks go to Dr. Barbara Grosz, then the Science Dean of Radcliffe and now Dean; to Dr. Drew Gilpin Faust, then the Dean of the Radcliffe Institute and now the President of Harvard University; and to Dr. Judith Vichniac, the Director of the Fellowship Program at the Radcliffe Institute, for their support of interdisciplinary research.

Anna Nagurney acknowledges the support of the Fulbright Program through its Senior Specialist Program. Nagurney was a Fulbright Senior Specialist at the University of Catania in Italy for two weeks in March 2008 and conducted a workshop, co-organized with Professor Patrizia Daniele of the University of Catania, titled, "Equilibrium and Vulnerability Analysis with Applications." Nagurney's Fulbright project theme was "Complex Networks and Vulnerability Analysis." The hospitality of her host institution plus extraordinary additional support provided by Professor Antonino Maugeri and the Department of Mathematics and Computer Science at the University of Catania were greatly appreciated. Results in Chapters 2, 3, 4, 7, and 9 are evolutions, in part, from this Fulbright experience.

The support of the Rockefeller Foundation is acknowledged for its Bellagio Center Program, which sponsored the conference "Humanitarian Logistics: Networks for Africa," organized by Anna Nagurney, which took place May 5-9, 2008 at the center in Italy. The discussions and presentations among the participants inspired Chapter 10 of this book. This Bellagio conference, which coincided with Cyclone Nargis and its immediate aftereffects, was a life-changing event for the organizer. The friendships made among participating conference practitioners and scholars, whose work in the field and in scholarly writings makes a true difference, are treasured.

Thanks are also extended to the Isenberg School of Management and to the University of Massachusetts Amherst for additional support through the John F. Smith Memorial Fund. Special thanks go to Mr. John F. Smith Jr., the former CEO and Chairman of the Board of General Motors, for endowing Anna Nagurney's chaired professorship. Additional thanks are extended to the University of Massachusetts Amherst and its Graduate School for support provided for Qiang Qiang's 2008-2009 Graduate Fellowship.

The authors are indebted to their colleagues at the Isenberg School of Management and at the Virtual Center for Supernetworks, to their collaborators, and to their students for their collegiality and continual support.

The authors acknowledge the support of their Wiley editor, Ms. Susanne Steitz-Filler, throughout the preparation and writing of this book, and the assistance of Ms. Jacqueline Palmieri and Ms. Kristen Parrish of Wiley during the book's production.

Finally, the authors are indebted to their families, who enabled the completion of this book project.

CHAPTER 2

FUNDAMENTAL METHODOLOGIES, NETWORK MODELS, AND ALGORITHMS

Networks are complex, typically, large-scale systems, and their formal study has attracted much interest from a plethora of scientific disciplines. The realities of networks today include such characteristics as congestion and distinct underlying behavioral principles as to the network users. Hence our focus throughout this book is on such networks.

In this chapter we review some of the fundamental methodologies, network models, and algorithms that are used as the foundations for the results in this book. The perspective that we take is that of *transportation* as a unifying theme, because networks are characterized by not only their underlying topologies but also by the associated flows, which are the results of behavior of the users, interpreted broadly, from travelers on urban transportation networks; to messages being routed in computer and communication networks; to products being produced, stored, and distributed through global supply chains, to financial products that are packaged and sold; and, of course, to electric power (and other forms of energy) being produced and transmitted according to supply and demand. All such problems may be interpreted as transportation problems given that specific flows are *transported* from origins to destinations in some manner. Moreover, many of the advances in transportation network modeling, analysis, and solution are directly applicable to other network domains.

A transportation approach to networks is conceptually rich, theoretically rigorous, and analytically sound, due to the decades of flow-based transportation network research. Moreover, network systems, as seemingly distinct as decentralized supply chain networks, financial networks with intermediation, and electric power supply chains have been recently shown to be transformable into transportation network problems, with the appropriate identification of origin/destination pairs of nodes, associated demands, and link cost functions [cf., respectively, Nagurney, Dong, and Zhang (2002), Liu and Nagurney (2007), and Nagurney et al. (2007)]. In addition, the study of the Internet, given its decentralized mode of operation and the analogies to transportation networks, from congestion and delay cost functions to the investigation of the Braess (1968) paradox in this setting, has benefited greatly from methodological tools originating in transportation [cf. Bertsekas and Gallager (1987), Papadimitrou (2001), Nagurney and Dong (2002a), Boyce, Mahmassani, and Nagurney (2005), and Roughgarden (2005)]. Finally, the representation/transformation of numerous problems in economics into network problems in general and into transportation network problems in particular, from Walrasian price equilibrium problems and classical spatial and aspatial oligopoly problems, to migration problems, further supports the relevance of such an approach to network systems [see, e.g., Nagurney (1999, 2006a)].

In this chapter, we provide a review of the fundamental methodologies and network models that are used in this book. The theory of finite-dimensional variational inequalities (VIs) is presented in Section 2.1, because it provides a unified mathematical formalism for optimization problems and equilibrium problems. Section 2.2 focuses on decentralized decision-making and user-optimizing behavior on networks, and Section 2.3 presents network models of centralized decision-making and system-optimizing behavior.

In Section 2.4 we review appropriate algorithms that are used to compute solutions to the models in this book, including the equilibration algorithms for classical user-optimized and system-optimized network problems, the projection method, the modified projection method, and the Euler method, all of which can be employed to solve, under appropriate assumptions, finite-dimensional variational inequality problems.

2.1 REVIEW OF VARIATIONAL INEQUALITY THEORY AND ITS RELATIONSHIPS TO OPTIMIZATION

In this section, we review the theory of (finite-dimensional) variational inequalities. All definitions and theorems, except where noted, are taken from Nagurney (1999), where proofs can also be found; see also Kinderlehrer and Stampacchia (1980). Throughout this book, we assume that vectors are column vectors. For basic definitions related to optimization theory and some mathematical fundamentals, we refer the reader to the appendix.

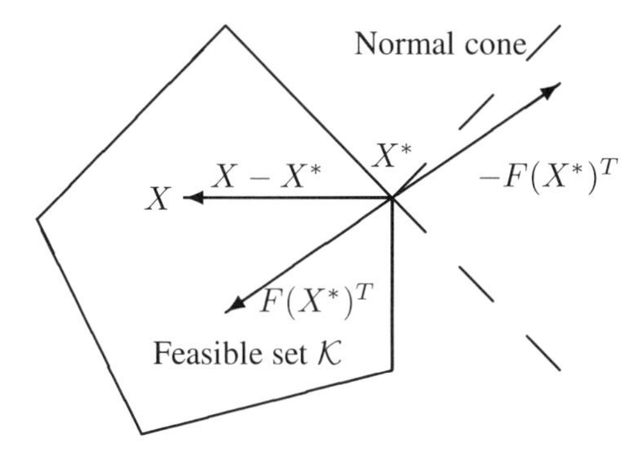

Figure 2.1: Geometric interpretation of VI$(F, \mathcal{K})$

Definition 2.1: The Finite-Dimensional Variational Inequality Problem
The finite-dimensional variational inequality problem, VI$(F, \mathcal{K})$, *is to determine a solution vector* $X^* \in \mathcal{K} \subset R^N$, *such that*

$$\langle F(X^*)^T, X - X^* \rangle \geq 0, \quad \forall X \in \mathcal{K}, \tag{2.1}$$

where F *is a given continuous function from* $\mathcal{K}$ *to* R^N, $\mathcal{K}$ *is a given closed convex set, and* $\langle \cdot, \cdot \rangle$ *denotes the inner product in* N*-dimensional space.*

In Figure 2.1, we provide a geometric interpretation of a variational inequality, which illustrates that $F(X^*)^T$ is "orthogonal" to the feasible set $\mathcal{K}$ at the point X^*.

It has been established that many constrained and unconstrained optimization problems can be formulated as VI problems [cf. Nagurney (1999)]. We now briefly discuss the relationship of VIs to such optimization problems.

Proposition 2.1
Let X^* *be a solution to the optimization problem*

$$\textit{Minimize} \quad f(X) \tag{2.2}$$

$$\textit{subject to: } X \in \mathcal{K},$$

where f *is continuously differentiable and* $\mathcal{K}$ *is closed and convex. Then* X^* *is a solution of the variational inequality problem*

$$\langle \nabla f(X^*)^T, X - X^* \rangle \geq 0, \quad \forall X \in \mathcal{K}, \tag{2.3}$$

where $\nabla f(X)$ *is the gradient vector of* f *with respect to X, that is,*

$$\nabla f(X) = (\frac{\partial f(X)}{\partial X_1}, \frac{\partial f(X)}{\partial X_2}, \ldots, \frac{\partial f(X)}{\partial X_N}).$$

Proposition 2.2
If $f(X)$ is a convex function and X^* is a solution to $\mathrm{VI}(\nabla f, \mathcal{K})$, then X^* is a solution to the optimization problem (2.2). In the case that the feasible set $\mathcal{K} = R^N$, then the unconstrained optimization problem is also a variational inequality problem.

The variational inequality problem can be reformulated as an optimization problem given certain symmetry conditions, which is rigorously shown in Theorem 2.1. First, however, we present the definitions of positive semidefiniteness, positive definiteness, and strongly positive definiteness.

Definition 2.2: Positive Semidefinite, Positive Definite, Strongly Positive Definite Matrix
An $N \times N$ matrix $M(X)$, whose elements $m_{ij}(X)$; $i, j = 1, ..., N$, are functions defined on the set $\mathcal{S} \subset R^N$, is said to be positive semidefinite on $\mathcal{S}$ if

$$v^T M(X) v \geq 0, \quad \forall v \in R^N, X \in \mathcal{S}. \tag{2.4}$$

It is said to be positive definite on $\mathcal{S}$ if

$$v^T M(X) v > 0, \quad \forall v \neq 0, v \in R^N, X \in \mathcal{S}. \tag{2.5}$$

It is said to be strongly positive definite on $\mathcal{S}$ if

$$v^T M(X) v \geq \alpha \|v\|^2, \text{ for some } \alpha > 0, \quad \forall v \in R^N, X \in \mathcal{S}. \tag{2.6}$$

Theorem 2.1
Assume that $F(X)$ is continuously differentiable on $\mathcal{K}$ and that the Jacobian matrix

$$\nabla F(X) = \begin{bmatrix} \frac{\partial F_1}{\partial X_1} & \cdots & \frac{\partial F_1}{\partial X_N} \\ \vdots & \cdots & \vdots \\ \frac{\partial F_N}{\partial X_1} & \cdots & \frac{\partial F_N}{\partial X_N} \end{bmatrix} \tag{2.7}$$

is symmetric and positive semidefinite. Then there is a real-valued convex function $f : \mathcal{K} \longmapsto R^1$ satisfying

$$\nabla f(X) = F(X) \tag{2.8}$$

with X^* the solution of $\mathrm{VI}(F, \mathcal{K})$ also being the solution of the mathematical programming (optimization) problem

$$\textit{Minimize} \quad f(X)$$

$$\textit{subject to: } X \in \mathcal{K},$$

where $f(X) = \int F(X)^T dx$, and $\int$ is a line integral.

Hence the variational inequality problem is a more general problem formulation and it can also handle a function $F(X)$ with an asymmetric Jacobian [see, e.g., Nagurney (1999)]. This feature of a VI formulation enables the more realistic modeling and ultimate solution of a spectrum of complex network problems. We now provide qualitative properties, including conditions for existence and uniqueness of a solution to $\mathrm{VI}(F, \mathcal{K})$.

Theorem 2.2: Existence of a Solution under Continuity and Compactness
If $\mathcal{K}$ is a compact convex set and $F(X)$ is continuous on $\mathcal{K}$, then $\mathrm{VI}(F, \mathcal{K})$ admits at least one solution X^*.

If the feasible set $\mathcal{K}$ is unbounded, Theorem 2.2 cannot be applied. However, existence of a solution to (2.1) may still be obtained provided that certain conditions are satisfied. For example, let $B_r(0)$ denote a closed ball with radius r centered at 0 and let $\mathcal{K}_r = \mathcal{K} \cap B_r(0)$. The set $\mathcal{K}_r$ is then bounded. Let VI_r denote the variational inequality problem: Determine $X_r^* \in \mathcal{K}_r$ such that

$$\langle F(X_r^*)^T, y - X_r^* \rangle \geq 0, \quad \forall y \in \mathcal{K}_r.$$

Then we have the following existence result.

Theorem 2.3
$\mathrm{VI}(F, \mathcal{K})$ admits a solution if and only if there exists an $r > 0$ and a solution of $\mathrm{VI}_r(F, \mathcal{K}_r)$, X_r^*, such that $\|X_r^*\| < r$.

The coercivity condition provides another guarantee of the existence of a solution to VIs, as summarized in the following theorem.

Theorem 2.4: Existence of a Solution under Coercivity
Suppose that $F(X)$ satisfies the coercivity condition

$$\frac{\langle (F(X) - F(X_0))^T, X - X_0 \rangle}{\|X - X_0\|} \to \infty \tag{2.9}$$

as $\|X\| \to \infty$ for $X \in \mathcal{K}$ and for some $X_0 \in \mathcal{K}$. Then $\mathrm{VI}(F, \mathcal{K})$ always has a solution.

Certain monotonicity conditions can be used to analyze the qualitative properties of the variational inequality problem. We first recall some basic definitions.

Definition 2.3: Monotonicity
$F(X)$ is monotone on $\mathcal{K}$ if

$$\langle (F(X^1) - F(X^2))^T, X^1 - X^2 \rangle \geq 0, \quad \forall X^1, X^2 \in \mathcal{K}. \tag{2.10}$$

Definition 2.4: Strict Monotonicity
$F(X)$ is strictly monotone on $\mathcal{K}$ if

$$\langle (F(X^1) - F(X^2))^T, X^1 - X^2 \rangle > 0, \quad \forall X^1, X^2 \in \mathcal{K},\ X^1 \neq X^2. \tag{2.11}$$

Definition 2.5: Strong Monotonicity
$F(X)$ is strongly monotone on $\mathcal{K}$ if, for some $\alpha > 0$

$$\langle (F(X^1) - F(X^2))^T, X^1 - X^2 \rangle \geq \alpha \|X^1 - X^2\|^2, \quad \forall X^1, X^2 \in \mathcal{K}. \tag{2.12}$$

Definition 2.6: Lipschitz Continuity
$F(X)$ is Lipschitz continuous on $\mathcal{K}$ if there exists an $L > 0$, such that

$$\langle (F(X^1) - F(X^2))^T, X^1 - X^2 \rangle \leq L \|X^1 - X^2\|^2, \quad \forall X^1, X^2 \in \mathcal{K}. \tag{2.13}$$

L is called the Lipschitz constant.

Theorem 2.5: Uniqueness of a Solution under Strict Monotonicity
Suppose that $F(X)$ is strictly monotone on $\mathcal{K}$. Then the solution to $\mathrm{VI}(F, \mathcal{K})$ is unique, if one exists.

Theorem 2.6: Existence and Uniqueness of a Solution under Strong Monotonicity
Suppose that $F(X)$ is strongly monotone on $\mathcal{K}$. Then there exists precisely one solution X^* to $\mathrm{VI}(F, \mathcal{K})$.

Based on these theorems, we know that if the feasible set $\mathcal{K}$ is unbounded, strong monotonicity of the function F guarantees both existence and uniqueness of a solution X^* to $\mathrm{VI}(F, \mathcal{K})$ due to the fact that the strong monotonicity condition implies coercivity and strict monotonicity. If the feasible set $\mathcal{K}$ is compact, that is, closed and bounded, then continuity of F alone guarantees the existence of a solution. Finally, strict monotonicity of F is then sufficient to ensure uniqueness. We will see how these theoretical results are applied to network models in the case of decentralized versus centralized decision-making and in the context of applications as varied as transportation, supply chains, finance, and humanitarian logistics.

Theorem 2.7
Suppose that $F(X)$ is continuously differentiable on $\mathcal{K}$ and that the Jacobian matrix

$$\nabla F(X) = \begin{bmatrix} \frac{\partial F_1}{\partial X_1} & \cdots & \frac{\partial F_1}{\partial X_N} \\ \vdots & \cdots & \vdots \\ \frac{\partial F_N}{\partial X_1} & \cdots & \frac{\partial F_N}{\partial X_N} \end{bmatrix},$$

which need not be symmetric, is positive semidefinite (positive definite). Then $F(X)$ is monotone (strictly) monotone.

2.2 DECENTRALIZED DECISION-MAKING AND USER-OPTIMIZATION

The fact that networks are used/operated in a decentralized vs. centralized manner would yield different flows, and that this would have associated different economic impacts was recognized as early as 1920 by Pigou [see also Kohl (1841) and Knight

(1924)]. In particular, Wardrop (1952) explicitly recognized alternative possible behaviors of users of transportation networks and stated two principles, which are commonly named after him. These principles correspond, in effect, to decentralized versus centralized behavior on networks and, although stated in a transportation context, have relevance to many different network systems. Hence, we now recall Wardrop's two principles:

First Principle: The journey times of all routes actually used are equal and less than those that would be experienced by a single vehicle on any unused route.

Second Principle: The average journey time is minimal.

The first principle corresponds to the behavioral principle in which travelers seek to (unilaterally) determine their minimal costs of travel whereas the second principle corresponds to the behavioral principle in which the total cost in the network is minimized.

Beckmann, McGuire, and Winsten (1956) were the first to rigorously formulate these conditions mathematically, as had Samuelson (1952) in the framework of spatial price equilibrium problems in which there were, however, no congestion effects. Specifically, Beckmann, McGuire, and Winsten (1956) established the equivalence between the *traffic network equilibrium* conditions, which state that all used paths connecting an origin/destination pair will have equal and minimal travel times (or costs); corresponding to Wardrop's first principle, and the Kuhn-Tucker conditions of an appropriately constructed optimization problem, under a symmetry assumption on the underlying functions. Hence, in this case, the equilibrium link and path flows could be obtained as the solution of a mathematical programming problem (as we show in this section).

Dafermos and Sparrow (1969) coined the terms *user-optimized* (U-O) and *system-optimized* (S-O) transportation networks to distinguish between the two distinct situations in which users act unilaterally, in their own self-interest in selecting their routes, and in which users select routes according to what is optimal from a societal point of view, in that the total cost in the system is minimized. In the latter problem, marginal (total) costs rather than average costs are equilibrated/equalized. The former problem coincides with Wardrop's first principle, and the latter with Wardrop's second principle. The authors also recognized the noncooperative game theoretic nature of the problem as in Nash (1950, 1951) equilibrium problems. In this book, we will often use the terms *user-optimization* and *network equilibrium* interchangeably.

The concept of *system-optimization* is also relevant to other types of routing models, as arise in transportation problems concerned with the routing of freight, in centralized supply chain and logistical networks, and in network systems in which a central controller has control as to how the flows can be routed (as, for example, in military operations or in certain humanitarian logistics operations).

Chapters 3 and 4 make substantive use of the fundamentals contained in this chapter. Chapters 5 and 6 develop decentralized network models, and Chapters 8 through 10 focus exclusively on centralized ones. Chapter 7 uses advanced material on evolutionary variational inequalities for dynamic networks. Chapter 11 develops game theoretic network models under oligopolistic competition.

2.2.1 The Network Equilibrium (U-O) Model with Fixed Demands

Consider a general network $\mathcal{G} = [\mathcal{N}, \mathcal{L}]$, where $\mathcal{N}$ denotes the set of nodes, and $\mathcal{L}$ the set of directed links. Let a denote a link of the network connecting a pair of nodes, and let p denote a path consisting of a sequence of links connecting an origin/destination (O/D) pair of nodes. The paths are assumed to be acyclic. In urban transportation networks, nodes correspond to origins and destinations as well as to intersections; links, on the other hand, correspond to roads/streets, whereas in freight networks, they would correspond to railroad lines or roads, as in the case of truck freight. A path in a transportation context thus can correspond to a sequence of roads or railroad links, which make up a route from an origin to a destination. In communications networks, on the other hand, nodes correspond to routers; in logistical and supply chain networks they can correspond to manufacturers, distributors, and demand markets; in financial and other economic networks, nodes can also correspond to decision-makers, from financial sources of funds to the ultimate consumers. Links in network models abstract physical connections (such as cables, transmission lines, manufacturing lines, etc.) or logical connections (such as economic transactions) in a graphical formalism. Of course, in the case of social networks, nodes correspond to individuals (or, in an aggregated case, to organizations) and links to social ties.

Let P_w denote the set of paths connecting the O/D pair of nodes w. Let P denote the set of all paths in the network and assume that there are n_W origin/destination pairs of nodes. We assume in all models in this chapter that the networks are (strongly) connected, that is, that there is at least one path connecting each pair of O/D nodes.

Let x_p represent the nonnegative flow on path p and let f_a denote the flow on link a. These flows, in different settings, would correspond to vehicles, commodities or products, computer messages, financial flows, electric power or energy flows, etc. The path flows on the network are grouped into the vector $x \in R^{n_P}_+$, where n_P denotes the number of paths in the network. The link flows, in turn, are grouped into the vector $f \in R^n_+$, where n denotes the number of links in the network.

The cost experienced by a user traversing link a is denoted by c_a, for all links $a \in \mathcal{L}$, and these costs are assumed to be continuous and nonnegative. The user cost on a path p is denoted by C_p for all paths $p \in P$. We group the user link costs and the user path costs into the vectors $c \in R^n_+$ and $C \in R^{n_P}_+$, respectively.

Let d_w denote the demand associated with O/D pair w, for all $w \in W$, assumed, for now, as being fixed and known. The following conservation of flow of equations must hold

$$d_w = \sum_{p \in P_w} x_p, \quad \forall w \in W, \tag{2.14}$$

that is, the sum of the path flows on paths connecting each O/D pair w must be equal to the given demand d_w. We denote the disutility associated with O/D pair w by λ_w.

Also, the path flows must be nonnegative, that is

$$x_p \geq 0, \quad \forall p \in P. \tag{2.15}$$

The following conservation of flow equations relate the link flows to the path flows

$$f_a = \sum_{p \in P} x_p \delta_{ap}, \quad \forall a \in \mathcal{L}, \tag{2.16}$$

where $\delta_{ap} = 1$, if path p contains link a, and $\delta_{ap} = 0$, otherwise. Hence, the flow on a link is equal to the sum of the flows on paths that contain that link.

The user cost on a path is equal to the sum of user costs on links the path consists of, which can be represented by the following expression

$$C_p = \sum_{a \in \mathcal{L}} c_a \delta_{ap}, \quad \forall p \in P. \tag{2.17}$$

For the sake of generality, the user cost on a link is allowed to depend on the entire vector of link flows, so that

$$c_a = c_a(f), \quad \forall a \in \mathcal{L}. \tag{2.18}$$

The following definition is based on Wardrop's (1952) first principle [see also Beckmann, McGuire, and Winsten (1956), Smith (1979), and Dafermos (1980)]. Note that, in view of (2.16), (2.17), and (2.18), one may express the cost on a path p as a function of the path flow variables.

Definition 2.7: Network Equilibrium – Fixed Demands
A path flow pattern $x^* \in \mathcal{K}^1$, where $\mathcal{K}^1 \equiv \{x | x \in R_+^{n_P}$ and (2.14) holds$\}$, is said to be a network equilibrium in the case of fixed demands if the following conditions hold for each O/D pair $w \in W$ and every path $p \in P_w$

$$C_p(x^*) - \lambda_w^* \begin{cases} = 0, & \text{if} \quad x_p^* > 0, \\ \geq 0, & \text{if} \quad x_p^* = 0. \end{cases} \tag{2.19}$$

Conditions (2.19) state that the user costs of all utilized paths joining an O/D pair are equal and minimal. Hence, in equilibrium, no user has any incentive to switch his or her path. As described in Dafermos (1980) and Smith (1979) the network equilibrium pattern according to conditions (2.19) coincides with the solution to the following finite-dimensional variational inequalities.

Theorem 2.8: Variational Inequality Formulations of Network Equilibrium with Fixed Demands
A path flow pattern is a network equilibrium according to Definition 2.7 if and only if it satisfies the following variational inequality in path flows: Determine $x^* \in \mathcal{K}^1$ such that

$$\sum_{w \in W} \sum_{p \in P_w} C_p(x^*) \times (x_p - x_p^*) \geq 0, \quad \forall x \in \mathcal{K}^1. \tag{2.20}$$

Equivalently, a link flow pattern is a network equilibrium according to Definition 2.7 if and only if it satisfies the variational inequality problem: Determine $f^* \in \mathcal{K}^2$ satisfying

$$\sum_{a \in \mathcal{L}} c_a(f^*) \times (f_a - f_a^*) \geq 0, \quad \forall f \in \mathcal{K}^2, \tag{2.21}$$

where $\mathcal{K}^2 \equiv \{f \in R^n_+ |$ *there exists an* x *satisfying* (2.15) *and* (2.16)$\}$.

If we define $X \equiv x$ and $F(X) \equiv C(x)$, then variational inequality (2.20) can be rewritten in standard form as in (2.1). Similarly, if we define $X \equiv f$ and $F(X) \equiv c(f)$, then the variational inequality (2.21) can be transformed into standard form (2.1).

The continuity of the link cost functions and the compactness of the feasible sets $\mathcal{K}^1$ and $\mathcal{K}^2$ guarantee, according to Theorem 2.2, the existence of solutions to both (2.20) and (2.21). Moreover, according to Theorem 2.5, if the vector of user link cost functions c is strictly monotone, then the solution f^* to (2.21) is unique. In reality, and hence in practical applications, this is not an unrealistic assumption; on the other hand, the vector of user path cost functions C can be expected to be monotone, but it is unlikely, except in very specially structured networks, that the vector C will be strictly monotone. Hence uniqueness of path flow solutions is much less likely in practice.

Beckmann, McGuire, and Winsten (1956) [see also Dafermos and Sparrow (1969)] established that the solution to the network equilibrium problem, in the case of separable user link cost functions, that is, when the user link cost functions (2.18) take on the form $c_a = c_a(f_a)$, $\forall a \in \mathcal{L}$, could be obtained by solving the following convex optimization problem

$$\text{Minimize} \sum_{a \in \mathcal{L}} \int_0^{f_a} c_a(x)dx \tag{2.22}$$

subject to: (2.14) through (2.16) or, simply, $f \in \mathcal{K}^2$, provided that $c(f)$ is continuously differentiable and positive semidefinite. Clearly, in this case, $\nabla c(f)$ is symmetric because it is of diagonal form due to the separability of the user link cost functions (see also Theorem 2.1).

Hence in applications in which the separability assumption of the user link cost functions can be expected to hold, one can appeal, if the resulting objective function in (2.22) is convex, which would also be the case if c is monotone, to convex optimization algorithms for the solution of the associated U-O (or network equilibrium) problem with fixed demands. If c is strictly monotone, then the objective function in (2.22) would be strictly convex and the corresponding solution f^*, as expected, unique, that is, no other link flow would satisfy the network equilibrium conditions.

Example 2.1: A U-O Fixed Demand Example with Separable User Link Cost Functions

Consider the network depicted in Figure 2.2 in which there are two nodes 1, 2; three links a, b, and c, and a single O/D pair $w_1 = (1, 2)$. Let path $p_1 = a$, path $p_2 = b$, and path $p_3 = c$.

The user link cost functions are

$$c_a(f_a) = 2f_a + 4, \quad c_b(f_b) = f_b + 10, \quad c_c(f_c) = f_c + 8,$$

and the demand

$$d_{w_1} = 10.$$

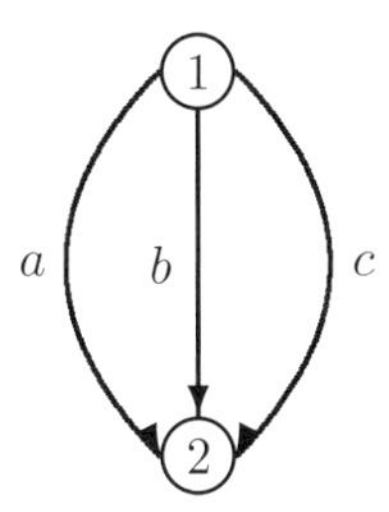

Figure 2.2: Network topology for Example 2.1

Decision-makers (or travelers) operating in a user-optimized manner will select their paths or links as follows

$$x_{p_1}^* = f_a^* = 4, \quad x_{p_2}^* = f_b^* = 2, \quad x_{p_3}^* = f_c^* = 4,$$

with incurred user costs

$$C_{p_1} = c_a(f_a^*) = C_{p_2} = c_b(f_b^*) = C_{p_3} = c_c(f_c^*) = 12.$$

Indeed, this path (and link flow) pattern satisfies the network equilibrium conditions (2.19) because all used paths connecting the O/D pair w_1 have equal and minimal user costs; in this example $\lambda_{w_1}^* = 12$. The equilibrium solution is unique.

Of course, in the case of *extended* user link cost functions, as in (2.18), in which the symmetry assumption: $\frac{\partial c_a}{\partial c_b} = \frac{\partial c_b}{\partial c_a}$ holds for all links $a, b \in \mathcal{L}$, then the solution to the network equilibrium conditions can be obtained by solving an associated optimization problem (cf. Theorem 2.1). However, in the case where the user link cost functions are no longer symmetric, one cannot compute the solution to the U-O problem using standard optimization algorithms. *Asymmetric* cost functions are very important from an application standpoint because they allow for asymmetric interactions on the network. For example, allowing for asymmetric cost functions permits one to handle the situation when the flow on a particular link affects the cost on another link in a different way than the cost on the particular link is affected by the flow on the other link.

Example 2.2: A U-O Fixed Demand Example with Asymmetric User Link Cost Functions

Returning to the network depicted in Figure 2.2, an example of an asymmetric user link cost structure would be

$$c_a(f) = 2f_a + f_b + 4, \quad c_b(f) = f_b + 10, \quad c_c(f) = f_c + .5f_b + 8,$$

because $\frac{\partial c_a}{\partial f_b} = 1 \neq \frac{\partial c_b}{\partial f_a} = 0$, and $\frac{\partial c_c}{\partial f_b} = .5 \neq \frac{\partial c_b}{\partial f_c} = 0$. The U-O solution for this example, with demand as in Example 2.1, is unique and given by

$$x_{p_1}^* = f_a^* = 3, \quad x_{p_2}^* = f_b^* = 3\frac{1}{3}, \quad x_{p_3}^* = f_c^* = 3\frac{2}{3},$$

with incurred user costs

$$C_{p_1} = c_a(f^*) = C_{p_2} = c_b(f^*) = C_{p_3} = c_c(f^*) = 13\frac{1}{3}.$$

Hence the network equilibrium conditions (2.19) are satisfied and variational inequality (2.21) is satisfied because

$$\sum_{a \in \mathcal{L}} c_a(f^*) \times (f_a - f_a^*) = 13\frac{1}{3} \times (f_a - 3) + 13\frac{1}{3} \times (f_b - 3\frac{1}{3}) + 13\frac{1}{3} \times (f_c - 3\frac{2}{3})$$

$$= 13\frac{1}{3} \times (f_a + f_b + f_c - 10) = 13\frac{1}{3}(10 - 10) = 0.$$

The allowance for such asymmetric interactions also enables the more realistic modeling of multimodal/multiclass networks in which a particular flow of a mode/class affects the costs of other modes/classes in a different manner from how it is affected by the other modes/classes. Hence, in such models, heterogeneity of the costs and flows is explicitly considered and handled.

Remark 2.1
It is also worth recognizing and emphasizing that multimodal and multiclass network equilibrium problems with fixed demands can be transformed into single-class, but *extended* models, as discussed in this section, by making as many copies of the network as there are classes/modes and by redefining the demands, costs, and flows on these multicopy networks [cf. Dafermos (1972)]. Such a technique is now often used in practice in both transportation and telecommunication settings and applications. In Section 10.4, we discuss an application of this technique to multiproduct supply chain networks.

In Section 2.4, we provide algorithms for the solution of these network equilibrium models, in both their variational inequality and optimization realizations.

2.2.2 Network Equilibrium (U-O) Models with Elastic Demands

In this section, two versions of network equilibrium models with elastic demands are discussed, one with given demand functions [see Dafermos and Nagurney (1984)], and the other with given disutility functions [see Dafermos (1982)].

2.2.2.1 Network Equilibrium Model with Given Demand Functions The notation in this section follows that in Section 2.2.1. We assume that equations (2.14) through (2.16) still hold. However, now, rather than assuming that the demands are fixed and given as in (2.14), we assume that the demand functions, $d_w(\lambda)$, $w \in W$, are given, where λ is the vector of disutilities and the demands are now variables in (2.14). Furthermore, now d denotes the column vector of the demands and $d(\lambda)$ the vector of demand functions.

The following definition is based on Dafermos and Nagurney (1984) [see also Aashtiani and Magnanti (1981), Fisk and Boyce (1983), Nagurney and Zhang (1996), and Nagurney (1999)].

Definition 2.8: Network Equilibrium – Elastic Demands with Given Demand Functions

A path flow and disutility pattern $(x^*, d^*, \lambda^*) \in \mathcal{K}^3$, *where* $\mathcal{K}^3 \equiv \{(x, d, \lambda) | (x, d, \lambda) \in R_+^{n_P + 2n_W}$ *and* $d_w = \sum_{p \in P_w} x_p, \forall w\}$, *is said to be a network equilibrium, in the case of elastic demands with known demand functions, if, once established, no user has any incentive to alter his or her path decisions. The state is expressed by the following conditions, which must hold for each O/D pair* $w \in W$ *and every path* $p \in P_w$

$$C_p(x^*) - \lambda_w^* \begin{cases} = 0, & \text{if} \quad x_p^* > 0, \\ \geq 0, & \text{if} \quad x_p^* = 0, \end{cases} \tag{2.23}$$

and

$$d_w^* \begin{cases} = d_w(\lambda^*), & \text{if} \quad \lambda_w^* > 0, \\ \geq d_w(\lambda^*), & \text{if} \quad \lambda_w^* = 0. \end{cases} \tag{2.24}$$

Conditions (2.23) state that all used paths connecting an O/D pair have equal and minimal costs that are equal to the disutility associated with "traveling" between that O/D pair. Conditions (2.24) state that the market clears for each O/D pair under a positive price or disutility. As described in Dafermos and Nagurney (1984) the network equilibrium conditions (2.23) and (2.24) can be expressed as the following variational inequalities.

Theorem 2.9: Variational Inequality Formulations of Network Equilibrium with Given Demand Functions

A path flow, demand, and disutility pattern is a network equilibrium according to Definition 2.8 if and only if it satisfies the variational inequality: Determine $(x^*, d^*, \lambda^*) \in \mathcal{K}^3$ *such that*

$$\sum_{w \in W} \sum_{p \in P_w} C_p(x^*) \times (x_p - x_p^*) - \sum_{w \in W} \lambda_w^* \times (d_w - d_w^*)$$

$$+ \sum_{w \in W} (d_w^* - d_w(\lambda^*)) \times (\lambda_w - \lambda_w^*) \geq 0, \quad \forall (x, d, \lambda) \in \mathcal{K}^3. \tag{2.25}$$

Equivalently, a link flow pattern and associated demand and disutility pattern is a network equilibrium according to Definition 2.8 if and only if it satisfies the variational inequality problem: Determine $(f^*, d^*, \lambda^*) \in \mathcal{K}^4$ *satisfying*

$$\sum_{a \in \mathcal{L}} c_a(f^*) \times (f_a - f_a^*) - \sum_{w \in W} \lambda_w^* \times (d_w - d_w^*) + \sum_{w \in W} (d_w^* - d_w(\lambda^*)) \times (\lambda_w - \lambda_w^*) \geq 0,$$

$$\forall (f, d, \lambda) \in \mathcal{K}^4, \tag{2.26}$$

where $\mathcal{K}^4 \equiv \{(f, d, \lambda) \in R_+^{n_L + 2n_W} \,|\, \text{there exists an } x \text{ satisfying } (2.14) - (2.16)\}$.

Let $X \equiv (x^T, d^T, \lambda^T)^T$ and $F(X) \equiv (C(x)^T, -\lambda^T, (d - d(\lambda))^T)^T$, which allows one to put variational inequality (2.25) into standard form (2.1).

Letting now $X \equiv (f^T, d^T, \lambda^T)^T$ and $F(X) \equiv (c(f)^T, -\lambda^T, (d - d(\lambda))^T)^T$, one can easily also rewrite variational inequality (2.26), which is in link flows, in standard form (2.1).

2.2.2.2 Network Equilibrium Model with Given Disutility Functions The network equilibrium model with known disutility functions is now recalled [cf. Dafermos (1982)]. The notation is the same as in the model with known demand functions. However, it is now assumed that the disutility functions are given as follows

$$\lambda_w = \lambda_w(d), \quad \forall w \in W, \tag{2.27}$$

where d is the vector of demands with demand associated with O/D pair w being denoted by d_w.

Definition 2.9: Network Equilibrium – Elastic Demands with Given Disutility Functions
A path flow and disutility pattern $(x^*, d^*) \in \mathcal{K}^5$, where $\mathcal{K}^5 \equiv \{(x,d)|(x,d) \in R_+^{n_P+n_W}$ and $d_w = \sum_{p \in P_w} x_p, \forall w\}$, is said to be a network equilibrium, in the case of elastic demands with known disutility functions, if, once established, no user has any incentive to alter his or her path selection decisions. The state is expressed by the following condition, which must hold for each O/D pair $w \in W$ and each path $p \in P_w$

$$C_p(x^*) - \lambda_w(d^*) \begin{cases} = 0, & if \quad x_p^* > 0, \\ \geq 0, & if \quad x_p^* = 0. \end{cases} \tag{2.28}$$

The interpretation of conditions (2.28) is the same as that of conditions (2.23). As proved in Dafermos (1982), the network equilibrium conditions (2.28) are equivalent to the following variational inequalities.

Theorem 2.10: Variational Inequality Formulations of Network Equilibrium with Given Disutility Functions
A path flow and demand pattern is a network equilibrium according to Definition 2.9 if and only if it satisfies the following variational inequality (in path flows): Determine $(x^*, d^*) \in \mathcal{K}^5$ such that

$$\sum_{w \in W} \sum_{p \in P_w} C_p(x^*) \times (x_p - x_p^*) - \sum_{w \in W} \lambda_w(d^*) \times (d_w - d_w^*) \geq 0, \quad \forall (x,d) \in \mathcal{K}^5. \tag{2.29}$$

Equivalently, a link flow pattern and associated demand pattern is a network equilibrium according to Definition 2.9 if and only if it satisfies the variational inequality problem (in link flows): Determine $(f^*, d^*) \in \mathcal{K}^6$ satisfying

$$\sum_{a \in \mathcal{L}} c_a(f^*) \times (f_a - f_a^*) - \sum_{w \in W} \lambda_w(d^*) \times (d_w - d_w^*) \geq 0, \quad \forall (f,d) \in \mathcal{K}^6, \tag{2.30}$$

where $\mathcal{K}^6 \equiv \{(f,d) \in R_+^{n+n_W} \mid$ there exists an x satisfying $(2.14) - (2.16)\}$.

Let $X \equiv (x^T, d^T)^T$ and $F(X) \equiv (C(x)^T, -\lambda(d)^T)^T$. We can then rewrite VI (2.29) in standard form (2.1).

By defining $X \equiv (f^T, d^T)^T$ and $F(X) \equiv (c(f)^T, -\lambda(d)^T)^T$, we can also put VI (2.30) into standard form (2.1).

It is interesting that Beckmann, McGuire, and Winsten (1956) formulated the elastic demand network equilibrium problem [rather than the fixed one that was studied by Dafermos and Sparrow (1969)] and considered both separable O/D pair disutility functions and user link cost functions. Under such an assumption, Theorem 2.1 can be applied. In particular, Beckmann, McGuire, and Winsten (1956) demonstrated that the network equilibrium conditions as in (2.28) could be reformulated and solved as the following convex optimization problem

$$\text{Minimize} \quad \sum_{a \in \mathcal{L}} \int_0^{f_a} c_a(x)dx - \sum_{w \in W} \int_0^{d_w} \lambda_w(y)dy \qquad (2.31)$$

subject to: $(f, d) \in \mathcal{K}^6$, provided that $c(f)$ and $\lambda(d)$ are continuously differentiable and that $\nabla c(f)$ and $-\nabla \lambda$ are each positive semidefinite. Because both $c(f)$ and $\lambda(d)$ are separable functions in the standard model, we know that their respective Jacobian matrices are always symmetric since they are diagonal matrices.

Note that in the elastic demand models (unlike in the fixed demand models), we no longer have a compact feasible set; hence to obtain existence (and uniqueness) results stronger assumptions are needed (cf. Theorems 2.3, 2.4, and 2.6).

Remark 2.2

It is interesting that Gartner (1980a, b) considered the separable elastic network equilibrium problem satisfying conditions (2.28) and proved that such elastic demand problems could be transformed into fixed demand problems using an excess overflow formulation. The formulation consists of adding to the existing network an additional path for each O/D pair $w \in W$ consisting of the single link a_w with an associated cost of $c_{a_w} = \lambda_w(\bar{d}_w - f_{a_w})$, where $\bar{d}_w$ is a fixed upper bound for the demand for O/D pair w.

We now present an example of an elastic demand network equilibrium problem, which we transform into a fixed demand problem.

Example 2.3: A U-O Elastic Demand Example with Separable User Link Cost and Disutility Functions

Consider the first network depicted in Figure 2.3 in which there are three nodes 1, 2, and 3; three links a, b, and c; and a single O/D pair $w_1 = (1, 3)$. Let path $p_1 = (a, b)$ and path $p_2 = (a, c)$.

Assume that the user link cost functions are

$$c_a(f_a) = 5f_a + 5, \quad c_b(f_b) = f_b + 7, \quad c_c(f_c) = f_c + 5$$

and that the disutility function is

$$\lambda_{w_1}(d_{w_1}) = -2d_{w_1} + 101.$$

It is easy to see that the unique path flow and associated link flow and demand patterns satisfying the network equilibrium conditions (2.28) are

$$x_{p_1}^* = 5, \quad x_{p_2}^* = 7; \quad f_a^* = 12, \quad f_b^* = 5, \quad f_c^* = 7; \quad d_{w_1}^* = 12;$$

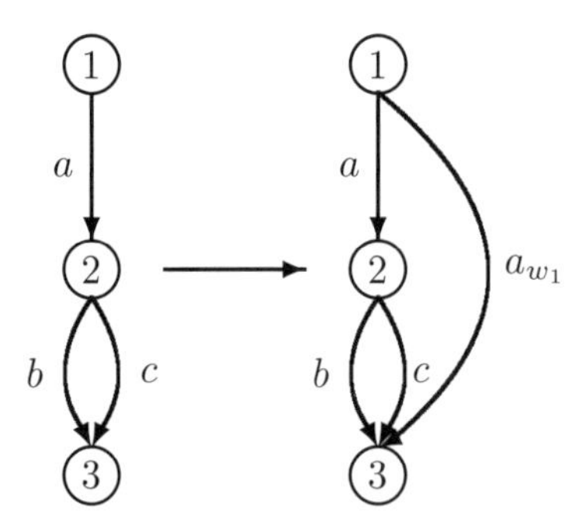

Figure 2.3: Network topology for Example 2.3 with elastic demand and its transformation into a fixed demand problem

with incurred user path costs and disutility

$$C_{p_1}(x^*) = c_a(f_a^*) + c_b(f_b^*) = C_{p_2}(x^*) = c_a(f_a^*) + c_a(f_c^*) = 77 = \lambda_{w_1}(d_{w_1}^*).$$

Of course, as noted, this elastic demand network equilibrium problem can be transformed into a fixed demand network equilibrium problem. Refer to the second network in Figure 2.3. Given that $\lambda_{w_1}(d_{w_1}) = -2d_{w_1} + 101$, it is reasonable to assume that $\bar{d}_{w_1} = 50\frac{1}{2}$, in which case $c_{a_{w_1}} = \lambda_{w_1}(\bar{d}_{w_1} - f_{a_{w_1}}) = 2f_{a_{w_1}}$. If we let path p_3 denote the excess overflow path in the second network in Figure 2.3 and assign a cost to its sole link a_{w_1} of $c_{a_{w_1}} = 2f_{a_{w_1}}$, we can see that the network equilibrium flow pattern for the fixed demand problem for the second network is

$$x_{p_1}^* = 5, \quad x_{p_2}^* = 7, \quad x_{p_3}^* = 38\frac{1}{2};$$

$$f_a^* = 12, \quad f_b^* = 5, \quad f_c^* = 7, \quad f_{a_{w_1}}^* = 38\frac{1}{2},$$

and $C_{p_1} = C_{p_2} = C_{p_3} = 77$.

Example 2.4: A U-O Elastic Demand Example with Asymmetric User Link Cost and Disutility Functions

We now present an elastic demand network equilibrium problem with asymmetric user link cost functions and disutility functions (see Figure 2.4). There are three nodes 1, 2, and 3, and three links a, b, and c. The O/D pairs are $w_1 = (1, 2)$ and $w_2 = (1, 3)$. The paths are for O/D pair w_1: $p_1 = a$, and for O/D pair w_2: $p_2 = b$ and $p_3 = (a, c)$.

The user link cost functions are

$$c_a(f) = 5f_a + 2f_b + 5, \quad c_b(f) = f_b + .5f_c + 7, \quad c_c(f) = 4f_c + f_b + 10,$$

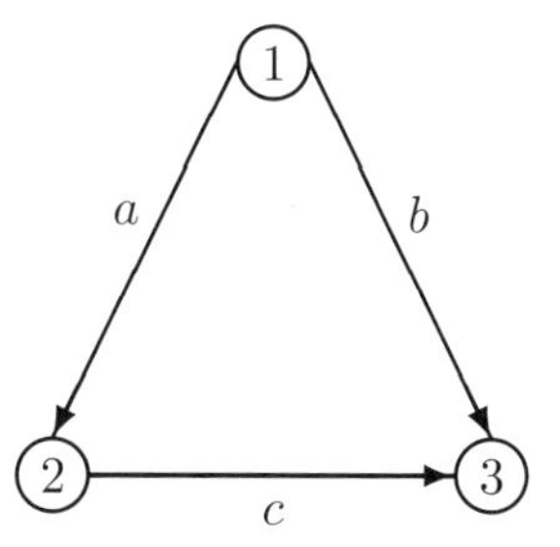

Figure 2.4: Network topology for Example 2.4

and the disutility functions are

$$\lambda_{w_1}(d) = -2d_{w_1} - d_{w_2} + 50, \quad \lambda_{w_2}(d) = -3d_{w_2} - 2d_{w_1} + 57.$$

The unique network equilibrium solution, satisfying conditions (2.28), is

$$x^*_{p_1} = 5, \quad x^*_{p_2} = 10, \quad x^*_{p_3} = 0,$$
$$f^*_a = 5, \quad f^*_b = 10, \quad f^*_c = 0,$$
$$d^*_{w_1} = 5, \quad d^*_{w_2} = 10.$$

Indeed

$$C_{p_1}(x^*) = c_a(f^*) = \lambda_{w_1}(d^*) = 30$$

and

$$C_{p_2}(x^*) = c_b(f^*) = \lambda_{w_2}(d^*) = 17 \leq C_{p_3}(x^*) = c_a(f^*) + c_c(f^*) = 50.$$

Observe that this link flow and demand pattern satisfies variational inequality (2.30). If we substitute the corresponding values for this example into (2.30), we obtain

$$30(f_a - 5) + 17(f_b - 10) + 20(f_c - 0) - 30(d_{w_1} - 5) - 17(d_{w_2} - 10),$$

which, because of the conservation of flow equations, simplifies to

$$30(d_{w_1} - 5) + 17(f_b - 10) + 20(f_c - 0) - 30(d_{w_1} - 5) - 17(f_b + f_c - 10) = 3f_c \geq 0$$

because the path flows and hence the link flows, are always nonnegative.

The variational inequality in path flows [cf. (2.29)] for this problem is also satisfied by the path flow and demand pattern because substitution of the corresponding values into (2.29) yields

$$30(x_{p_1} - 5) + 17(x_{p_2} - 10) + 50(x_{p_3} - 0) - 30(d_{w_1} - 5) - 17(d_{w_2} - 10),$$

which, with use of the conservation of flow equations, simplifies to

$$30(d_{w_1} - 5) + 17(x_{p_2} - 10) + 50x_{p_3} - 30(d_{w_1} - 5) - 17(x_{p_2} + x_{p_3} - 10)$$
$$= 33x_{p_3} \geq 0.$$

2.3 CENTRALIZED DECISION-MAKING AND SYSTEM-OPTIMIZATION

We now consider centralized decision-making by which a central controller seeks to determine the optimal nonnegative flows between O/D pairs of nodes that satisfy the demands, but at minimal total cost. In such problems, there is a single objective function to be optimized, subject to the constraints. As noted earlier, applications of the system-optimized (S-O) problem occur not only in transportation and telecommunication but also in a variety of logistics settings, including military operations, centralized supply chains, and even in the context of humanitarian logistics. We develop specific S-O models, including multiproduct and multicriteria ones, in Chapters 8 through 10.

2.3.1 The System-Optimization (S-O) Models

As in Section 2.2.1, the network $\mathcal{G} = [\mathcal{N}, \mathcal{L}]$, the demands associated with the O/D pairs and the user link cost functions are assumed as given. We first focus on the S-O problem based on separable user link cost functions, that is, $c_a = c_a(f_a)$, $\forall a \in \mathcal{L}$, for definiteness, and then consider more general user link cost functions as in (2.18) and the associated total cost functions as described next.

The total cost on link a, denoted by $\hat{c}_a(f_a)$, is given by

$$\hat{c}_a(f_a) = c_a(f_a) \times f_a, \quad \forall a \in \mathcal{L}, \tag{2.32}$$

that is, the total cost on a link is equal to the user link cost on the link times the flow on the link. As noted earlier, in the system-optimized problem, there exists a central controller who seeks to minimize the total cost in the network system, where the total cost is expressed as

$$\sum_{a \in \mathcal{L}} \hat{c}_a(f_a) \tag{2.33}$$

and the total cost on a link is given by expression (2.32).

The S-O problem is, thus, given by

$$\text{Minimize}_{f \in \mathcal{K}^2} \quad \sum_{a \in \mathcal{L}} \hat{c}_a(f_a). \tag{2.34}$$

The total cost on a path, denoted by $\hat{C}_p$, is the user cost on a path times the flow on a path, that is

$$\hat{C}_p = C_p x_p, \quad \forall p \in P, \tag{2.35}$$

where the user cost on a path, C_p, is given by (2.17). An alternative version of the S-O problem can be stated in path flow variables only, where one has now the problem

$$\text{Minimize}_{x \in \mathcal{K}^1} \quad \sum_{p \in P} C_p(x) x_p. \tag{2.36}$$

2.3.2 System-Optimality Conditions

Under the assumption of convex (strictly convex) cost functions, the objective function (2.34) in the S-O problem is convex (strictly convex). The feasible set $\mathcal{K}^2$ is also convex. Therefore, the optimality conditions, that is, the Kuhn-Tucker conditions (see Dafermos and Sparrow (1969) and the appendix) are: For each O/D pair $w \in W$ and each path $p \in P_w$, the flow pattern x (and the corresponding link flow pattern f), satisfying constraints (2.14) through (2.16), must satisfy

$$\hat{C}'_p(x) \begin{cases} = \mu_w, & \text{if} \quad x_p > 0, \\ \geq \mu_w, & \text{if} \quad x_p = 0, \end{cases} \tag{2.37}$$

where $\hat{C}'_p(x)$ denotes the marginal of the total cost on path p, given by

$$\hat{C}'_p(x) = \sum_{a \in \mathcal{L}} \frac{\partial \hat{c}_a(f_a)}{\partial f_a} \delta_{ap} \tag{2.38}$$

and evaluated in (2.37) at the solution. The term μ_w denotes the Lagrange multiplier associated with constraint (2.14) for that O/D pair w.

Observe that conditions (2.37) may be rewritten so that there exists an ordering of paths for each O/D pair whereby all used paths (i.e., those with positive flow) have equal and minimal marginal total costs and the unused paths (i.e., those with zero flow) have higher (or equal) marginal total costs than those of the used paths. Hence, in the S-O problem, according to the optimality conditions (2.37), it is the marginal of the total cost on each used path connecting an O/D pair that is equalized and minimal.

Example 2.5: A S-O Fixed Demand Example with Separable User Link Cost Functions

We now return to Example 2.1, but we determine the S-O pattern.

The system-optimized path flow pattern satisfying conditions (2.37) is given by $x_{p_1} = 3$, $x_{p_2} = 3$, and $x_{p_3} = 4$, which corresponds to the link flow pattern $f_a = 3$, $f_b = 3$, and $f_c = 4$. The marginals of the total costs on the paths [cf. (2.38)] are

$$\hat{C}'_{p_1} = \hat{c}'_a = 16, \quad \hat{C}'_{p_2} = \hat{c}'_b = 16, \quad \hat{C}'_{p_3} = \hat{c}'_c = 16.$$

Observe that the U-O flow pattern for this problem, which is $f_a^* = 4$, $f_b^* = 2$, and $f_c^* = 4$, is distinct from the S-O problem.

Consider now user link cost functions that are of the general form [refer to (2.18)], where the cost on a link may depend also on the flow on this as well as other flows on the network, that is

$$c_a = c_a(f), \quad \forall a \in \mathcal{L}. \tag{2.39}$$

The system-optimization problem in the case of nonseparable user link cost functions becomes

$$\text{Minimize}_{f \in \mathcal{K}^2} \quad \sum_{a \in \mathcal{L}} \hat{c}_a(f), \tag{2.40}$$

where $\hat{c}_a(f) = c_a(f) \times f_a, \forall a \in \mathcal{L}$.

The system-optimality conditions remain as in (2.37), but now the marginal of the total cost on a path becomes, in the more general case

$$\hat{C}'_p = \sum_{a,b \in \mathcal{L}} \frac{\partial \hat{c}_b(f)}{\partial f_a} \delta_{ap}, \quad \forall p \in P. \tag{2.41}$$

Of course, if the total link cost functions are strictly convex, then there is a unique S-O link flow pattern.

Remark 2.3

One question that naturally arises is, under which user link cost functions, if any, is the S-O solution the same as the U-O solution? This question is interesting because in networks with such cost functions users would behave *individually* in a way that is also *optimal from a societal perspective.* In a general network, for user link cost functions given by

$$c_a(f_a) = t^0_a f^\beta_a, \quad \forall a \in \mathcal{L}, \tag{2.42}$$

with $t^0_a > 0$, for all links $a \in \mathcal{L}$ and with β being a nonnegative constant, the U-O link solution coincides with the S-O link solution. A simple proof follows. The variational inequality formulation of the U-O problem for a network with such user link cost functions (see Theorem 2.8) would take the form: Determine $f^* \in \mathcal{K}^2$, such that

$$\sum_{a \in \mathcal{L}} (t^0_a f^{*\beta}_a) \times (f_a - f^*_a) \geq 0, \quad \forall f \in \mathcal{K}^2. \tag{2.43}$$

Because the total link cost functions are hence given by

$$\hat{c}_a(f_a) = c_a(f_a) \times f_a = t^0_a f^\beta_a \times f_a = t^0_a f^{\beta+1}_a, \quad \forall a \in \mathcal{L},$$

we know that the corresponding variational inequality for the S-O problem (see Proposition 2.2) would be: Determine $f' \in \mathcal{K}^2$, such that

$$\sum_{a \in \mathcal{L}} (\beta + 1)(t^0_a f'^\beta_a) \times (f_a - f'_a) \geq 0, \quad \forall f' \in \mathcal{K}^2. \tag{2.44}$$

Dividing inequality (2.44) through by $(\beta + 1)$, we obtain: Determine $f' \in \mathcal{K}^2$ satisfying

$$\sum_{a \in \mathcal{L}} (t^0_a f'^\beta_a) \times (f_a - f'_a) \geq 0, \quad \forall f' \in \mathcal{K}^2, \tag{2.45}$$

which implies [see also (2.43)] that $f^*_a = f'_a$, for all links $a \in \mathcal{L}$, and the conclusion follows.

Of course, for networks with special structure, the class of functions can be broadened for the above equivalence to hold.

It is also interesting and relevant to observe that networks with user link cost functions of the form (2.42) with $\beta = 0$ are *uncongested* networks because the user cost on each of their links is independent of the flow on the link. Hence, in uncongested networks, the user link cost on any link is not an increasing function of the flow but, rather, is fixed because $c_a = t_a^0$, for all links $a \in \mathcal{L}$. In view of the result given earlier, we know that, in uncongested networks, the S-O link flow pattern coincides with the U-O link flow pattern.

2.4 ALGORITHMS

In this section, we present equilibration algorithms [cf. Dafermos and Sparrow (1969) and Nagurney (1999)]; the projection method [Dafermos (1983)]; the modified projection method due to Korpelevich (1977); and the Euler-type method, which is based on the general iterative scheme of Dupuis and Nagurney (1993).

2.4.1 Equilibration Algorithms

Equilibration algorithms for the solution of network problems with separable and linear "user" link cost functions are presented in this section. We begin with the equilibration algorithm for the U-O single O/D pair problem with fixed demand, and then derive the analogue for the S-O problem. Both are, subsequently, generalized to n_W O/D pairs. The notation for the network problems is as in Sections 2.2.1 and 2.3.1.

The equilibration algorithms identify the most "expensive" used path for an O/D pair and the "cheapest" path and equilibrate the cost for the two paths by reassigning a portion of the flow from the most expensive path to the cheapest path. This process continues until the solution is achieved to a prespecified tolerance. In the case of linear and separable user link cost functions, that is, when the user cost on link a is given by

$$c_a(f_a) = g_a f_a + h_a, \quad g_a, h_a > 0, \quad \forall a \in \mathcal{L}, \tag{2.46}$$

this reassignment or reallocation process can be computed in closed form.

Assume for the time being that there is only a single O/D pair w_i on a given network. A U-O equilibration algorithm is now presented for the computation of the equilibrium path and link flows satisfying conditions (2.19), where the conservation of flow equations (2.14) through (2.16) are also satisfied by the solution pattern.

2.4.1.1 U-O Single O/D Pair Equilibration The statement of this algorithm is as follows.

Step 0: Initialization
Construct an initial feasible path flow pattern, that is, a path flow pattern satisfying (2.14) and (2.15) that induces a feasible link flow pattern through (2.16). Set the iteration counter $k = 1$.

Step 1: Selection and Convergence Verification
Determine

$$r = \{p | \max_p C_p, \quad x_p^{k-1} > 0\};$$

$$q = \{p | \min_p C_p\}.$$

If $|C_r - C_q| \leq \epsilon$, for a prespecified tolerance $\epsilon > 0$, then stop; else, go to Step 2.

Step 2: Computation
Compute the following

$$\Delta' = \frac{[C_r - C_q]}{\sum_{a \in \mathcal{L}} g_a(\delta_{aq} - \delta_{ar})^2} \tag{2.47}$$

$$\Delta = \min\{\Delta', x_r^{k-1}\}.$$

Set

$$x_r^k = x_r^{k-1} - \Delta; \quad x_q^k = x_q^{k-1} + \Delta;$$

$$x_p^k = x_p^{k-1}, \quad \forall p \neq q \cup r.$$

Let $k = k + 1$ and return to Step 1.

2.4.1.2 S-O Single O/D Pair Equilibration In a similar manner, we can construct a S-O analogue as follows. The notation for the most expensive and cheapest path is now within the context of the S-O problem. Hence, w.l.o.g., we retain the path delimiters r and q.

Step 0: Initialization
Construct an initial feasible path flow pattern, that is, a path flow pattern satisfying (2.14) and (2.15) that induces a feasible link flow pattern through (2.16). Set the iteration counter $k = 1$.

Step 1: Selection and Convergence Verification
Determine

$$r = \{p | \max_p \hat{C}'_p, \quad x_p^{k-1} > 0\};$$

$$q = \{p | \min_p \hat{C}'_p\}.$$

If $|\hat{C}'_r - \hat{C}'_q| \leq \epsilon$, for a prespecified tolerance $\epsilon > 0$, then stop; else, go to Step 2.

Step 2: Computation
Compute the following

$$\Delta' = \frac{[\hat{C}'_r - \hat{C}'_q]}{\sum_{a \in \mathcal{L}} 2g_a(\delta_{aq} - \delta_{ar})^2} \tag{2.48}$$

$$\Delta = \min\{\Delta', x_r^{k-1}\}.$$

Set:

$$x_r^{\mathcal{T}} = x_r^{k-1} - \Delta; \quad x_q^k = x_q^{k-1} + \Delta;$$

$$x_p^k = x_p^{k-1}, \quad \forall p \neq q \cup r.$$

Let $k = k + 1$ and return to Step 1.

2.4.1.3 U-O Multiple O/D Pair Equilibration On a network in which there are multiple O/D pairs, the single O/D pair U-O equilibration procedure (see Section 2.4.1.1) is applicable as well. We term Step 1 (without the convergence check) and Step 2 in Section 2.4.1.1 as the equilibration operator $E^{\mathrm{U-O}}_{w_i}$ for a fixed O/D pair w_i.

U-O General Equilibration

Let $E^{\mathrm{U-O}} \equiv (E^{\mathrm{U-O}}_{w_{n_W}} \circ (\ldots (E^{\mathrm{U-O}}_{w_{n_W}}))) \circ \ldots \circ (E^{\mathrm{U-O}}_{w_1} \circ (\ldots \circ (E^{\mathrm{U-O}}_{w_1})))$.

Step 0: Initialization

Construct a feasible path and link flow pattern. Set $\mathcal{T} = 1$, where $\mathcal{T}$ represents an iteration counter.

Step 1: U-O Equilibration

Apply $E^{\mathrm{U-O}}$. Note that $E^{\mathrm{U-O}}$ equilibrates only one pair of paths for an O/D pair at a time and continues working on a particular O/D pair until convergence is achieved for that pair. One then moves to the next O/D pair, and so on.

Step 2: Convergence Verification

If $|C_{r_{w_i}} - C_{q_{w_i}}| \leq \epsilon$; $i = 1, ..., n_W$, for a prespecified tolerance, $\epsilon > 0$, where $r_{w_i} = \{p|\max_{p \in P_{w_i}} C_p, \ x^{\mathcal{T}}_p > 0\}$; $i = 1, ..., n_W$, and $q_{w_i} = \{p|\min_{p \in P_{w_i}} C_p\}$; $i = 1, ..., n_W$, then stop; else, let $\mathcal{T} = \mathcal{T} + 1$ and return to Step 1.

2.4.1.4 S-O Multiple O/D Pair Equilibration On a network in which there are multiple O/D pairs, the single O/D pair S-O equilibration procedure (see Section 2.4.1.2) is also applicable. We term the S-O version of Step 1 (without the convergence check) and Step 2 in Section 2.4.1.2 as the S-O equilibration operator $E^{S-O}_{w_i}$ for a fixed O/D pair w_i.

S-O General Equilibration

Let $E^{\mathrm{S-O}} \equiv (E^{\mathrm{S-O}}_{w_{n_W}} \circ (\ldots (E^{\mathrm{S-O}}_{w_{n_W}}))) \circ \ldots \circ (E^{\mathrm{S-O}}_{w_1} \circ (\ldots \circ (E^{\mathrm{S-O}}_{w_1})))$.

Step 0: Initialization

Initialize the feasible flow pattern as in Step 0 for U-O General Equilibration. Set $\mathcal{T} = 1$, where $\mathcal{T}$ represents an iteration counter.

Step 1: S-O Equilibration

Apply $E^{\mathrm{S-O}}$. Note that $E^{\mathrm{S-O}}$ equilibrates one O/D pair before proceeding to the next O/D pair, and so on.

Step 2: Convergence Verification

If $|\hat{C}'_{r_{w_i}} - \hat{C}'_{q_{w_i}}| \leq \epsilon$; $i = 1, ..., n_W$, for a prespecified tolerance, $\epsilon > 0$, where $r_{w_i} = \{p|\max_{p \in P_{w_i}} \hat{C}'_p, \ x^{\mathcal{T}}_p > 0\}$; $i = 1, ..., n_W$, and $q_{w_i} = \{p|\min_{p \in P_{w_i}} \hat{C}'_p\}$; $i = 1, ..., n_W$, then stop; else, let $\mathcal{T} = \mathcal{T} + 1$ and return to Step 1.

The following convergence theorem is due to Dafermos and Sparrow (1969).

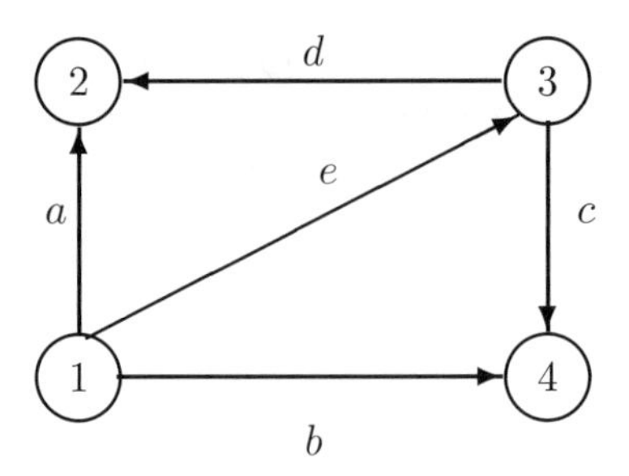

Figure 2.5: Network topology for Example 2.6

Theorem 2.11: Convergence of the Equilibration Algorithms
The above equilibration algorithms are guaranteed to converge, respectively, to the unique U-O flow solution, and to the unique S-O link flow solution, for the given user link cost functions (2.46).

Example 2.6
We now present an example for which we compute both the U-O and the S-O flow patterns using the respective general equilibration algorithm just given. The network topology is depicted in Figure 2.5. The network consists of four nodes and five links. The user link cost functions are

$$c_a(f_a) = 6f_a + 1, \quad c_b(f_b) = f_b + 4, \quad c_c(f_c) = 2f_c + 3,$$

$$c_d(f_d) = 3f_d + 1, \quad c_e(f_e) = 2f_e + 1.$$

The O/D pairs are $w_1 = (1,2)$ and $w_2 = (1,4)$, with demands $d_{w_1} = 40$ and $d_{w_2} = 80$. The paths are $p_1 = a$, $p_2 = (e,d)$, $p_3 = b$, and $p_4 = (e,c)$.

Both the U-O and the S-O general equilibration algorithms were implemented in Fortran and the system used was a Unix system at the University of Massachusetts Amherst.

The computed U-O solution obtained via the U-O general equilibration algorithm: the equilibrium path flows were

$$x^*_{p_1} = 19.7059, \quad x^*_{p_2} = 20.2941, \quad x^*_{p_3} = 72.1176, \quad x^*_{p_4} = 7.8824,$$

and the equilibrium link flows were

$$f^*_a = 19.7059, \quad f^*_b = 72.1176, \quad f^*_c = 7.8824, \quad f^*_d = 20.2941, \quad f^*_e = 28.1765.$$

The user path costs for paths connecting the first O/D pair were $C_{p_1} = C_{p_2} = 119.2353$. The user path costs for paths connecting the second O/D pair were $C_{p_3} = C_{p_4} = 76.1176$.

The S-O solution, in turn, computed via the S-O general equilibration algorithm was: the optimal path flows were

$$x_{p_1} = 19.6569, \quad x_{p_2} = 20.3431, \quad x_{p_3} = 72.1373, \quad x_{p_4} = 7.8627,$$

and the optimal link flows were

$$f_a = 19.6569, \quad f_b = 72.1373, \quad f_c = 7.8627, \quad f_d = 20.3431, \quad f_e = 28.2059.$$

The marginal total costs evaluated at the S-O solution were as follows: For O/D pair w_1

$$\hat{C}'_{p_1} = \hat{C}'_{p_2} = 236.8824;$$

for O/D pair w_2

$$\hat{C}'_{p_3} = \hat{C}'_{p_4} = 148.2745.$$

2.4.2 The Projection Method

The projection method is a special case of the general iterative scheme for the solution of the VI problem (2.1) developed by Dafermos (1983). Its statement is as follows, where $\mathcal{T}$, again, denotes an iteration counter. Note that variational inequality problems are, typically, solved as series of optimization problems.

Step 0: Initialization
Start with an $X^0 \in \mathcal{K}$. Set $\mathcal{T} = 1$.

Step 1: Construction and Computation
Compute $X^{\mathcal{T}}$ by solving the VI subproblem

$$\langle g(X^{\mathcal{T}}, X^{\mathcal{T}-1})^T, X - X^{\mathcal{T}} \rangle \geq 0, \ \forall X \in \mathcal{K}, \tag{2.49}$$

where $g(X, y) = F(y) + \frac{1}{\rho}G(X - y)$, $\rho > 0$, and G is a fixed $(N \times N)$ symmetric positive definite matrix.

Note that at each step $\mathcal{T}$ of the projection method the subproblem (2.49) is equivalent to

$$\text{Minimize}_{X \in \mathcal{K}} \ \frac{1}{2}\langle X^T, GX \rangle + \langle (\rho F(X^{\mathcal{T}-1}) - GX^{\mathcal{T}-1})^T, X \rangle. \tag{2.50}$$

In particular, if G is selected to be a diagonal matrix, then (2.50) is a separable quadratic programming problem.

Step 2: Convergence Verification
If max $|X_l^{\mathcal{T}} - X_l^{\mathcal{T}-1}| \leq \epsilon$, for all l, with $\epsilon > 0$, a prespecified tolerance, then stop; else, set $\mathcal{T} = \mathcal{T} + 1$, and return to Step 1.

Theorem 2.12: Convergence of the Projection Method
Assume that

$$|||I - \rho G^{-\frac{1}{2}} \nabla_X F(X) G^{-\frac{1}{2}}||| \leq 1, \ \forall X \in \mathcal{K}, \tag{2.51}$$

where $0 < \rho \leq 1$ *and is fixed. Then the sequence generated by the projection method (2.49) converges to the solution of variational inequality (2.1).*

2.4.3 The Modified Projection Method

A necessary condition for the convergence of the projection method is that $F(X)$ is strictly monotone. However, such a condition may not be met. Of course, if the projection method converges, then it converges to a solution of (2.1). The modified projection method can be applied to solve VI (2.1), if the function F that enters the variational inequality problem satisfies monotonicity and Lipschitz continuity conditions (provided that a solution exists).

We now present the modified projection method, where $\mathcal{T}$ denotes an iteration counter.

Step 0: Initialization
Set $X^0 \in \mathcal{K}$. Let $\mathcal{T} = 1$ and let α be a scalar such that $0 < \alpha \leq \frac{1}{L}$, where L is the Lipschitz continuity constant [cf. (2.13)].

Step 1: Computation
Compute $\bar{X}^{\mathcal{T}}$ by solving the VI subproblem

$$\langle \bar{X}^{\mathcal{T}} + \alpha F(X^{\mathcal{T}-1}) - X^{\mathcal{T}-1}, X - \bar{X}^{\mathcal{T}} \rangle \geq 0, \quad \forall X \in \mathcal{K}. \tag{2.52}$$

Step 2: Adaptation
Compute $X^{\mathcal{T}}$ by solving the VI subproblem

$$\langle X^{\mathcal{T}} + \alpha F(\bar{X}^{\mathcal{T}}) - X^{\mathcal{T}-1}, X - X^{\mathcal{T}} \rangle \geq 0, \quad \forall X \in \mathcal{K}. \tag{2.53}$$

Step 3: Convergence Verification
If $\max |X_l^{\mathcal{T}} - X_l^{\mathcal{T}-1}| \leq \epsilon$, for all l, with $\epsilon > 0$, a prespecified tolerance, then stop; else, set $\mathcal{T} = \mathcal{T} + 1$, and return to Step 1.

Theorem 2.13: Convergence of the Modified Projection Method
If $F(X)$ is monotone and Lipschitz continuous (and a solution exists), the modified projection algorithm converges to a solution of variational inequality (2.1).

2.4.4 The Euler Method

In this section, we recall the Euler method, which can be applied to solve finite-dimensional projected dynamic systems (PDSs) [see Dupuis and Nagurney (1993) and Nagurney and Zhang (1996); see Sandholm, Dokumaci, and Lahkar (2008) for applications of PDSs in evolutionary game theory]. Due to the equivalence between the set of equilibria of a finite-dimensional PDS and the set of solutions to the corresponding finite-dimensional VI problem [cf. Dupuis and Nagurney (1993)], the algorithm can also be used to compute a solution that satisfies VI (2.1). We now recall the algorithm.

Step 0: Initialization
Set $X^0 \in \mathcal{K}$. Let $\mathcal{T} = 1$ and set the sequence $\{\alpha_{\mathcal{T}}\}$ so that $\sum_{\mathcal{T}=1}^{\infty} \alpha_{\mathcal{T}} = \infty$, $\alpha_{\mathcal{T}} > 0$ for all $\mathcal{T}$, and $\alpha_{\mathcal{T}} \to 0$ as $\mathcal{T} \to \infty$.

Step 1: Computation
Compute $X^{\mathcal{T}} \in \mathcal{K}$ by solving the VI subproblem

$$\langle X^{\mathcal{T}} + \alpha_{\mathcal{T}} F(X^{\mathcal{T}-1}) - X^{\mathcal{T}-1}, X - X^{\mathcal{T}} \rangle \geq 0, \qquad \forall X \in \mathcal{K}. \tag{2.54}$$

Step 2: Convergence Verification
If $|X^{\mathcal{T}} - X^{\mathcal{T}-1}| \leq \epsilon$, with $\epsilon > 0$, a prespecified tolerance, then stop; otherwise, set $\mathcal{T} = \mathcal{T} + 1$, and return to Step 1.

It is worth comparing the iterative step of the Euler method (2.54) with the corresponding step(s) of the modified projection method (2.52) and (2.53). First, the modified projection method has a fixed step size α, whereas the Euler method uses a diminishing step size $\alpha_{\mathcal{T}}$. In addition, the modified projection method focuses on the solution of the VI problem, while the Euler method also provides an alternative discrete-time approximation for the continuous-time PDSs. Moreover, note that the subproblems (2.52), (2.53), and (2.54) are actually optimization problems similar to (2.50). Note that the projection method (2.49) also has what may be interpreted as a fixed step size.

For definiteness, we now recall the definition of a projection, along with a theoretical result, which helps fix the motivation behind the above projection methods.

Proposition 2.3
Let $\mathcal{K}$ be a closed set in R^N. Then for each $X \in R^N$, there is a unique point $y \in \mathcal{K}$, such that

$$\|X - y\| \leq \|X - z\|, \quad \forall z \in \mathcal{K}, \tag{2.55}$$

and y is known as the orthogonal projection of X on the set $\mathcal{K}$ with respect to the Euclidean norm, that is

$$y = P_{\mathcal{K}} X = arg\ min_{z \in \mathcal{K}} \|X - z\|. \tag{2.56}$$

Theorem 2.14: Relationship between the Variational Inequality Problem and a Fixed Point Problem
Assume that $\mathcal{K}$ is closed and convex. Then $X^* \in \mathcal{K}$ is a solution of the variational inequality problem VI $(F, \mathcal{K})$ *if and only if, for any $\gamma > 0$, X^* is a fixed point of the map*

$$P_{\mathcal{K}}(I - \gamma G) : \mathcal{K} \mapsto \mathcal{K},$$

that is

$$X^* = P_{\mathcal{K}}(X^* - \gamma F(X^*)). \tag{2.57}$$

In the subsequent theorem, due to Nagurney and Zhang (1997), the realization of the Euler method [cf. (2.54)] for the solution of the fixed demand network equilibrium problem, along with convergence results, is given.

Theorem 2.15: Convergence of the Euler Method for Fixed Demand Network Equilibrium Problems
Suppose that the user link cost functions c are strictly monotone (increasing) and that the sequence $\alpha_{\mathcal{T}}$ satisfies the conditions in the statement of the Euler method above. Then the Euler method given by

$$x^{\mathcal{T}} = P_{\mathcal{K}^1}(x^{\mathcal{T}-1} - \alpha_{\mathcal{T}} C(x^{\mathcal{T}-1})) \tag{2.58}$$

converges to some network equilibrium path flow pattern satisfying variational inequality (2.20).

For the elastic demand network equilibrium problem with known disutility functions, we have the following result, due to Nagurney and Zhang (1996).

Theorem 2.16: Convergence of the Euler Method for Elastic Demand Network Equilibrium Problems with Given Disutility Functions
Suppose that the user link cost functions c are regular, that is, for every link $a \in \mathcal{L}$

$$c_a(f) \to \infty, \text{ as } f_a \to \infty \tag{2.59}$$

holds uniformly true for all link flow patterns and that the vector c is also strictly monotone (increasing) as is the vector $-\lambda$. Then with the sequence $\{a_{\mathcal{T}}\}$, as in the statement of the Euler method, the realization of the Euler method (due to the simplicity of the underlying feasible set) given by

$$x_p^{\mathcal{T}} = \max\{0, x_p^{\mathcal{T}-1} + \alpha_{\mathcal{T}}(\lambda_w(d^{\mathcal{T}-1}) - C_p(x^{\mathcal{T}-1}))\}, \quad \forall p \in P \tag{2.60}$$

converges to some network equilibrium path flow pattern satisfying variational inequality (2.29).

We now present an elastic demand network equilibrium numerical example, which is solved by the Euler method using the explicit formulae for the path flows at each iteration given by (2.60).

Example 2.7
Consider the network given in Figure 2.6 consisting of three nodes and four links. There is a single O/D pair $w_1 = (1, 3)$ with four paths $p_1 = (a, c)$, $p_2 = (a, d)$, $p_3 = (b, d)$, and $p_4 = (b, c)$. The user link cost functions are

$$c_a(f) = 5f_a + 2f_b + 5, \quad c_b(f) = 4f_b + f_a + 9,$$

$$c_c(f) = 7f_c + f_d + 4, \quad c_d(f) = 5f_d + 3f_c + 12,$$

and the disutility function is

$$\lambda_{w_1}(d) = -2d_{w_1} + 107.$$

The Euler method [cf. (2.60)] was implemented in Fortran and a Unix system at the University of Massachusetts Amherst used for the computations. The algorithm was initialized with all path flows and demand equal to zero. We set the sequence

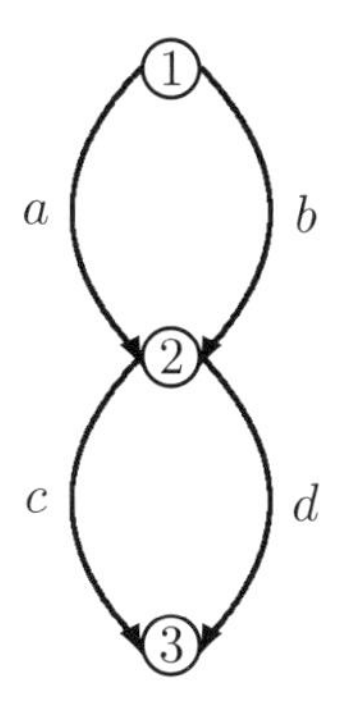

Figure 2.6: Network topology for Example 2.7

$\{\alpha_\mathcal{T}\}$=$.1\{1, \frac{1}{2}, \frac{1}{2}, \frac{1}{3}, \frac{1}{3}, \frac{1}{3}, ...\}$. The convergence tolerance $\epsilon = .0001$. The Euler method converged in 65 iterations, yielding the following equilibrium path flow pattern

$$x^*_{p_1} = 2.50, \quad x^*_{p_2} = 1.50, \quad x^*_{p_3} = 3.50, \quad x^*_{p_4} = 2.50,$$

with the corresponding equilibrium link flow pattern, which is unique

$$f^*_a = 4.00, \quad f^*_b = 6.00, \quad f^*_c = 6.00, \quad f^*_d = 4.00,$$

and unique equilibrium demand $d^*_{w_1} = 10.00$. The user path costs are $C_{p_1} = C_{p_2} = 87.00$, and these costs are equal to the disutility λ_{w_1}. Hence the network equilibrium conditions (2.28) are, indeed, satisfied.

For completeness, in the next theorem, we provide the realization of the Euler method for the solution of the network equilibrium problem with known demand functions, along with convergence results. The theorem is due to Zhang and Nagurney (1997).

Theorem 2.17: Convergence of the Euler Method for Elastic Demand Network Equilibrium Problems with Given Demand Functions
Suppose that the link costs c are strictly monotone increasing in link flows and that the demands d are strictly monotone decreasing in the disutilities λ. In addition, assume the following: There exist sufficiently large constants: M_d, M_x, and M_λ, such that, for any w and p

$$d_w(\lambda) \leq M_d, \quad \forall \lambda \in R^{nW}_+,$$

$$\lambda_w \leq C_p(x), \quad \textit{if} \quad x_p \geq M_x,$$

$$d_w(\lambda) \leq \sum_{p \in P_w} x_p, \quad \textit{if} \quad \lambda_w \geq M_\lambda.$$

Then the Euler method, given by

$$x^\mathcal{T}_p = \max\{0, x^{\mathcal{T}-1}_p + \alpha_\mathcal{T}(\lambda^{\mathcal{T}-1}_w - C_p(x^{\mathcal{T}-1}))\}, \quad \forall p \in P, \tag{2.61}$$

$$\lambda_w^{\mathcal{T}} = \max\{0, \lambda_w^{\mathcal{T}-1} + \alpha_{\mathcal{T}}(d_w(\lambda^{\mathcal{T}-1}) - \sum_{p \in P_w} x_p^{\mathcal{T}-1})\}, \quad \forall w \in W, \tag{2.62}$$

produces sequences $\{x^{\mathcal{T}}, \lambda^{\mathcal{T}}\}$*, which converge to some network equilibrium path flow, associated demand, and disutility pattern satisfying variational inequality (2.25).*

Example 2.7 Revisited

We return now to Example 2.7 but we considered that the demand function was given. Hence we inverted the disutility function for w_1 and obtained the corresponding demand function

$$d_{w_1}(\lambda) = -.5\lambda_{w_1} + 53.50.$$

We retained the user link cost functions as in Example 2.7 and implemented the Euler method in Fortran for the network equilibrium model with known demand functions [cf. (2.61) and (2.62)]. We initialized the algorithm with path flows set to zero and the initial disutility was also set to zero. The convergence tolerance ϵ was as in Example 2.6 but note that now successive computed path flows and the disutility had to satisfy this convergence tolerance for convergence to be achieved. We used the same $\{\alpha_{\mathcal{T}}\}$ sequence as in Example 2.6. As expected, given the above theoretical results, the Euler method converged to the identical equilibrium link flow pattern as obtained for Example 2.6 by the Euler method for known disutility functions.

Remark 2.4

Note that the realizations of the Euler method for the computation of solutions to the fixed demand and elastic demand network equilibrium problems can be interpreted as discrete-time adjustment processes.

2.4.4.1 An Exact Equilibration Algorithm For completeness and because it will also be utilized in Chapter 6 for the solution of financial network subproblems, we now provide an explicit statement of an exact equilibration algorithm [see Dafermos and Sparrow (1969) and Nagurney (1999)]. It can be used to solve the subproblems with a special network structure encountered in (2.58). Indeed, expression (2.58), due to the definition of the projection in (2.56), is equivalent to the following quadratic programming problem: Determine the vector of path flows $x^{\mathcal{T}}$ according to

$$x^{\mathcal{T}} = \min_{x \in \mathcal{K}^1} \frac{1}{2}\langle x^T, x\rangle - \langle (x^{\mathcal{T}-1} - \alpha_{\mathcal{T}} C(x^{\mathcal{T}-1}))^T, x\rangle. \tag{2.63}$$

In view of the feasible set $\mathcal{K}^1$, problem (2.63), in turn, can be decomposed into as many subproblems as there are O/D pairs in the network problem, each of which is a quadratic programming problem with a special structure that can be solved exactly and in closed form using exact equilibration, which was proposed by Dafermos and Sparrow (1969) and also noted by Nagurney and Zhang (1996). In particular, problem (2.63) is equivalent to solving the following: For each O/D pair w, compute

$$\text{Minimize} \quad \frac{1}{2}\sum_{p \in P_w} x_p^2 + \sum_{p \in P_w} h_p^{\mathcal{T}} x_p \tag{2.64}$$

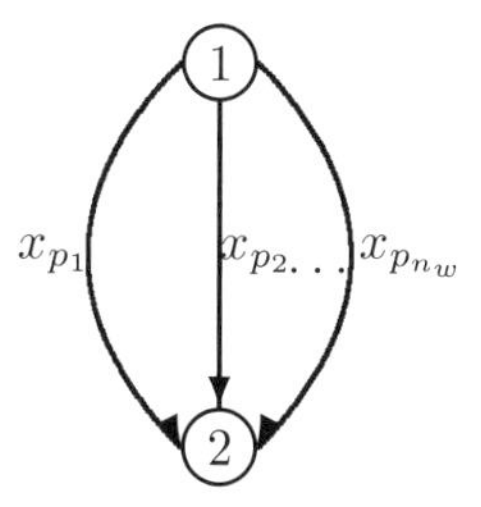

Figure 2.7: Network with special structure

subject to

$$d_w = \sum_{p \in P_w} x_p, \tag{2.65}$$

$$x_p \geq 0, \quad \forall p \in P_w, \tag{2.66}$$

where

$$h_p^{\mathcal{T}} = \alpha_{\mathcal{T}} C_p(x^{\mathcal{T}-1}) - x_p^{\mathcal{T}-1}. \tag{2.67}$$

Observe that this problem for each O/D pair w is over a network of special structure, in view of constraints (2.65) and (2.66). In particular, the specially structured network has disjoint paths, that is, the paths have no links in common. This is a notable feature of the Euler method for fixed demand network problems in path flow variables.

We now state the exact equilibration algorithm for which the iteration counter $\mathcal{T}$ is suppressed, for simplicity. It can be used to solve (2.64) through (2.67) and can hence be embedded in the Euler method for fixed demand network problems (discussed earlier) to compute network equilibrium solutions. The special network structure, in turn, is depicted in Figure 2.7.

An Exact Equilibration Algorithm for O/D pair w

Step 0: Sort

Sort the fixed cost terms, $h_p; p \in P_w$, in nondescending order, and relabel accordingly. Assume, from this point on, that they are relabeled. Set $h_{p_{n_w}+1} = \infty$, where n_w denotes the number of paths in O/D pair w. Set $p = 1$.

Step 1: Computation

Compute

$$\lambda_w^p = \frac{\sum_{i=1}^{p} h_p + d_w}{p}. \tag{2.68}$$

Step 2: Evaluation

If $h_p < \lambda_w^p \leq h_{p+1}$, then stop; set $q = p$, and go to Step 3. Otherwise, let $p = p + 1$, and return to Step 1.

Step 3: Update
Set

$$x_p = \lambda_w^q - h_p, \quad p = 1, \dots, q; \tag{2.69}$$

$$x_p = 0, \quad p = q+1, \dots, n_w. \tag{2.70}$$

Throughout this book we include, as appropriate, instances of the implementation of specific algorithms, whether done in Fortran or MATLAB (www.mathworks.com), to demonstrate the flexibility of the various algorithmic schemes and the ease of their implementation. In addition, for a wide range of numerical examples we also provide information about the initialization of the algorithms and the convergence tolerances, for reproducibility purposes.

2.5 SOURCES AND NOTES

In this chapter we laid the foundations for the subsequent chapters in the book. The appendix contains additional mathematical fundamentals with a focus on optimization theory. Proofs of the variational inequality theorems in this chapter can be found in Kinderlehrer and Stampacchia (1980) and Nagurney (1999). This chapter contains examples to help the reader familiarize himself or herself with the material. The explicated algorithms have the advantage of being easy to implement, have been used to solve numerous network problems, and are also used in this book.

Clearly, the overall efficiency of a given variational inequality algorithm will depend on the optimization algorithm used to solve the encountered subproblems [cf. (2.50), (2.52), (2.53), and (2.54)]; hence it is important that the network structure of the underlying problems be exploited for efficiency purposes. Of course, if the feasible set $\mathcal{K}$ underlying the variational inequality problem is simply the nonnegative orthant, then the solutions to (2.50), (2.52), (2.53), and (2.54) can be obtained via explicit formulae (as was done in (2.60), and in (2.61) and (2.62), for the Euler method for the respective network equilibrium model with elastic demands).

Indeed, in practice, and as mentioned earlier, network problems may be (very) large-scale. For example, in the context of transportation networks, the subject of Chapter 4, available datasets for urban transportation networks may consist of tens of thousands of nodes and tens of thousands of links; see the datasets maintained and made available by Bar-Gera (2008). For such large-scale networks, paths joining a given O/D pair may be generated as needed using, for example, column generation techniques [Leventhal, Nemhauser, and Trotter (1973)].

Remark 2.1 merits further elaboration. We have chosen to present, for simplicity and clarity of exposition, the network models in this chapter for single-modal, single-class networks. In the context of transportation networks, for example, single-modal would correspond to a single mode of transportation and, similarly, single-class would correspond to a single class of network user. However, we emphasize that the framework is sufficiently general to also capture multimodal/multiclass interactions on networks. In multimodal networks, there would be different modes of transport, for example, or different classes of users, with each one perceiving the cost on the links in a particular way. Hence in multimodal/multiclass networks, each mode or class of user would be characterized by mode or class specific link cost functions to reflect the perception of cost associated with traversing a link. One could explicitly state the models in such a format, which would be done, for example, in the elastic demand network equilibrium case [see, e.g., Dafermos (1982)] by constructing multidimensional vectors of the notation to correspond to multiple modes or classes. Of course, the network equilibrium conditions would have to hold then for each class or mode of user in the network. As Remark 2.1 states, one can transform multimodal and multiclass networks into single-modal/single-class, but, extended, networks by making as many copies of the network as there are modes or classes and by defining the demands, costs, and flows accordingly on the expanded network(s). The cost on a link would then generally depend not just on its own flow.

The result in Remark 2.3 was originally noted in Dafermos and Sparrow (1969). Here we provide a proof using variational inequality theory, which makes the result more transparent.

Those interested in further background reading on the traffic assignment or transportation network equilibrium problem, along with additional models and algorithms, are referred to Patriksson (1994). Additional references to and background on projected dynamical systems and variational inequality problems can be found in Nagurney and Zhang (1996). For those interested in a focused book on dynamic transportation networks, see Ran and Boyce (1996). For a historical perspective on user-optimized problems and algorithms, see Boyce, Mahmassani, and Nagurney (2005).

CHAPTER 3

NETWORK PERFORMANCE MEASUREMENT AND ROBUSTNESS ANALYSIS

Who is the most important individual in an organization? Which manufacturing plant is the most critical to a supply chain? Which bank, if it were to fail, would cause the greatest economic impact? Which bridge or road if it were to deteriorate or become untraversable would affect the vehicular flows in the immediate region (and beyond) the most? Which Internet link, if destroyed, would disrupt communications in a major way? Such questions in seemingly disparate systems actually have much in common, and their answers become possible once one identifies not only the underlying network structure of the given system, but, also, the appropriate network performance measure. Indeed, it is only when the performance of a network can be adequately assessed that can one begin to identify which nodes and links truly matter in that their deterioration or outright removal will make the greatest impact. The importance identification and the rankings of nodes and links in a network provide essential and invaluable information to decision-makers, including executives, managers, planners, government officials, network designers, engineers, and policymakers. Such crucial information, obtained in a theoretically rigorous and quantitative way, allows decision-makers to determine a priori which network components should be better maintained and secured because their reduction in service (or outright elimination due to structural failures, natural disasters, planned attacks, etc.) will affect network performance to a greater degree than that of other network components.

In this chapter, we provide a unified approach to network performance and efficiency measurement. Our perspective explicitly acknowledges that different networks may operate under different decision-making constructs, as revealed in Chapter 2. In networks that are under centralized operation, one has outright control of the flows. This may be the underlying behavioral mechanism, for example, in specific freight networks and in supply chain networks in which a firm owns its own manufacturing plants and distribution centers, and supplies the particular retailers with its products. In the former application, the central controller may route the freight flows through the network in a system-optimizing fashion. In the latter application, the product flows may be allocated/routed through the various economic link activities to minimize total cost while satisfying the demand for the products.

On the other hand, in congested urban transportation networks, travelers select their routes of travel between origin and destination pairs of nodes unilaterally in a decentralized manner until they, acting independently, cannot improve on their situation, as measured by the travel time or cost on used paths. The underlying behavioral principle, as discussed in Chapter 2, would hence be that of user-optimization (U-O) or network equilibrium, and the network would operate under a decentralized decision-making principle. Furthermore, in the case of economic decision-makers interacting with one another, there may be competition across a tier of the network (as in financial networks) but cooperation between tiers of the network; similar behavior may be associated with decision-makers in supply chains who own either manufacturing plants or distribution centers, for example, and must interact with one another so that the consumers at the demand markets are ultimately supplied with the products.

Consequently, in network systems, there may exist 1). a single, centralized decision-maker; 2). numerous decision-makers, acting individually in a decentralized manner, as in large-scale urban transportation networks; or 3). a smaller, finite number of decision-makers, such as firms, interacting, and competing and cooperating with one another, as appropriate, in a given network setting. The decision-makers or agents are characterized by their own respective objectives, their respective constraints, and their behavior results in flows on the networks with associated incurred costs, profits, etc. Hence, to appropriately assess the performance of a network one must capture not only the underlying network topology, that is, the nodes and links and their connections, but also the underlying behavioral and economic constructs and the induced flows. The graph alone, consisting solely of nodes and links and representing an organization, a transportation, or logistical network such as a supply chain, a telecommunication network, or even an electric power network or financial network is not sufficient to capture the performance of a complex network.

This chapter uses the fundamental models of network decision-making described in Chapter 2. In Section 3.1, we present some preliminaries in terms of network performance measurement and network component importance identification. In Section 3.2, we introduce a novel unified (in that it can handle either fixed demands or elastic demands) network performance measure, based on decentralized decision-making under network equilibrium that can be used to assess the efficiency/performance of a network in the case of either fixed or elastic demands. Such a measure is needed for many different applications, ranging from transportation networks, which we explore in Chapter 4, to decentralized supply chain networks, which we study in Chapter 5, and even in financial networks, the subject of Chapter 6. Again, we emphasize that only when the performance of a network can be quantifiably measured can the network be appropriately managed. Moreover, as we demonstrate in this chapter, the unified network performance measure, which captures flow information and behavior, allows one to determine the criticality of various nodes (as well as links) through the identification of their importance and ranking, under decentralized decision-making behavior. Furthermore, we present specific networks for which the network efficiency is computed along with the importance rankings of the nodes and links. The ranking results are compared to those using the measure of Latora and Marchiori (2001, 2002, 2003, 2004), who investigated the MBTA subway system in Boston and the Internet, and the advantages demonstrated.

We then turn, in Section 3.3, to the examination of *network robustness*, within the framework of the performance measure introduced in Section 3.2. The concept of *system robustness* has been studied in both computer science and in engineering. According to the Institute of Electrical and Electronic Engineers (1990), robustness can be defined as "the degree to which a system or component can function correctly in the presence of invalid inputs or stressful environmental conditions." Gribble (2001) defined system robustness as "the ability of a system to continue to operate correctly across a wide range of operational conditions, and to fail gracefully outside of that range." Ali et al. (2003) considered an allocation mapping to be robust if it "guarantees the maintenance of certain desired system characteristics despite fluctuations in the behavior of its component parts or its environment." Schillo et al. (2001) argued that robustness has to be studied "in relation to some definition of performance measure." According to Holmgren (2007): "Robustness signifies that the system will retain its system structure (function) intact (remain unchanged or nearly unchanged) when exposed to perturbations."

In addition, the physics research on complex networks has also examined network robustness according to different network measures and the accompanying degradation of network performance in the presence of attacks on the network; see, for example, Albert, Jeong, and Barabási (2000). However, the focus of that research has been on the impact of the removal of nodes on networks, whereas in this chapter, in terms of network robustness, we focus on the degradation of links through reductions in their capacities and the effects on the induced costs under different functional forms for the links. Hence, when we formalize the measures for network robustness, we are not concerned directly with extreme events that may lead to the removal of nodes and links from the network but, rather, with the deterioration of the network infrastructure, such as roads, Internet links, electric power transmission lines, etc., through changes in the link practical capacities. Of course, we also assess and quantify, in Section 3.2, the outright removal of nodes and links, which may occur because of extreme events, and the associated impacts in terms of performance degradation.

It is worth noting that there is a literature in robust optimization, which is a mathematical approach to deal with uncertainty and, in particular, when a problem's data may be known only within certain bounds. Robustness is a well-known concept in control, and the subject of robust optimization dates to the pioneering work of Soyster (1973) and Ben-Tal and Nemirovsky (1998, 1999). Here, however, we deal with specific link functional forms, which are assumed to be known, and we vary the link capacities.

Specifically, we propose, in Section 3.3, a network robustness measure based on the decentralized network performance measure of Section 3.2, by assuming that links in the particular network are characterized by user link cost functions in which the capacities of each link are embedded in the link cost function. Such link cost functions are used both in transportation networks and in telecommunication networks. We also derive some lower bounds for the robustness of certain networks of special structure.

Subsequently, in Section 3.4, we study network robustness from a different perspective, that of *total network cost*. In particular, we construct relative total cost indices that can be used to assess network robustness, in this context, and which can be evaluated at either U-O or S-O solution flows. For completeness, we also identify the relationship between the ratio of the proposed indices and the *price of anarchy* [cf. Roughgarden (2005)], which has received much attention in both the computer science and transportation science literature, in which networks play a pivotal role in problem formulation and solution.

Throughout this chapter, we provide illustrative examples, including variants of the classical Braess paradox (1968) example, to cement the concepts and tools.

3.1 SOME PRELIMINARIES AND NETWORK CENTRALITY MEASURES

As noted in the complex network literature by Barrat et al. (2004), "The identification of the most central nodes in the system is a major issue in network characterization." The centrality property of a network has important implications for determining network vulnerability. Only when we are furnished with a better understanding of the critical network components can we protect the network more effectively.

Many studies have addressed centrality in networks beginning with contributions in sociology. Specifically, the *degree, shortest-path betweenness*, and *closeness* have been widely applied node centrality measures [cf. Freeman (1979)]. The node degree is the number of links connecting with the node. The shortest-path betweenness of a particular node, in turn, is the ratio between the number of the shortest paths that use that particular node and the total number of shortest paths in a network. The larger the degree or the betweenness of a node, it has been argued, the more important that particular node is to the network. However, the degree centrality is considered to be a local measure because it is determined only by the number of its neighbors [Koschützki et al. (2005)]. The closeness of a node, on the other hand, is the reciprocal of the mean geodesic distance (i.e., the shortest path) between a node and all other nodes reachable from that node. A small value of the closeness indicates the central role of a node in a network. Another interpretation of this type of closeness is that, given shorter distances from a node to other nodes, the easier it is for communication purposes.

Besides these commonly used centrality measures, there are other centrality measures that focus on different aspects of a network. According to *eigenvector centrality* [Bonacich (1972)], a node acquires a high eigenvector centrality either by being connected to a lot of others (as with degree centrality) or by being connected to others that themselves are highly central. *Flow centrality* [cf. Freeman, Borgatti, and White (1991)] and *betweenness centrality using flows* [see Izquierdo and Hanneman (2006)] evaluate the importance of a node by measuring the amount of flow passing through the node, assuming a maximum flow network model. Although the flow centrality measure takes into consideration network flows, it does not capture behavior and link flow adjustment *after* the removal of network components (as our network measure does). Moreover, to extend the centrality study to edges/links in networks, *edge betweenness* has been applied to study social and biological networks [see Girvan and Newman (2002) and Holme and Kim (2002)]. Similar to node betweenness, edge betweenness is the number of geodesic shortest paths between nodes that pass along the edge of the network.

All the above noted centrality measures are for nonweighted networks, that is, networks that use solely binary variables to indicate the existence of a connection between a pair of nodes with no associated weights on the links. Recently, some complex network researchers have argued that different weights should be assigned on each link joining a pair of nodes to capture the relationships and flows between different nodes [cf. Yook et al. (2001), Barrat et al. (2004), Newman (2004), and Barrat, Barthélémy, and Vespignani (2005)]. These types of networks are referred to as *weighted networks*. As indicated by Barrat et al. (2004), "networks are specified not only by their topology but also by the dynamics of information or traffic flow taking place on the structure. The amount of traffic characterizing the connections in communication systems and large transport infrastructures is fundamental for a full description of these networks."

Nevertheless, there exist very few centrality measures for weighted networks. Newman (2004) proposed a simple mapping method to adapt the existing centrality measures for nonweighted networks to that of weighted networks. Dall'Asta et al. (2006) defined the weighted betweenness centrality measure by including such information as the traffic between each pair of nodes in a network. However, although claimed by the authors that the weighted betweenness centrality consists of economic factors, the measure does not include decision-maker behavior and cost information, which are the important properties of many real-world large-scale networks. Because our unified network measure will be compared to the network efficiency measure of Latora and Marchiori (2001, 2002, 2003, 2004), we now briefly review their measure.

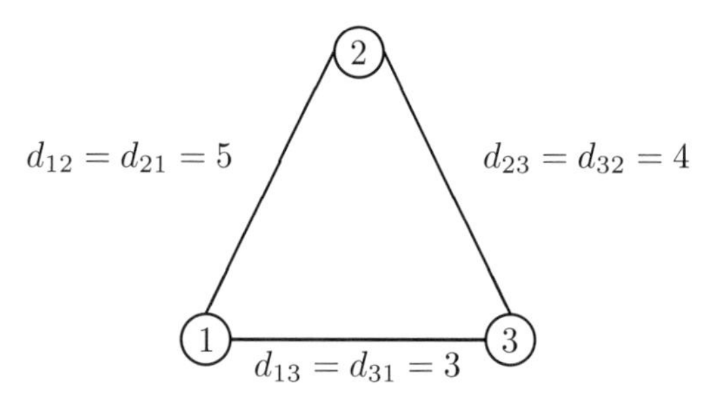

Figure 3.1: Network topology for Example 3.1

The Latora and Marchiori Network Measure

The Latora and Marchiori network efficiency measure, $E(\mathcal{G})$, of a network $\mathcal{G}$ is defined as

$$E = E(\mathcal{G}) = \frac{1}{n(n-1)} \sum_{i \neq j \in \mathcal{G}} \frac{1}{d_{ij}}, \tag{3.1}$$

where n is the number of nodes in $\mathcal{G}$ and d_{ij} is the shortest path length (the geodesic distance) between nodes i and j. For simplicity, in this chapter we refer to the Latora and Marchiori measure as E.

The E measure is a network centrality measure that examines how the function of a network deteriorates once a certain network component is removed from the network. Note that according to (3.1) a network is considered to be more efficient and hence to have a better performance if the distances between pairs of nodes are smaller. Typically, a given link need not be directed, that is, represented by an arrow, and hence the distance $d_{ij} = d_{ji}$ for all pairs of distinct nodes (i, j).

Example 3.1

As an illustration, we now present a small example. For the network in Figure 3.1 (with distances denoted above the particular links), the efficiency $E = \frac{\frac{1}{5}+\frac{1}{5}+\frac{1}{3}+\frac{1}{3}+\frac{1}{4}+\frac{1}{4}}{6} = \frac{47}{180} = .2611$. The degree of each node in this network is 2 because each node has two links connected to it. Therefore, all three nodes are equally important in this network in terms of the degree measure. The shortest-path betweenness for each node is 0 because each node is directly connected to every other node (without using any intermediate nodes). All three nodes hence are also equally important when evaluated by the shortest-path betweenness measure. The closeness measure, in turn, for the nodes is $\frac{5+3}{2} = 4$ for node 1; $\frac{5+4}{2} = 4.5$ for node 2, and $\frac{3+4}{2} = 3.5$ for node 3. Node 3 has the closest average distance to the other nodes in the network; consequently, it is the most important node if the closeness measure is used.

Besides using network efficiency to assess the performance of a network, closeness can also be used to evaluate the functionality of a network [cf. Koschützki et al. (2005)]. However, such an approach is based on the assumption that the network is connected after the removal of a network component. We do not make such a restrictive assumption in the definition of our unified network performance measure for decentralized networks. Indeed, following a disaster, be it a natural one or not, one may expect that pairs of origin and destination nodes may very well become disconnected.

3.2 A UNIFIED NETWORK PERFORMANCE MEASURE BASED ON DECENTRALIZED DECISION-MAKING

Here we propose a new, unified network performance/efficiency measure that, as we will demonstrate, is applicable and relevant to the case of either fixed demands or elastic demands. This measure assumes decentralized decision-making behavior on the network.

Definition 3.1: A Unified Network Performance Measure
The network performance/efficiency measure, $\mathcal{E}(\mathcal{G}, d)$, for a given network topology $\mathcal{G}$ and the fixed (or, in the elastic case, equilibrium) demand vector d, is defined as follows

$$\mathcal{E} = \mathcal{E}(\mathcal{G}, d) = \frac{\sum_{w \in W} \frac{d_w}{\lambda_w}}{n_W}, \tag{3.2}$$

where n_W is the number of O/D pairs in the network, and d_w and λ_w denote, for simplicity, the fixed (or equilibrium) demand and the equilibrium disutility for O/D pair w, respectively (see Section 2.2).

The equilibrium disutilities λ_w, $w \in W$, are assumed to be positive in both fixed-demand and elastic-demand networks, which is reasonable for most networks with commonly used link cost functions. Therefore, the network performance measure in (3.2) is well-defined.

For concreteness, we now interpret the measure $\mathcal{E}$ given by (3.2) in terms of transportation networks. The demand d_w is measured over a period of time, such as an hour, whereas λ_w is the minimum equilibrium travel cost (or time) associated with the O/D pair w. Suppose that we have only a single O/D pair w in the network, and that the $d_w = 100$ vehicles with $\lambda_w = .5$ hour. Then $\mathcal{E} = 200$ (vehicles/hour). Consequently, the network can process, in effect, 200 vehicles in the hour. If λ_w was, instead, 1 hour, then the efficiency $\mathcal{E}$ would be 100 (vehicles/hour), and this network would be half as efficient as the original network. Depending on the network under consideration, the unit of measurement would correspond to the type of flow on the network. For general networks, the performance/efficiency measure $\mathcal{E}$ defined in (3.2) is actually the average demand to price ratio. When $\mathcal{G}$ and d are fixed, a network is more efficient if it can satisfy a higher demand at a lower price.

It is interesting that we demonstrate in the following theorem that, under certain assumptions, the measure $\mathcal{E}$ collapses to the E measure (3.1), which, however, considers neither explicit demands nor flows!

Theorem 3.1

If positive demands exist for all pairs of nodes in the network $\mathcal{G}$, and each of these demands is equal to 1 *and if* d_{ij} *is set equal to* λ_w, *where* $w = (i, j)$, *for all* $w \in W$ *then the* $\mathcal{E}$ *measure (3.2) and the* E *measure (3.1) are one and the same.*

Proof: Let n be the number of nodes in $\mathcal{G}$. Hence the total number of O/D pairs, n_W, is equal to $n(n-1)$ given the assumption that there exist positive demands for all pairs of nodes in $\mathcal{G}$. Furthermore, by assumption, we have $d_w = 1$, $\forall w \in W$, $w = (i, j)$, and $d_{ij} = \lambda_w$, where $i \neq j$, $\forall i, j \in \mathcal{G}$. Also, an implicit assumption is that each pair of nodes is connected by a directed link and $d_{ij} = d_{ji}; i \neq j$, $\forall i, j \in \mathcal{G}$. Then the E measure becomes as follows.

$$E = E(\mathcal{G}) = \frac{1}{n(n-1)} \sum_{i \neq j \in \mathcal{G}} \frac{1}{d_{ij}} = \frac{\sum_{i \neq j \in \mathcal{G}} \frac{1}{d_{ij}}}{n_W} = \frac{\sum_{w \in W} \frac{d_w}{\lambda_w}}{n_W} = \mathcal{E}. \tag{3.3}$$

The conclusion, thus, follows.

Note that [see (2.19), (2.23), and (2.28)] according to its definition λ_w is the equilibrium disutility or, given the network equilibrium conditions, in effect, the value of the "shortest path" for O/D pair w (assuming that at least one path is used), and d_{ij} is the shortest path length (the geodesic distance) between nodes i and j. Therefore, the assumption of d_{ij} being equal to λ_w is not unreasonable. The $\mathcal{E}$ measure, however, is a more general measure because it also captures the flows on the network through the disutilities, costs, and the demands.

Furthermore, we note that in the E measure, there is no information regarding the demand for each O/D pair. Therefore, $n(n-1)$ can be interpreted as the total possible number of O/D pairs regardless of whether there exists a demand for a pair of nodes or not. However, because our measure $\mathcal{E}$ is an average network efficiency measure, it does not make sense to count a pair of nodes that has no associated demand in the computation of the network efficiency. Therefore, the number of O/D pairs, n_W, is more appropriate as a divisor in the $\mathcal{E}$ measure than $n(n-1)$. Of course, if there is a positive associated demand between all pairs of nodes in the network then $n_W = n(n-1)$.

In the $\mathcal{E}$ measure (3.2), the elimination of a link is treated by removing that link from the network while the removal of a node is managed by removing the links entering or exiting that node. If the removal results in no path connecting an O/D pair, we simply assign the demand for that O/D pair (either fixed or elastic) to an abstract path with a cost of infinity.

3.2.1 A Desirable Property of the Network Performance Measure

Before we present the definition of importance of a network component, we state a desirable and important property of any network performance measure in the case of elastic (or fixed) demands and decentralized, user-optimizing behavior.

Desirable Network Performance Property
The performance/efficiency measure for a given network should be nonincreasing with respect to the equilibrium disutility for each O/D pair, holding the equilibrium disutilities for the other O/D pairs constant.

In a network with elastic demands, when there is a disconnected O/D pair w, we have, from the above discussion, that the associated "path cost" of the abstract path, say, r, $C_r(x^*)$, is equal to infinity. If the disutility functions are known, as discussed in Section 2.2 of Chapter 2, according to equilibrium condition (2.28), we then have $C_r(x^*) > \lambda_w(d^*)$, and hence $x_r^* = 0$, so that $d_w^* = 0$, which leads to the conclusion $d_w^*/\lambda_w = 0$. Therefore, the disconnected O/D pair w makes a zero "contribution" to the efficiency measure and the $\mathcal{E}$ measure is well-defined in both the fixed and elastic demand cases. The above procedure(s) for handling disconnected O/D pairs will be illustrated by examples in Sections 3.2.4 and 3.2.5, when we compute the importance of the network components and their rankings.

We believe that this feature of the unified performance measure is essential. In reality, it is relevant to investigate the efficiency of a large-scale network even in the case of disconnected O/D pairs. A measure with such adaptability and flexibility can enable the study of the performance of a wider range of networks, especially when evaluating networks under disruptions. Moreover, it also allows us to investigate the criticality of various network components without worrying about the connectivity assumption. Notably, Latora and Marchiori (2001) also mentioned this important characteristic, which gives their measure an attractive property over the measure used for the small-world model [cf. Watts and Strogatz (1998)].

A network, be it a transportation network or a supply chain network or an economic/financial network, is characterized by its topology, its vector of demands (either fixed or elastic), and associated costs. To evaluate the importance of nodes and links of a network, the examination of only the topology of the network is insufficient. A reasonable measure should capture the efficiency deterioration with the increase of path costs in a network. Let us examine whether the measure in (3.2) has such a feature. Assume that the disutility functions are, again, known. The disutility function for each $w \in W$ is assumed to depend, for the sake of generality, on the entire demand vector. With the assumption of the equilibrium disutilities for all the other O/D pairs being held constant, the partial derivative of $\mathcal{E}(\mathcal{G}, d)$ in (3.2) in regard to λ_w for the network with elastic demands is then given as follows

$$\frac{\partial \mathcal{E}(\mathcal{G}, d)}{\partial \lambda_w} = \frac{\frac{-d_w}{(\lambda_w(d))^2} + \sum_{v \in W} \frac{(\frac{\partial \lambda_w(d)}{\partial d_v})^{-1}}{\lambda_v(d)}}{n_W}. \tag{3.4}$$

Given the assumptions that $d_w \geq 0$, $\lambda_w > 0$, and $\frac{\partial \lambda_w}{\partial d_v} < 0$, $\forall v \in W$, it is obvious that $\mathcal{E}(\mathcal{G}, d)$ in (3.2) is a nonincreasing function of λ_w, $\forall w \in W$.

Hence the unified network performance/efficiency measure $\mathcal{E}$ has the desirable property discussed earlier.

3.2.2 The Importance of Network Components

With the network performance/efficiency measure $\mathcal{E}$, we are ready to investigate the importance of network components by studying their effect on network efficiency through their removal. Network efficiency can be expected to deteriorate when a critical network component is eliminated from the network. Such a component can include a link or a node or a subset of nodes and links, depending on the network problem under study. Furthermore, the removal of a critical network component will cause a more severe impact than that of a trivial one. Hence similar to the definition of importance of network components in Latora and Marchiori (2001), we define the importance of a network component as follows.

Definition 3.2: Importance of a Network Component
The importance of a network component $g \in \mathcal{G}$, $I(g)$, is measured by the relative network efficiency drop after g is removed from the network

$$I(g) = \frac{\triangle \mathcal{E}}{\mathcal{E}} = \frac{\mathcal{E}(\mathcal{G}, d) - \mathcal{E}(\mathcal{G} - g, d)}{\mathcal{E}(\mathcal{G}, d)} \tag{3.5}$$

where $\mathcal{G} - g$ is the resulting network after component g is removed from network $\mathcal{G}$.

Recall that in the $\mathcal{E}$ measure the elimination of a link is treated by removing that link from the network while the removal of a node is managed by removing the links entering or exiting that node. If the removal results in no path connecting an O/D pair, we simply assign the demand for that O/D pair (either fixed or elastic) to an abstract path with a cost of infinity. The upper bound of the importance of a network component is 1. The higher the importance value (3.5), the more important that network component is.

3.2.3 Numerical Examples

Two network examples are now presented for which the unified network performance/efficiency measure is computed. The first example, reported in Section 3.2.3.1, is a fixed demand example, whereas the second example, given in Section 3.2.3.2, is an elastic demand example. Moreover, the importance of individual nodes and links are determined, ranked, and compared by using both the $\mathcal{E}$ measure and the E measure. In the following examples, we assume that d_{ij} in the E measure is equal to λ_w, where $w = (i, j)$ for $w \in W$. Note that if a pair of nodes (i, j) becomes disconnected, then according to the E measure, $d_{ij} = \infty$ and hence $\frac{1}{d_{ij}} = 0$, in that case. We also, for definiteness, use $\frac{1}{n(n-1)}$ as in (3.1) when applying the E measure rather than n_W as in (3.2).

To further illustrate that the amount of flow has an effect on the importance rankings of network components, in Section 3.2.4, we apply the $\mathcal{E}$ network performance measure to study the link/node importance in the Braess network with varying demand.

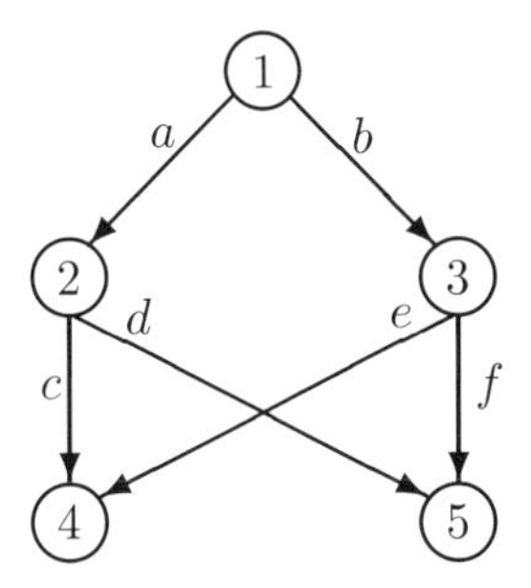

Figure 3.2: Network for Examples 3.2 and 3.3

3.2.3.1 A Fixed Demand Example We now present a fixed demand network example (see Section 2.2.1) for which we determine the network efficiency using the measures (3.2) and (3.1).

Example 3.2: A Network with Fixed Demands

Example 3.2 is a fixed demand network equilibrium problem with the topology given in Figure 3.2.

There are two O/D pairs in the network given by $w_1 = (1, 4)$ and $w_2 = (1, 5)$. There are two paths connecting each O/D pair:
for O/D pair w_1

$$p_1 = (a, c),\ p_2 = (b, e),$$

for O/D pair w_2

$$p_3 = (a, d),\ p_4 = (b, f).$$

The user link cost functions are as follows

$$c_a(f_a) = f_a,\ c_b(f_b) = f_b,\ c_c(f_c) = f_c,\ c_d(f_d) = f_d,\ c_e(f_e) = f_e,\ c_f(f_f) = f_f.$$

The fixed demands for the O/D pairs w_1 and w_2 are $d_{w_1} = 100$ and $d_{w_2} = 20$. The equilibrium path flow solution for this network is

$$x^*_{p_1} = 50,\ x^*_{p_2} = 50,\ x^*_{p_3} = 10,\ x^*_{p_4} = 10,$$

with the equilibrium link flow solution being

$$f^*_a = 60, \quad f^*_b = 60, \quad f^*_c = 50, \quad f^*_d = 10, \quad f^*_e = 50, \quad f^*_f = 10,$$

and the associated disutilities

$$\lambda_{w_1} = 110,\ \lambda_{w_2} = 70.$$

The network performance/efficiency using $\mathcal{E}$ for Example 3.2 is then given by

$$\mathcal{E}(\mathcal{G}, d) = \frac{1}{n_W}\left[\frac{d_{w_1}}{\lambda_{w_1}} + \frac{d_{w_2}}{\lambda_{w_2}}\right] = \frac{\frac{100}{110} + \frac{20}{70}}{2} = .5974.$$

Table 3.1: Importance and ranking of links for Example 3.2

Link	$\mathcal{E}$ Measure		E Measure	
	Importance Value	Importance Ranking	Importance Value	Importance Ranking
a	.5000	1	—	—
b	.5000	1	—	—
c	.1630	2	—	—
d	.0422	3	.5119	1
e	.1630	2	—	—
f	.0422	3	.5119	1

Table 3.2: Importance and ranking of nodes for Example 3.2

Node	$\mathcal{E}$ Measure		E Measure	
	Importance Value	Importance Ranking	Importance Value	Importance Ranking
1	1.0000	1	—	—
2	.5000	2	.7303	1
3	.5000	2	.7303	1
4	.1630	3	-.5166	3
5	.1630	3	.6967	2

The efficiency according to the E measure for Example 3.2 is

$$E(\mathcal{G}) = \frac{1}{n(n-1)}\left[\frac{1}{d_{12}} + \frac{1}{d_{13}} + \frac{1}{d_{14}} + \frac{1}{d_{15}} + \frac{1}{d_{24}} + \frac{1}{d_{25}} + \frac{1}{d_{34}} + \frac{1}{d_{35}}\right]$$

$$= \frac{1}{20}\left[\frac{1}{60} + \frac{1}{60} + \frac{1}{110} + \frac{1}{70} + \frac{1}{50} + \frac{1}{10} + \frac{1}{50} + \frac{1}{10}\right] = .0148.$$

The importance of links and nodes and their rankings for Example 3.2 are reported, respectively, in Tables 3.1 and 3.2; see also Latora and Marchiori (2004). Note that Latora and Marchiori (2004) defined the importance of a network component as $I(g) = E(\mathcal{G}) - E(\mathcal{G} - g) = \Delta E$, but they use $I(g) = \frac{\Delta E}{E}$ in their calculations, which we do as well when we compare the $\mathcal{E}$ measure to the E measure.

Moreover, note that in the above link and node ranking results the importance values (and hence their rankings) of links a, b, c, and e, and node 1 are not defined for the E measure. This is because the cost functions of all the links depend solely on the flow on the respective link (and do not have any fixed cost terms). Take link a for example: the removal of link a is treated by removing it from the network (see Section 3.1). But the cost on links c and d will be zero because of the cost structure on the link, which makes the E measure undefined. However, our measure $\mathcal{E}$ is still well-defined with the removal of link a.

Table 3.3: Importance and ranking of links for Example 3.3

	$\mathcal{E}$ Measure		E Measure	
Link	Importance Value	Importance Ranking	Importance Value	Importance Ranking
a	.5327	1	—	—
b	.5327	1	—	—
c	.1475	2	—	—
d	.0533	3	.4516	1
e	.1475	2	—	—
f	.0533	3	.4516	1

Table 3.4: Importance and ranking of nodes for Example 3.3

	$\mathcal{E}$ Measure		E Measure	
Node	Importance Value	Importance Ranking	Importance Value	Importance Ranking
1	1.0000	1	—	—
2	.5327	2	.2775	2
3	.5327	2	.2775	2
4	.1475	3	.3509	1
5	.1475	3	.3509	1

3.2.3.2 An Elastic Demand Network Example As discussed in Section 3.2.1, the network measure $\mathcal{E}$ can also be used to gauge the efficiency/performance of networks with elastic demands. This is illustrated by Example 3.3.

Example 3.3: A Network with Elastic Demands

We return now to Example 3.2. However, we now let the demand for O/D pairs w_1 and w_2 be elastic, so that the problem is as described in Section 2.2.2, with the following disutility functions

$$\lambda_{w_1}(d_{w_1}) = -d_{w_1} + 100, \quad \lambda_{w_2}(d_{w_2}) = -d_{w_2} + 40.$$

It is easy to calculate the equilibrium path flow solution

$$x^*_{p_1} = 24, \; x^*_{p_2} = 24, \; x^*_{p_3} = 4, \; x^*_{p_4} = 4,$$

the equilibrium link flow solution

$$f^*_a = 28, \quad f^*_b = 28, \quad f^*_c = 24, \quad f^*_d = 4, \quad f^*_e = 24, \quad f^*_f = 4,$$

and the equilibrium demands

$$d^*_{w_1} = 48, \; d^*_{w_2} = 8,$$

so that the incurred disutilities are

$$\lambda_{w_1} = 52, \ \lambda_{w_2} = 32.$$

The network performance/efficiency for Example 3.3, according to $\mathcal{E}$ is hence

$$\mathcal{E}(\mathcal{G}, d) = \frac{1}{n_W}\left[\frac{d_{w_1}}{\lambda_{w_1}} + \frac{d_{w_2}}{\lambda_{w_2}}\right] = \frac{\frac{48}{52} + \frac{8}{32}}{2} = .5865.$$

The efficiency according to the E measure is

$$E(\mathcal{G}) = \frac{1}{n(n-1)}\left[\frac{1}{d_{12}} + \frac{1}{d_{13}} + \frac{1}{d_{14}} + \frac{1}{d_{15}} + \frac{1}{d_{24}} + \frac{1}{d_{25}} + \frac{1}{d_{34}} + \frac{1}{d_{35}}\right]$$

$$= \frac{1}{20}\left[\frac{1}{28} + \frac{1}{28} + \frac{1}{52} + \frac{1}{32} + \frac{1}{24} + \frac{1}{4} + \frac{1}{24} + \frac{1}{4}\right] = .0353.$$

In Tables 3.3 and 3.4, the importance of links and nodes and their rankings using our measure $\mathcal{E}$ and the E measure are given for this example.

As discussed in Section 3.2, by adding an abstract (and infinite cost) path to a disconnected O/D pair, the $\mathcal{E}$ measure can be used to study networks with disconnected O/D pairs. This feature enables us to investigate the importance of nodes 4 and 5 in Examples 3.2 and 3.3.

3.2.4 An Application of the New Network Measure to the Braess Network with Varying Demands

We now consider the Braess paradox example after the addition of a new link e and as depicted in Figure 3.3 [see also Braess (1968) and Braess, Nagurney, and Wakolbinger (2005)]. There are four nodes: 1, 2, 3, and 4; five links: a, b, c, d, and e; and a single O/D pair $w = (1, 4)$. There are hence three paths connecting the single O/D pair, which are denoted, respectively, by $p_1 = (a, c)$, $p_2 = (b, d)$, and $p_3 = (a, e, d)$.

The user link cost functions are

$$c_a(f_a) = 10f_a, \ \ c_b(f_b) = f_b + 50,$$

$$c_c(f_c) = f_c + 50, \ \ c_d(f_d) = 10f_d, \ \ c_e(f_e) = f_e + 10.$$

We can also write the path cost functions, by making use of (2.14), (2.16), and (2.17) as follows

$$C_{p_1}(x) = 11x_{p_1} + 10x_{p_3} + 50, \ \ C_{p_2}(x) = 11x_{p_2} + 10x_{p_3} + 50,$$

$$C_{p_3}(x) = 10x_{p_1} + 21x_{p_3} + 10x_{p_2} + 10.$$

Recall that Braess (1968) demonstrated that for a fixed demand of $d_w = 6$ the addition of link e, which provides the users with the new path p_3 as in the network

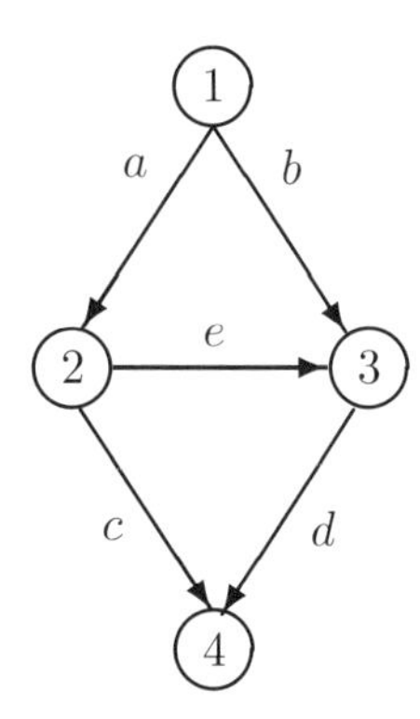

Figure 3.3: The Braess network with the addition of the new link

in Figure 3.3 actually makes all users worse off because without the link e, the path costs (and O/D pair disutility) are 83, whereas with the new link/path, the path costs (and disutility) increase for all users to 92! The Braess paradox has achieved great attention both in theoretical research and in practice. It has fascinated researchers in networks including those in operations research and management science, in transportation science, and, more recently, in computer science. Its relevance in a variety of congested networks operating in a decentralized manner ranges from urban transportation networks (the original application setting) to, more recently, the Internet; see also, e.g., Korilis, Lazar, and Orda (1999), Boyce, Mahmassani, and Nagurney (2005), Boyce and Nagurney (2005), and Roughgarden (2005).

However, a natural question arises as to what happens to the equilibrium flows (and the incurred path costs) as demand varies. Nagurney (2006a) and Nagurney, Parkes, Daniele (2007) explored this question (see also Chapter 7) and provided an evolutionary variational inequality formulation of the time-dependent (demand-varying) Braess paradox, which was formulated in a static setting, but without any qualitative analysis, by Pas and Principio (1997). It was found that different paths are used (i.e., have positive flow in equilibrium) in three different demand ranges as the demand increases. Therefore, the importance and the ranking of individual nodes and links can be expected to be different, depending on which demand range the demand of concern is in.

Furthermore, because the Braess paradox occurs in a certain part of Demand Range I below (as referred to in the cited references), in the following analysis, we discuss the importance and the ranking of the network components in four (rather than three) different demand ranges. It is notable that we are able to derive explicit formulae for both the network efficiency and the importance of network components as a function of the demand d_w.

3.2.4.1 Demand Range I: $d_w \in [0, 2\frac{18}{31})$

Assume that the demand $d_w \in [0, 2\frac{18}{31})$. In this demand range, only path p_3 is used in equilibrium and the Braess paradox does not occur. The equilibrium path flow pattern is $x^*_{p_1} = x^*_{p_2} = 0$ and

Table 3.5: Link results for Section 3.2.4.1 network using the $\mathcal{E}$ measure

Link	Importance Value	Importance Ranking
a	$\frac{10(4-d_w)}{11d_w+50}$	1
b	.00	3
c	.00	3
d	$\frac{10(4-d_w)}{11d_w+50}$	1
e	$\frac{(80-31d_w)}{(11d_w+100)}$	2

Table 3.6: Nodal results for Section 3.2.4.1 network using the $\mathcal{E}$ measure

Node	Importance Value	Importance Ranking
1	1.00	1
2	$\frac{10(4-d_w)}{11d_w+50}$	2
3	$\frac{10(4-d_w)}{11d_w+50}$	2
4	1.00	1

Table 3.7: Link results for Section 3.2.4.2 network using the $\mathcal{E}$ measure

Link	Importance Value	Importance Ranking
a	$\frac{10(4-d_w)}{11d_w+50}$	1
b	.00	2
c	.00	2
d	$\frac{10(4-d_w)}{11d_w+50}$	1
e	$\frac{(80-31d_w)}{(11d_w+100)}$	3

$x^*_{p_3} = d_w$. The equilibrium disutility is $\lambda_w = 21d_w + 10$. The network performance according to (3.2) for this range of demand is $\mathcal{E} = \frac{d_w}{21d_w+10}$.

The importance and the rankings of the links and the nodes are given, respectively, in Tables 3.5 and 3.6.

3.2.4.2 Demand Range II: $d_w \in [2\frac{18}{31}, 3\frac{7}{11}]$

Assume now that the demand $d_w \in [2\frac{18}{31}, 3\frac{7}{11}]$. Similar to the results for Demand Range I, only path p_3 is used in equilibrium but now the Braess paradox occurs. Hence the equilibrium solution in this demand range is $x^*_{p_1} = x^*_{p_2} = 0$ and $x^*_{p_3} = d_w$. The equilibrium disutility is $\lambda_w = 21d_w + 10$. The efficiency is now: $\mathcal{E} = \frac{d_w}{21d_w+10}$. The importance ranking of the links and the nodes are given, respectively, in Tables 3.7 and 3.8.

Table 3.8: Nodal results for Section 3.2.4.2 network using the $\mathcal{E}$ measure

Node	Importance Value	Importance Ranking
1	1.00	1
2	$\frac{10(4-d_w)}{11d_w+50}$	2
3	$\frac{10(4-d_w)}{11d_w+50}$	2
4	1.00	1

Table 3.9: Link results for Section 3.2.4.3 network using the $\mathcal{E}$ measure

Link	Importance Value	Importance Ranking
a	$\frac{8(14d_w-45)}{13(11d_w+50)}$	1
b	$\frac{121(11d_w-40)}{13(131d_w+560)}$	2
c	$\frac{121(11d_w-40)}{13(131d_w+560)}$	2
d	$\frac{8(14d_w-45)}{13(11d_w+50)}$	1
e	$\frac{9(9d_w-80)}{13(11d_w+100)}$	3

Table 3.10: Nodal results for Section 3.2.4.3 network using the $\mathcal{E}$ measure

Node	Importance Value	Importance Ranking
1	1.00	1
2	$\frac{8(14d_w-45)}{13(11d_w+50)}$	2
3	$\frac{8(14d_w-45)}{13(11d_w+50)}$	2
4	1.00	1

3.2.4.3 *Demand Range III:* $d_w \in (3\frac{7}{11}, 8\frac{8}{9}]$ Assume now that the demand $d_w \in (3\frac{7}{11}, 8\frac{8}{9}]$. Please note that, in this range of demand, all three paths are used in equilibrium and the Braess paradox still occurs. We now have $x^*_{p_1} = x^*_{p_2} = \frac{11}{13}d_w - \frac{40}{13}$ and $x^*_{p_3} = -\frac{9}{13}d_w + \frac{80}{13}$. The equilibrium disutility is now $\lambda_w = \frac{31d_w+1010}{13}$. The network performance measure is now $\mathcal{E} = \frac{13d_w}{31d_w+1010}$. The importance rankings of links and nodes are given, respectively, in Tables 3.9 and 3.10.

3.2.4.4 *Demand Range IV:* $d_w \in (8\frac{8}{9}, \infty)$ Assume now that $d_w \in (8\frac{8}{9}, \infty)$. Only paths p_1 and p_2 are now used in equilibrium and the Braess paradox vanishes. Hence we have $x^*_{p_1} = x^*_{p_2} = \frac{d_w}{2}$ and $x^*_{p_3} = 0$. The equilibrium disutility is now given by the expression $\lambda_w = \frac{11}{2}d_w + 50$. The efficiency is now $\mathcal{E} = \frac{2d_w}{(11d_w+100)}$.

Table 3.11: Link results for Section 3.2.4.4 network using the $\mathcal{E}$ measure

Link	Importance Value	Importance Ranking
a	$\frac{11d_w}{2(11d_w+50)}$	1
b	$\frac{5(13d_w-8)}{(131d_w+560)}$	2
c	$\frac{5(13d_w-8)}{(131d_w+560)}$	2
d	$\frac{11d_w}{2(11d_w+50)}$	1
e	.00	3

Table 3.12: Nodal results for Section 3.2.4.4 network using the $\mathcal{E}$ measure

Node	Importance Value	Importance Ranking
1	1.00	1
2	$\frac{11d_w}{2(11d_w+50)}$	2
3	$\frac{11d_w}{2(11d_w+50)}$	2
4	1.00	1

The importance and the rankings of the links and the nodes are given, respectively, in Tables 3.11 and 3.12.

3.2.4.5 Discussion Note that this example demonstrates that the importance ranking of a link may be different in different demand ranges. For example, links b and c are less important in Demand Range I than in Demand Ranges III and IV. This is due to the fact that in Demand Range I, links b and c carry zero flows; therefore, they are not critical links in evaluating network performance in those ranges. However, in Demand Ranges III and IV, links b and c carry positive flows; thus, in these ranges of demand, the removal of these links will deteriorate the network performance/efficiency.

The importance rankings of the nodes in this network example remain the same across all the demand ranges. This feature is quite interesting and reasonable.

The different ranking results for links b and c clearly explain why "flow matters" and why an appropriate network performance/efficiency measure for congested networks should capture not only costs and distances but also flows and the behavior of the network users.

To show the advantage of the $\mathcal{E}$ measure, we give a comparison of the importance rankings for links and nodes in the Braess network using the measure $\mathcal{E}$ and the E measure in Tables 3.13 and 3.14 when d_w is fixed and equal to 6.

It is interesting that we see that the links identified as the most important ones according to our measure $\mathcal{E}$, that is, links a and d, are ranked the least important according to the E measure. On the other hand, link e, which is ranked least

Table 3.13: Link results for the Braess network for d_w=6

	$\mathcal{E}$ Measure		E Measure	
Link	Importance Value	Importance Ranking	Importance Value	Importance Ranking
a	.2069	1	.1056	3
b	.1794	2	.2153	2
c	.1794	2	.2153	2
d	.2069	1	.1056	3
e	-.1084	3	.3616	1

Table 3.14: Nodal results for the Braess network for d_w=6

	$\mathcal{E}$ Measure		E Measure	
Node	Importance Value	Importance Ranking	Importance Value	Importance Ranking
1	1.0000	1	—	—
2	.2069	2	.7635	1
3	.2069	2	.7635	1
4	1.0000	1	—	—

important according to the $\mathcal{E}$ measure, is ranked as most important according to the E measure. Because the addition of link e causes the Braess paradox when demand is equal to 6 [cf. Braess (1968)], it will, obviously, be detrimental to the network performance, which is clearly shown by the negative importance value of link e obtained via the $\mathcal{E}$ measure. The fact that link e is ranked as the most important link in the E measure is unreasonable.

3.3 A NETWORK ROBUSTNESS MEASURE UNDER DECENTRALIZED DECISION-MAKING BEHAVIOR

In this section, we introduce a network robustness measure based on the network performance measure $\mathcal{E}$ introduced in Section 3.2. The network robustness measure assumes, however, that the underlying user link cost functions have embedded, explicit link capacities. The robustness of a network is then evaluated, in the case of decentralized decision-making behavior, as the practical capacity varies. Specifically, if we let u denote the vector of link capacities, then we evaluate the robustness of a given network as the relative performance retained under a given uniform capacity retention ratio γ with $\gamma \in (0, 1]$. The network robustness measure is relevant in the case of either fixed or elastic demands, as is the network performance/efficiency measure. After stating the network robustness measure, we identify several specific link functional forms in practice that occur in transportation and in telecommunications and that have explicit embedded link capacities.

We define the robustness measure of a network under decentralized decision-making behavior as follows

Definition 3.3: Network Robustness Measure under Decentralized Decision-Making Behavior
The robustness measure $\mathcal{R}^\gamma$ for a network $\mathcal{G}$ with the vector of user link cost functions c, the vector of link capacities u, the vector of demands d (either fixed or elastic) is defined as the relative performance retained under a given uniform capacity retention ratio γ with $\gamma \in (0,1]$ so that the new capacities [see, e.g., (3.7$a-c$)] are given by γu. Its mathematical definition is given as

$$\mathcal{R}^\gamma = \mathcal{R}(\mathcal{G}, c, \gamma, u) = \frac{\mathcal{E}^\gamma}{\mathcal{E}} \times 100\% \tag{3.6}$$

where $\mathcal{E}$ [cf. (3.2)] and $\mathcal{E}^\gamma$ are the network performance measures with the original capacities and the remaining capacities, respectively.

For example, if $\gamma = .9$, this means that the user link cost functions now have the link capacities given by $.9u_a$ for $a \in \mathcal{L}$; if $\gamma = .7$, then the link capacities become $.7u_a$ for all links $a \in \mathcal{L}$, and so on.

From Definition 3.3, a network under a given level of capacity retention or deterioration is considered to be robust if the network performance stays close to the original level. Next we identify several explicit user link cost functions with embedded capacities for which the robustness measure can be applied, with the proviso that the link flows in (3.7b) and (3.7c) do not attain their capacities. We return to this issue in Section 3.4.

Bureau of Public Roads (BPR) Function
The functional form of the Bureau of Public Roads (1964) link cost functions is

$$c_a(f_a) = t_a^0[1 + k(\frac{f_a}{u_a})^\beta], \quad \forall a \in \mathcal{L}, \tag{3.7a}$$

where f_a is the flow on link a; u_a is the "practical" capacity on link a, which also has the interpretation of the level-of-service flow rate; t_a^0 is the free-flow travel time or cost on link a; k and β are the model parameters and both take on positive values. Often in applications $k = .15$ and $\beta = 4$. In this context $c_a(f_a)$ measures the travel time on the link.

This function, as we shall also see in Chapter 4, is widely used in transportation planning practice.

The Davidson Function
The Davidson (1966) user link cost function is based on queueing theory considerations and is given by

$$c_a(f_a) = t_a^0[1 + J\frac{f_a}{u_a - f_a}], \quad \forall a \in \mathcal{L}. \tag{3.7b}$$

As in the case of the BPR function, t_a^0 is the free-flow speed (the speed at zero flow), and u_a denotes the road's (link's) capacity. The term J controls the shape of the associated curve.

The M/M/1 Delay Function

In telecommunication networks, the M/M/1 delay function is widely used. It also is based on queueing considerations of messages. Its form [see also Bertsekas and Gallager (1987) and Roughgarden (2005)] is

$$c_a(f_a) = \frac{1}{[u_a - f_a]}, \quad \forall a \in \mathcal{L}. \tag{3.7c}$$

Remark 3.1

By applying a similar concept, we can also study network robustness when the network capacities are *enhanced*. We believe that this type of analysis can provide implications for network capacity investments to assess the effect on network performance. However, because in such situations $\gamma \geq 1$, we refer to the network robustness measure in this context as the "capacity increment ratio."

3.3.1 Some Theoretical Results for the Network Robustness Measure under BPR Functions and Fixed Demands

In this section, we consider networks with special structure for which we can obtain some theoretical results in terms of robustness. In particular, we first consider a very simple network with BPR functions given by (3.7a) for any β. We then consider networks of special topology consisting of a single O/D pair with parallel links for which the associated BPR link cost functions have $\beta = 1$. In this section, we focus on fixed demands.

Theorem 3.2

Consider a network consisting of two nodes 1 *and* 2 *as in Figure 3.4, which are connected by a single link* a *and with a single O/D pair* $w_1 = (1, 2)$. *Assume that the user link cost function associated with link* a *is of the BPR form given by (3.7a). Then the network robustness measure (3.6) is given by the explicit formula*

$$\mathcal{R}^\gamma = \frac{\gamma^\beta [u_a^\beta + k d_{w_1}^\beta]}{[\gamma^\beta u_a^\beta + k d_{w_1}^\beta]} \times 100\%, \tag{3.8}$$

where d_{w_1} *is the given demand for O/D pair* $w_1 = (1, 2)$.

Moreover, the network robustness $\mathcal{R}^\gamma$ *is bounded from below by* $\gamma^\beta \times 100\%$.

Proof: Clearly, because there is only a single O/D pair and a single path, we have that

$$\mathcal{R}^\gamma = \frac{\mathcal{E}^\gamma}{\mathcal{E}} \times 100\% = \frac{\lambda_{w_1}}{\lambda_{w_1}^\gamma} \times 100\%,$$

where $\lambda_{w_1}^\gamma$ denotes the incurred disutility between O/D pair w_1 with the capacity γu_a on link a.

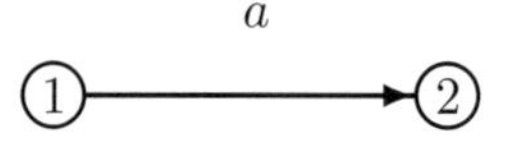

Figure 3.4: Network for Theorem 3.2

We can write $\lambda_{w_1}^{\gamma}$ and λ_{w_1} explicitly for this simple network, which yields

$$\mathcal{R}^{\gamma} = \frac{t_a^0[1 + k(\frac{d_{w_1}}{u_a})^{\beta}]}{t_a^0[1 + k(\frac{d_{w_1}}{\gamma u_a})^{\beta}]} \times 100\%.$$

After simplification, we obtain

$$\mathcal{R}^{\gamma} = \frac{\gamma^{\beta}[u_a^{\beta} + kd_{w_1}^{\beta}]}{[\gamma^{\beta}u_a^{\beta} + kd_{w_1}^{\beta}]} \times 100\%,$$

which is exactly the form of (3.8).

To determine the lower bound of $\mathcal{R}^{\gamma}$, we can rearrange (3.8) and get the following form

$$\mathcal{R}^{\gamma} = \frac{\gamma^{\beta}[1 + k(\frac{d_{w_1}}{u_a})^{\beta}]}{[\gamma^{\beta} + k(\frac{d_{w_1}}{u_a})^{\beta}]} \times 100\%.$$

Because $\gamma \in (0, 1]$, we have that $\gamma^{\beta} \in (0, 1], \forall \beta > 0$. Hence we have the following

$$\mathcal{R}^{\gamma} \geq \frac{\gamma^{\beta}[1 + k(\frac{d_{w_1}}{u_a})^{\beta}]}{[1 + k(\frac{d_{w_1}}{u_a})^{\beta}]} \times 100\% = \gamma^{\beta} \times 100\%,$$

which completes the proof.

Now let us consider a network with a special topology as depicted in Figure 3.5. The network consists of a single O/D pair that is connected by parallel links. In the following theorem, we give the general form of the network robustness as well as its lower bound for the network just given.

Theorem 3.3
Consider a network consisting of two nodes 1 and 2 as in Figure 3.5, which are connected by a set of parallel links. Assume that the associated BPR link cost functions have $\beta = 1$ [cf. (3.7a)]. Furthermore, assume that there are positive flows on all the links at both the original and partially degraded capacity levels. Then the network robustness measure (3.6) is given by the explicit formula

$$\mathcal{R}^{\gamma} = \frac{\gamma U + k\gamma d_{w_1}}{\gamma U + kd_{w_1}} \times 100\%, \tag{3.9}$$

where d_{w_1} is the given demand for O/D pair $w_1 = (1, 2)$ and $U \equiv u_a + u_b + \ldots + u_n$.

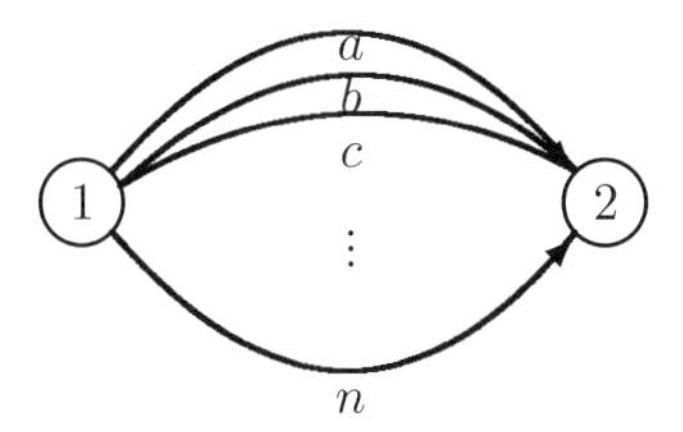

Figure 3.5: A network with special structure

Moreover, the network robustness $\mathcal{R}^\gamma$ is bounded from below by $\gamma \times 100\%$.

Proof: Clearly, because there is only a single O/D pair we have that

$$\mathcal{R}^\gamma = \frac{\mathcal{E}^\gamma}{\mathcal{E}} \times 100\% = \frac{\lambda_{w_1}}{\lambda_{w_1}^\gamma} \times 100\%,$$

where $\lambda_{w_1}^\gamma$ denotes the incurred disutility between O/D pair w_1 under the capacity retention ratio γ.

Due to the special structure of the network and the assumption that there are positive flows on all the links before and after the capacity reduction, by referring to the network equilibrium conditions (cf. (2.19) in Section 2.2), we can write λ_{w_1} and $\lambda_{w_1}^\gamma$ explicitly as follows

$$\lambda_{w_1} = t_a^0(1 + k\frac{f_a^*}{u_a}) = t_b^0(1 + k\frac{f_b^*}{u_b}) = \ldots = t_n^0(1 + k\frac{f_n^*}{u_n})$$

where f_a^*, $f_b^* \ldots f_n^*$ are the equilibrium link flows under the link capacities: u_a, u_b,. . .,u_n, respectively, and

$$\lambda_{w_1}^\gamma = t_a^0(1 + k\frac{f_a^{**}}{\gamma u_a}) = t_b^0(1 + k\frac{f_b^{**}}{\gamma u_b}) = \ldots = t_n^0(1 + k\frac{f_n^{**}}{\gamma u_n})$$

where f_a^{**}, f_b^{**},. . .,f_n^{**} are the equilibrium link flows under the link capacities: γu_a, γu_b,. . .,γu_n, respectively.

Hence we have

$$\mathcal{R}^\gamma = \frac{\lambda_{w_1}}{\lambda_{w_1}^\gamma} \times 100\% = \frac{t_a^0(1 + k\frac{f_a^*}{u_a})}{t_a^0(1 + k\frac{f_a^{**}}{\gamma u_a})} \times 100\%$$

$$= \frac{t_b^0(1 + k\frac{f_b^*}{u_b})}{t_b^0(1 + k\frac{f_b^{**}}{\gamma u_b})} \times 100\% = \ldots = \frac{t_n^0(1 + k\frac{f_n^*}{u_n})}{t_n^0(1 + k\frac{f_n^{**}}{\gamma u_n})} \times 100\%,$$

which yields

$$\mathcal{R}^\gamma = \frac{(1 + k\frac{f_a^*}{u_a}) + (1 + k\frac{f_b^*}{u_b}) + \ldots + (1 + k\frac{f_n^*}{u_n})}{(1 + k\frac{f_a^{**}}{\gamma u_a}) + (1 + k\frac{f_b^{**}}{\gamma u_b}) + \ldots + (1 + k\frac{f_n^{**}}{\gamma u_n})} \times 100\%.$$

After some simplification and from the fact that $f_a^* + f_b^* + \ldots + f_n^* = f_a^{**} + f_b^{**} + \ldots + f_n^{**} = d_{w_1}$, we have

$$\mathcal{R}^\gamma = \frac{\gamma U + k\gamma d_{w_1}}{\gamma U + k d_{w_1}} \times 100\%,$$

which is exactly (3.9).

To determine the lower bound of the network robustness, we can rearrange (3.9) to get

$$\mathcal{R}^\gamma = \frac{\gamma(1 + k\frac{d_{w_1}}{U})}{\gamma + k\frac{d_{w_1}}{U}} \times 100\%.$$

Because $\gamma \in (0, 1]$, we have the following

$$\mathcal{R}^\gamma \geq \frac{\gamma(1 + k\frac{d_{w_1}}{U})}{(1 + k\frac{d_{w_1}}{U})} \times 100\% = \gamma \times 100\%,$$

which completes the proof.

3.3.2 A Braess-Inspired Network

We now consider a Braess-inspired network after the addition of a new link e and as depicted in Figure 3.3.

Instead of using the original link cost functions, however, because they are not in explicit BPR form, we construct a set of BPR functions under which the Braess paradox still occurs (without any capacity reduction). We assume that $k = 1$. Let $t_a^0 = t_d^0 = 1$, $t_b^0 = t_c^0 = 50$, and $t_e^0 = 10$. Furthermore, let $u_a = u_d = 20$, $u_b = u_c = 50$, and $u_e = 100$. The user link cost functions are hence given by

$$c_a(f_a) = 1 + (\frac{f_a}{20})^\beta, \quad c_b(f_b) = 50(1 + (\frac{f_b}{50})^\beta),$$

$$c_c(f_c) = 50(1 + (\frac{f_b}{50})^\beta), \quad c_d(f_d) = 1 + (\frac{f_d}{20})^\beta,$$

$$c_e(f_e) = 10(1 + (\frac{f_e}{100})^\beta).$$

The demand $d_{w_1} = 110$. Figure 3.6 shows the network robustness R^γ for the Braess-inspired network under β values equal to 1, 2, 3, and 4, respectively.

From this example, we see that, for a given capacity retention ratio, when the value of β is small, the robustness of the network drops less severely than when β is large. This is due to the fact that β indicates, in part, the effect of congestion on links in a network. Therefore, for a certain capacity reduction, a "less congestion-sensitive" network can keep its efficiency closer to the original value.

Moreover, as discussed earlier, we can also study the robustness of a network when its link capacities are enhanced. Figure 3.7 presents the network robustness for

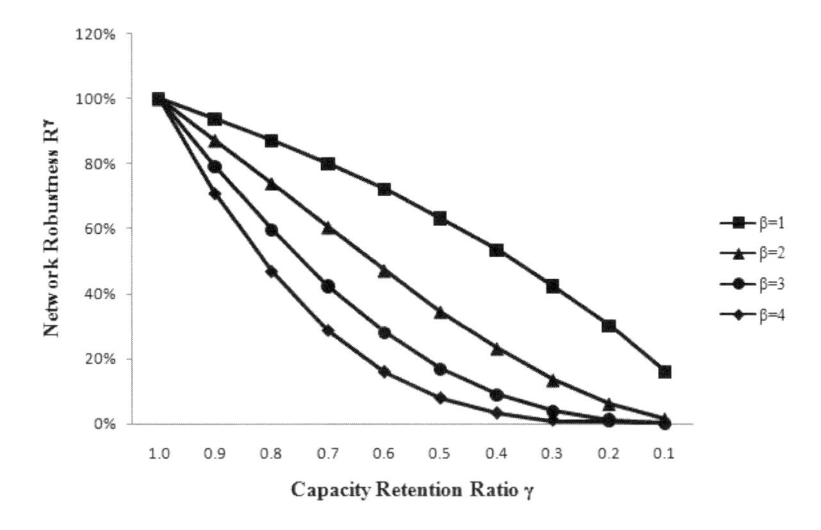

Figure 3.6: Robustness vs. capacity retention ratio for different β values

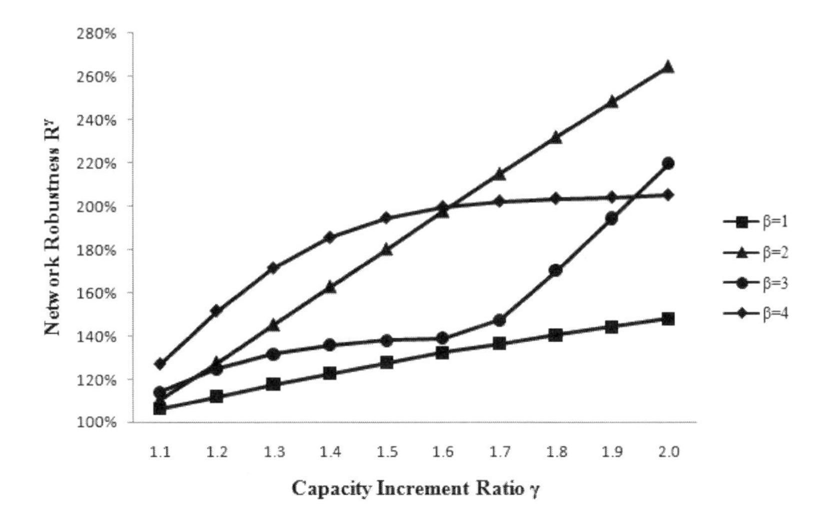

Figure 3.7: Robustness vs. capacity increment ratio for different β values

the Braess-inspired network with capacity enhancement for β values equal to 1, 2, 3, and 4, respectively.

From Figure 3.7, we can see that when $\gamma \leq 1.6$, the Braess-inspired network with $\beta = 4$ has the largest performance improvement under the same level of capacity increment ratio. However, when $\gamma \geq 1.6$, the Braess-inspired network with $\beta = 2$ has the largest performance improvement.

3.4 RELATIVE TOTAL COST INDICES FOR ASSESSING NETWORK ROBUSTNESS

We now propose indices based on the relative total cost that assess the robustness of a network based on the two behavioral solution concepts, namely, the total cost evaluated under the user-optimizing flow pattern, denoted by TC_{U-O}, and the system-optimizing flow pattern, denoted-by TC_{S-O}, in the case of fixed demands. In particular, TC_{U-O} denotes the total cost on the network as given by expression (2.33), where the vector f is the solution to the user-optimizing (or network equilibrium) conditions [cf. (2.19)]. On the other hand, TC_{S-O} is the total cost on the network as given also by expression (2.33) but now evaluated at the flow pattern given by the solution to the S-O problem [cf. (2.34)]. We believe that the total cost is also an appropriate measure because it represents the total cost to society associated with routing of flows on networks. Moreover, as the links degrade and the practical capacity of links decreases, the total cost is expected to increase and, hence, the relative total cost of a network reflects its robustness. Clearly, since we are considering the robustness relative to changes in link capacities, the functional forms in Section 3.3 are all relevant. Others may also be found in practice in particular network contexts; all we require is that the link capacity be explicitly captured within each link cost function. In Section 3.4.1 we then apply the index to networks with BPR functions and obtain theoretical results. Throughout this section, we assume that the demands are fixed and known.

The relative total cost index for a network $\mathcal{G}$ with the vector of fixed demands d, the vector of user link cost functions c, and the vector of link capacities u is defined as the relative total cost increase under a given uniform capacity retention ratio γ ($\gamma \in (0, 1]$) so that the new capacities are given by γu. Let c denote the vector of user link cost functions [cf. $(3.7a - c)$] and let d denote the vector of O/D demands. The mathematical definition of the index under the user-optimizing flow pattern, denoted by $\mathcal{I}^{\gamma}_{U-O}$, is then

$$\mathcal{I}^{\gamma}_{U-O} = \mathcal{I}_{U-O}(\mathcal{G}, c, d, \gamma, u) = \frac{TC^{\gamma}_{U-O} - TC_{U-O}}{TC_{U-O}} \times 100\%, \tag{3.10a}$$

where TC_{U-O} and TC^{γ}_{U-O} are the total network costs evaluated under the user-optimizing flow pattern with the original capacities and the remaining capacities (i.e., γu), respectively.

The mathematical definition of the index under the system-optimizing flow pattern is

$$\mathcal{I}^{\gamma}_{S-O} = \mathcal{I}_{S-O}(\mathcal{G}, c, d, \gamma, u) = \frac{TC^{\gamma}_{S-O} - TC_{S-O}}{TC_{S-O}} \times 100\%, \tag{3.10b}$$

where TC_{S-O} and TC^{γ}_{S-O} are the total network costs evaluated at the system-optimizing flow pattern with the original capacities and the remaining capacities (i.e., γu), respectively.

For example, if $\gamma = .8$ the total user link cost functions now have the link capacities given by $.8u_a$ for $a \in \mathcal{L}$; if $\gamma = .4$, then the link capacities become $.4u_a$ for all links $a \in \mathcal{L}$, and so on.

From these definitions, a network, under a given capacity retention/deterioration ratio γ (and either S-O or U-O behavior) is considered to be robust if the index $\mathcal{I}^\gamma$ is low. This means that the relative total cost does not change much; hence the network may be viewed as being more robust than if the relative total cost were large.

Remark 3.2

Similar to the discussion in Remark 3.1, we can also study the relative total cost improvement after capacity enhancement. In that case, because the relative total cost savings need to be computed, we reverse the order of substraction in $(3.10a)$ and $(3.10b)$ with $\gamma \geq 1$. Furthermore, γ is defined as the "capacity increment ratio." Therefore, the larger the relative total cost index is, the greater the expected total cost savings for a capacity enhancement plan for a specific γ.

Remark 3.3

It is worth noting the relationship between the *price of anarchy*, which is denoted here by $\mathcal{P}$ and is defined as

$$\mathcal{P} = \frac{TC_{U-O}}{TC_{S-O}} \tag{3.11}$$

and the ratio of the two proposed robustness indices. Observe that $\mathcal{P}$ captures the relationship between total costs *across* distinct behavioral principles, whereas the indices $(3.10a)$ and $(3.10b)$ are focused on the degradation of network performance *within* U-O or S-O behavior. Nevertheless, we have the following relationship between the ratio of the two indices and the price of anarchy

$$\frac{I^\gamma_{S-O}}{I^\gamma_{U-O}} = \frac{[TC^\gamma_{S-O} - TC_{S-O}]}{[TC^\gamma_{U-O} - TC_{U-O}]} \times \mathcal{P}. \tag{3.12}$$

The term preceding the price of anarchy in (3.12) may be less than 1, greater than 1, or equal to 1, depending on the network and data.

As established in Roughgarden (2003), the price of anarchy is bounded by $O(\frac{\beta}{log\beta})$, when the cost function (sometimes also referred to in the computer science literature as a *latency function*) on each link is a polynomial function with nonnegative coefficients and degree at most β. This is the form, for example, of the BPR function in $(3.7a)$. Specifically, when $\beta = 4$, we have that the price of anarchy $\mathcal{P}$ is bounded by $O(\frac{\beta}{log\beta}) = 6.6439$.

On the other hand, in the case of M/M/1 user link cost (or delay) functions, one must note that such cost functions are defined only for each link a, $\forall a \in \mathcal{L}$, on the set: $[0, u_a)$. Let d_{max} denote the maximum allowable amount of network traffic and let u_{min} denote the minimum allowable link capacity on the network. Further assume that $d_{max} < u_{min}$. Within this context [see also Roughgarden (2005)], the bound on the price of anarchy $\mathcal{P}$ for M/M/1 delay function networks is given by

$$\frac{1}{2}(1 + \sqrt{\frac{u_{min}}{u_{min} - d_{max}}}).$$

3.4.1 Theoretical Results for Networks with BPR Functions

In this section, we derive some theoretical results. In particular, we first prove that for certain networks of special structure, with links characterized by user link cost functions of BPR form and with identical free flow travel terms, the relative total cost index is identical under the U-O and the S-O flow patterns. We then show that for the same network topologies, with user link cost functions also of BPR form but linear, that the relative total cost index under the U-O flow pattern can be obtained via an explicit formula. In addition, we derive an upper bound for the relative total cost index under the U-O flow pattern. Finally, we derive an upper bound for the relative total cost index for network robustness, under the S-O flow pattern, for any network in the case of user link cost functions of BPR form.

Theorem 3.4
Consider a network consisting of two nodes 1 *and* 2 *as in Figure 3.5, which are connected by* n *parallel links. If the free-flow term,* t_a^0, $\forall a \in \mathcal{L}$, *is the same for all links* $a \in \mathcal{L}$ *in the BPR link cost function [cf. (3.7a)], the S-O flow pattern coincides with the U-O flow pattern and, therefore,* $\mathcal{I}_{U-O} = \mathcal{I}_{S-O}$.

Proof: Because the free-flow term t_a^0, $\forall a \in \mathcal{L}$, is the same for all the links, let us denote it as t^0. First, we show that any U-O flow pattern is also a S-O flow pattern. The reverse can be established in a similar manner. We distinguish between two cases.

Case 1: not all links are used
If not all links are used then we know that there is at least one link with zero flow.

Select any two links $a, b \in \mathcal{L}$ such that $f_a^* > 0$ and $f_b^* = 0$ under the U-O condition. We must have that $c_a(f_a^*) = \lambda_w \leq c_b(f_b^*)$, where λ_w is the equilibrium disutility for the O/D pair $w = (1, 2)$. Hence we have the following relationships

$$t_a^0[1 + (\frac{f_a^*}{u_a})^\beta] \leq t_b^0[1 + (\frac{f_b^*}{u_b})^\beta] \Longleftrightarrow t^0[1 + (\frac{f_a^*}{u_a})^\beta] \leq t^0[1 + (\frac{f_b^*}{u_b})^\beta] \Longleftrightarrow$$

$$(\frac{f_a^*}{u_a})^\beta \leq (\frac{f_b^*}{u_b})^\beta$$

$$\Longleftrightarrow t^0 + t^0(\beta + 1)(\frac{f_a^*}{u_a})^\beta \leq t^0 + t^0(\beta + 1)(\frac{f_b^*}{u_b})^\beta \Longleftrightarrow t_a^0 + t_a^0(\beta + 1)(\frac{f_a^*}{u_a})^\beta$$

$$\leq t_b^0 + t_b^0(\beta + 1)(\frac{f_b^*}{u_b})^\beta \tag{3.13}$$

where the last inequality is exactly the S-O condition [cf. (2.37)]. Therefore, f_a^* and f_b^* are also the S-O link flow patterns. Because links a and b were chosen arbitrarily, we can conclude that a U-O flow pattern is also a S-O flow pattern.

Case 2: all links are used
If all the links are used, under the U-O condition we must have

$$\lambda_w = t_a^0[1 + (\frac{f_a^*}{u_a})^\beta] = t_b^0[1 + (\frac{f_b^*}{u_b})^\beta] = \ldots = t_n^0[1 + (\frac{f_n^*}{u_n})^\beta]$$

$$\iff \lambda_w = t^0[1 + (\frac{f_a^*}{u_a})^\beta] = t^0[1 + (\frac{f_b^*}{u_b})^\beta] = \ldots = t^0[1 + (\frac{f_n^*}{u_n})^\beta]$$

$$\iff (\frac{f_a^*}{u_a})^\beta = (\frac{f_b^*}{u_b})^\beta = \ldots = (\frac{f_n^*}{u_n})^\beta$$

$$\iff t^0 + t^0(\beta+1)(\frac{f_a^*}{u_a})^\beta = t^0 + t^0(\beta+1)(\frac{f_b^*}{u_b})^\beta = \ldots = t^0 + t^0(\beta+1)(\frac{f_n^*}{u_n})^\beta$$

$$\iff t_a^0 + t_a^0(\beta+1)(\frac{f_a^*}{u_a})^\beta = t_b^0 + t_b^0(\beta+1)(\frac{f_b^*}{u_b})^\beta = \ldots = t_n^0 + t_n^0(\beta+1)(\frac{f_n^*}{u_n})^\beta \tag{3.14}$$

where the last line of (3.14) shows that f_a^*, f_b^*, ..., f_n^* are also S-O link flows [cf. (2.37)].

By combining the results of Case 1 and Case 2 and from the definitions of TC_{U-O}, TC_{U-O}^γ, TC_{S-O}, and TC_{S-O}^γ, we then have

$$TC_{U-O} = TC_{S-O} \quad \text{and} \quad TC_{U-O}^\gamma = TC_{S-O}^\gamma. \tag{3.15}$$

From the definitions of $\mathcal{I}_{U-O}^\gamma$ [cf. (3.10a)] and $\mathcal{I}_{S-O}^\gamma$ [cf. (3.10b)], we conclude that $\mathcal{I}_{U-O}^\gamma = \mathcal{I}_{S-O}^\gamma$, which completes the proof.

Theorem 3.5

Consider a network consisting of two nodes 1 and 2 as in Figure 3.5, which are connected by n parallel links. Assume that the associated BPR link cost functions [cf. (3.7a)] have $\beta = 1$. Furthermore, assume that there are positive flows on all the links at both the original and the partially degraded capacity levels given, respectively, by u and γu. Then the relative total cost index under the U-O flow pattern is given by the explicit formula

$$\mathcal{I}_{U-O}^\gamma = (\frac{\gamma U + kd_w}{\gamma U + k\gamma d_w} - 1) \times 100\%, \tag{3.16}$$

where d_w is the given demand for O/D pair $w = (1,2)$ and $U \equiv u_a + u_b + \ldots + u_n$. Moreover, the network robustness $\mathcal{I}_{U-O}^\gamma$ is bounded from above by $\frac{1-\gamma}{\gamma} \times 100\%$.

Proof: Clearly, because there is only a single O/D pair, we have

$$\frac{TC_{U-O}^\gamma}{TC_{U-O}} = \frac{d_w \times \lambda_w^\gamma}{d_w \times \lambda_w} = \frac{\lambda_w^\gamma}{\lambda_w}, \tag{3.17}$$

where λ_w^γ denotes the incurred equilibrium disutility for travelers between O/D pair w under the capacity retention ratio γ.

Due to the special structure of the network as well as the assumption that there are positive flows on all the links before and after the capacity reduction, by referring to the network equilibrium conditions [cf. (2.19)], we can write λ_w and λ_w^γ explicitly as follows

$$\lambda_w = t_a^0(1 + k\frac{f_a^*}{u_a}) = t_b^0(1 + k\frac{f_b^*}{u_b}) = \ldots = t_n^0(1 + k\frac{f_n^*}{u_n}), \tag{3.18}$$

where f_a^*, $f_b^* \ldots f_n^*$ are the equilibrium link flows under the link capacities: u_a, u_b,...,u_n, respectively, and

$$\lambda_w^\gamma = t_a^0(1 + k\frac{f_a^{**}}{\gamma u_a}) = t_b^0(1 + k\frac{f_b^{**}}{\gamma u_b}) = \ldots = t_n^0(1 + k\frac{f_n^{**}}{\gamma u_n}), \tag{3.19}$$

where f_a^{**}, f_b^{**},...,f_n^{**} are the equilibrium link flows under the link capacities: γu_a, γu_b,...,γu_n, respectively.

Hence we have

$$\frac{TC_{U-O}^\gamma}{TC_{U-O}} = \frac{\lambda_w^\gamma}{\lambda_w} = \frac{t_a^0(1 + k\frac{f_a^{**}}{\gamma u_a})}{t_a^0(1 + k\frac{f_a^*}{u_a})}$$

$$= \frac{t_b^0(1 + k\frac{f_b^{**}}{\gamma u_b})}{t_b^0(1 + k\frac{f_b^*}{u_b})} = \ldots = \frac{t_n^0(1 + k\frac{f_n^{**}}{\gamma u_n})}{t_n^0(1 + k\frac{f_n^*}{u_n})}, \tag{3.20}$$

which yields

$$\frac{TC_{U-O}^\gamma}{TC_{U-O}} = \frac{(1 + k\frac{f_a^{**}}{\gamma u_a}) + (1 + k\frac{f_b^{**}}{\gamma u_b}) + \ldots + (1 + k\frac{f_n^{**}}{\gamma u_n})}{(1 + k\frac{f_a^*}{u_a}) + (1 + k\frac{f_b^*}{u_b}) + \ldots + (1 + k\frac{f_n^*}{u_n})}. \tag{3.21}$$

After some simplification and from the fact that $f_a^* + f_b^* + \ldots + f_n^* = f_a^{**} + f_b^{**} + \ldots + f_n^{**} = d_w$, we obtain

$$\frac{TC_{U-O}^\gamma}{TC_{U-O}} = \frac{\gamma U + kd_w}{\gamma U + k\gamma d_w}. \tag{3.22a}$$

From the definition of $\mathcal{I}_{U-O}^\gamma$, we have

$$\mathcal{I}_{U-O}^\gamma = (\frac{TC_{U-O}^\gamma}{TC_{U-O}} - 1) \times 100\% = (\frac{\gamma U + kd_w}{\gamma U + k\gamma d_w} - 1) \times 100\%, \tag{3.22b}$$

which is exactly the form of (3.16).

To show the upper bound of $\mathcal{I}_{U-O}^\gamma$, we can rearrange (3.22*b*) to get the following form

$$\mathcal{I}_{U-O}^\gamma = [\frac{\gamma + k\frac{d_w}{U}}{\gamma(1 + k\frac{d_w}{U})} - 1] \times 100\%. \tag{3.23}$$

Because $\gamma \in (0, 1]$, we have the following

$$\mathcal{I}_{U-O}^\gamma \le [\frac{1 + k\frac{d_w}{U}}{\gamma(1 + k\frac{d_w}{U})} - 1] \times 100\% = \frac{1 - \gamma}{\gamma} \times 100\%, \tag{3.24}$$

which completes the proof.

Now we relax the assumptions in Theorem 3.5 in that

1. We do not require the network to be fully loaded before and after the capacity reduction.

2. We do not assume that $\beta = 1$.
3. We do not assume that the network has any special structure.

We now derive a general upper bound for $\mathcal{I}^{\gamma}_{S-O}$ in the following theorem under these relaxed assumptions.

Theorem 3.6
The upper bound for $\mathcal{I}^{\gamma}_{S-O}$ for a network with BPR link cost functions is $\frac{1-\gamma^{\beta}}{\gamma^{\beta}} \times 100\%$.

Proof: From the definition of TC_{S-O}, we have

$$TC_{S-O} = \sum_{a \in \mathcal{L}} t^0_a [1 + k(\frac{\hat{f}_a}{u_a})^{\beta}] \times \hat{f}_a, \tag{3.25}$$

where $\hat{f}_a, \quad \forall a \in \mathcal{L}$, is the S-O link flow pattern under the original link capacities u.

Because $\gamma \in (0, 1]$ and we assume that $t^0_a > 0, \quad \forall a \in \mathcal{L}$,

$$\frac{1}{\gamma^{\beta}} TC_{S-O} \geq \sum_{a \in \mathcal{L}} t^0_a [1 + k(\frac{\hat{f}_a}{\gamma u_a})^{\beta}] \times \hat{f}_a. \tag{3.26}$$

Furthermore, denote the S-O link flow pattern under the capacity retention ratio γ as $\widetilde{f}_a, \forall a \in \mathcal{L}$. Then, from the definition of TC^{γ}_{S-O}, we obtain

$$\sum_{a \in \mathcal{L}} t^0_a [1 + k(\frac{\hat{f}_a}{\gamma u_a})^{\beta}] \times \hat{f}_a \geq \sum_{a \in L} t^0_a [1 + k(\frac{\widetilde{f}_a}{\gamma u_a})^{\beta}] \times \widetilde{f}_a = TC^{\gamma}_{S-O}. \tag{3.27}$$

By combining (3.26) and (3.27), we have

$$\frac{1}{\gamma^{\beta}} TC_{S-O} \geq TC^{\gamma}_{S-O} \Longleftrightarrow \frac{TC^{\gamma}_{S-O}}{TC_{S-O}} \leq \frac{1}{\gamma^{\beta}}. \tag{3.28}$$

Hence

$$\mathcal{I}^{\gamma}_{S-O} \leq \frac{1 - \gamma^{\beta}}{\gamma^{\beta}} \times 100\%. \tag{3.29}$$

We have thus established the theorem.

3.4.2 A Network Example

We now present a numerical example for which the relative total cost indices are computed in the case of alternative decision-making behaviors.

Consider the following simple network as shown in Figure 3.8. There are two O/D pairs, namely, w_1=(1,3) and w_2=(1,4). The demands for the two O/D pairs are $d_{w_1} = 10$ and d_{w_2}=20. The paths connecting O/D pair w_1 are $p_1 = (a, b)$ and $p_2 = d$. The paths connecting O/D pair w_2 are $p_3 = (a, c)$ and $p_4 = e$. The capacities for links a, b, c, d, and e are 100, 50, 60, 10, and 20, respectively. Let t_0 and k [cf.

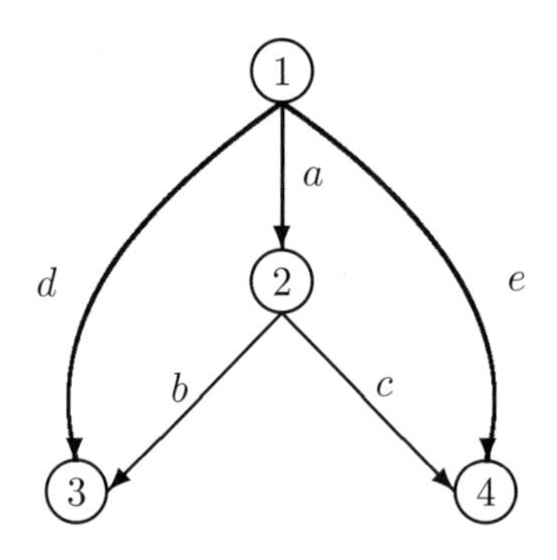

Figure 3.8: Network for the example in Section 3.4.2

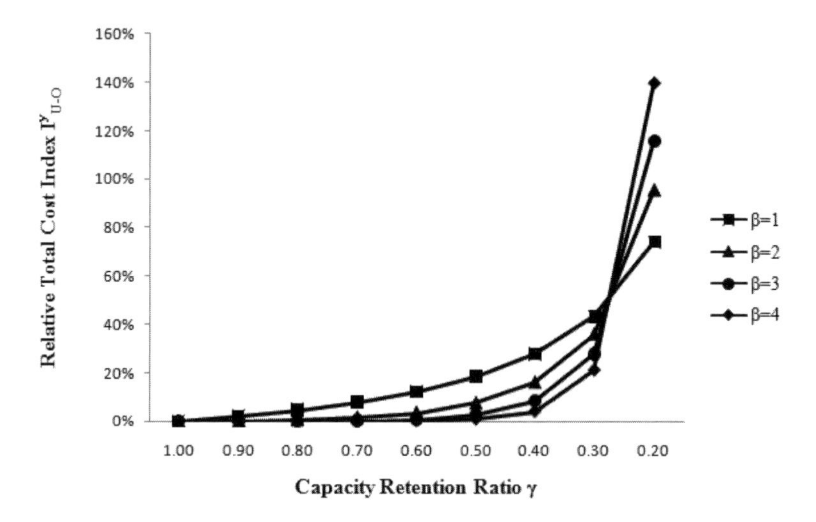

Figure 3.9: Relative total cost index under the U-O flow pattern for the example in Section 3.4.2

(3.7a)] be identical for all the links and equal to 10 and 1, respectively. The BPR link cost functions for the links in Figure 3.8 are hence given by

$$c_a(f_a) = 10[1 + (\frac{f_a}{100})^{\beta}], \quad c_b(f_b) = 10[1 + (\frac{f_b}{50})^{\beta}],$$

$$c_c(f_c) = 10[1 + (\frac{f_c}{60})^{\beta}], \quad c_d(f_d) = 10[1 + (\frac{f_d}{10})^{\beta}], \quad c_e(f_e) = 10[1 + (\frac{f_e}{20})^{\beta}].$$

Figures 3.9 and 3.11 present the relative total cost index for this example under the U-O flow pattern for β values equal to 1, 2, 3, and 4, respectively; γ in Figure 3.9 is the capacity retention ratio, and γ in Figure 3.11 is the capacity increment ratio.

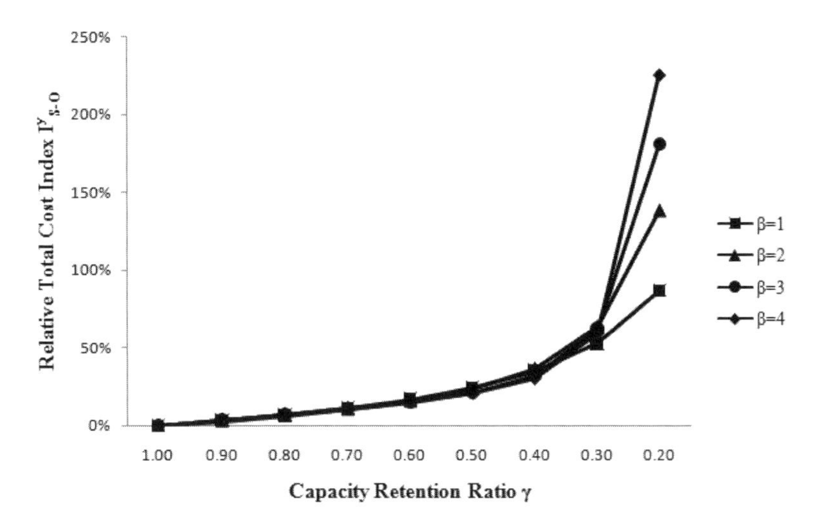

Figure 3.10: Relative total cost index under the S-O flow pattern for the example in Section 3.4.2

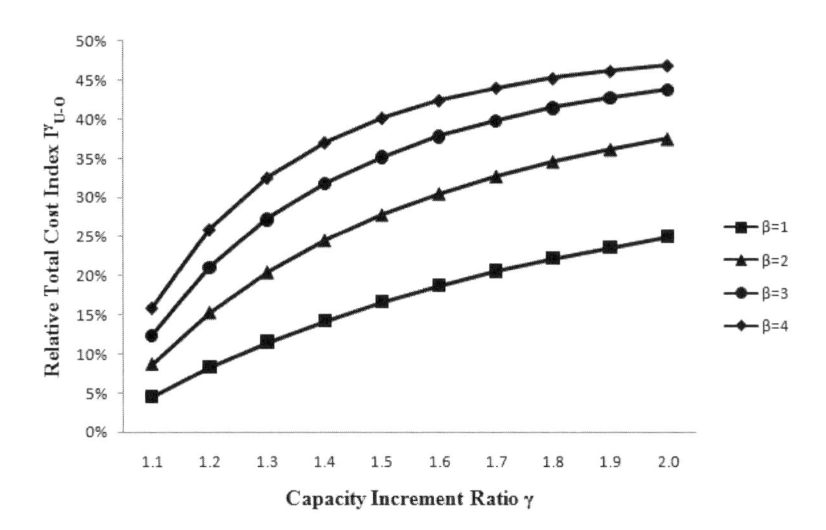

Figure 3.11: Relative total cost index under the U-O flow pattern for the example in Section 3.4.2 with capacity enhancement

Similarly, Figures 3.10 and 3.12 present the index for this example under the S-O flow pattern for β values equal to 1, 2, 3, and 4, respectively; γ in Figure 3.10 is the capacity retention ratio and γ in Figure 3.12 is the capacity increment ratio.

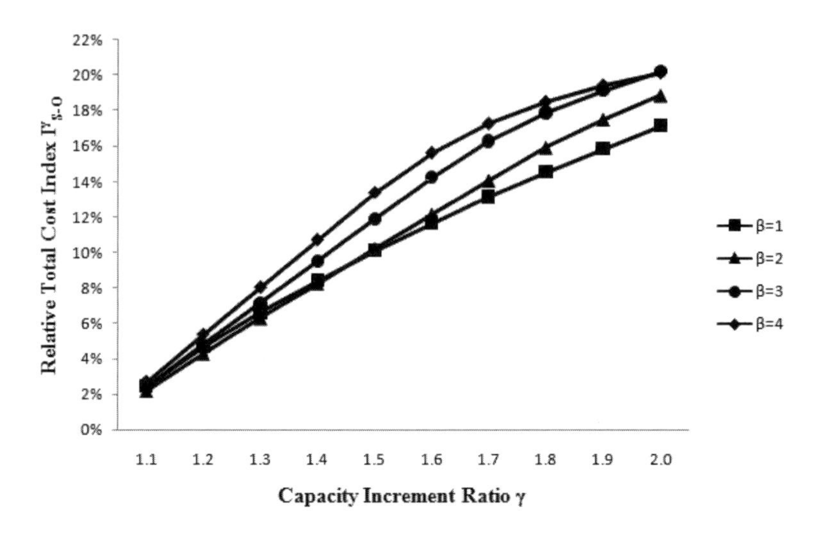

Figure 3.12: Relative total cost index under the S-O flow pattern for the example in Section 3.4.2 with capacity enhancement

It is interesting that from Figures 3.9 and 3.10 we can see that, although under the S-O flow pattern the total cost of the entire network is minimized, the relative total cost index under the U-O flow pattern is lower given the same capacity retention ratio γ. Therefore, this network under the U-O solution concept is more robust than under the S-O solution concept. Furthermore, from Figures 3.11 and 3.12, we can see that under the U-O flow pattern, the network has a larger relative total cost improvement as compared to that under the S-O flow pattern under the same capacity increment ratio.

3.5 SUMMARY AND CONCLUSIONS

In this chapter, we introduced a unified network efficiency/performance measure, denoted by $\mathcal{E}$, which can be applied to evaluate the network efficiency/performance of different types of networks, whether the demands on the network are fixed or elastic, and in the case of decentralized decision-making behavior. The measure assesses the network efficiency by incorporating flows and costs along with behavior, all important factors when dealing with network vulnerability. We also demonstrated that our network measure contains, as a special case, an existing widely used and applied efficiency measure in the complex network literature. Furthermore, the resulting importance definition of network components is well-defined even in the case of disconnected networks for our network efficiency measure.

In this chapter, we also compared our network performance measure $\mathcal{E}$ with several existing network centrality measures and explicated its advantages.

The advantages of the network performance measure $\mathcal{E}$ are summarized as follows:

- The network measure $\mathcal{E}$ is a weighted network measure that takes into account the information associated with the network flows, decentralized decision-making behavior/operation, and costs, all of which are the essential features of many network systems today, such as critical infrastructure networks from transportation to supply chains and the Internet. In particular, the measure $\mathcal{E}$ captures the readjustment of decision-makers on a network and the induced new flows, after the removal of network components.

- The network measure $\mathcal{E}$ can be used to analyze the importance of network components, that is, a node, a link, or a set of nodes and links, without assuming connectivity, after the network disruption, degradation, deterioration, or destruction.

- The importance measure $I(g)$ [cf. (3.5)], based on the network efficiency measure $\mathcal{E}$, is well-defined even when a given network becomes disconnected.

- The network measure $\mathcal{E}$ is applicable to networks with either fixed or elastic demands.

In addition, we presented a rigorous measure of network robustness [cf. (3.6)] based on the network performance measure $\mathcal{E}$ and derived some theoretical results. We also introduced relative total cost indices [cf. (3.10a) and (3.10b)] to study network robustness from the total network cost point of view under distinct decision-making behaviors, corresponding to either centralized (i.e., system-optimizing) behavior or decentralized (i.e., user-optimizing) behavior. Moreover, the results were applied to study numerical network examples, including several motivated by the classical Braess (1968) paradox.

3.6 SOURCES AND NOTES

This chapter is based on the papers by Nagurney and Qiang (2007a-c, 2008a) and on Qiang and Nagurney (2008). In particular, Section 3.2 is based on the paper by Qiang and Nagurney (2008), in which the unified network performance measure was first proposed. Section 3.3 is based on the paper by Nagurney and Qiang (2007b), in which we discussed the network robustness measure under decentralized decision-making behavior. The discussion on network robustness and capacity enhancement in this chapter has not been reported elsewhere. Section 3.4 is based on the paper by Nagurney and Qiang (2009) where we studied network robustness from the total network cost point of view.

It is important to recognize that other critical infrastructure networks, such as electric power supply chain networks can be transformed into transportation network equilibrium problems [cf. Nagurney (2006a)] as can decentralized supply chain networks, in general, and multitiered financial networks [see also, e.g., Nagurney (2006b) and Liu and Nagurney (2007)]. Hence the approach discussed in this chapter is also applicable to the formal and quantifiable analysis of the vulnerability and robustness of a variety of critical infrastructure networks. We further explore such networks in subsequent chapters of this book.

In this chapter, we made rigorous the important concept of *network robustness*, when links are faced with possible capacity deterioration. Such an approach can be applied to a spectrum of network systems with different link cost functions, ranging from transportation and logistics networks to telecommunication networks and the Internet. For example, Bertsekas and Gallager (1987) and Roughgarden (2005) discussed user link cost functions with embedded capacities [cf. (3.7c)] for queueing networks that arise in telecommunication networks and the Internet. Furthermore, we identified the relationship between the ratio of two proposed network robustness indices and the price of anarchy, a term coined by Papadimitrou (2001) for the ratio first proposed by Koutsoupias and Papadimitrou (1999). Subsequently, Perakis (2007) characterized the price of anarchy in the case of nonseparable, asymmetric user link cost functions, thus, generalizing earlier work that considered separable cost functions.

Zhu et al. (2006) introduced another measure, which we denote by $\hat{E}(\mathcal{G})$, which they then applied to gauge the efficiency of the Chinese airline transportation network with fixed demands. Their network performance measure is characterized by the average social travel cost, which is represented below (with their notation adapted to the one in this chapter, for clarity) as follows

$$\hat{E}(\mathcal{G}) = \frac{\sum_{w \in W} \lambda_w d_w}{\sum_{w \in W} d_w}.$$

The Zhu et al. (2006) measure is an average disutility weighted by the demands. It can be used for networks with fixed demands, provided that the network does not get disconnected, in which case the measure becomes undefined. Indeed, a very important feature of our network measure $\mathcal{E}$ is that there is no assumption made that the network needs to be connected. In contrast, the Zhu et al. (2006) measure requires such an assumption because, otherwise, their network performance measure becomes infinite-valued.

PART II

APPLICATIONS AND EXTENSIONS

CHAPTER 4

APPLICATION OF THE MEASURES TO TRANSPORTATION NETWORKS

Transportation networks are the pillars of a nation's economy, and a local transportation incident may have global consequences. Indeed, several recent disasters have caused severe damage to transportation networks and the impacts have propagated for many miles. For example, the infamous Hurricane Katrina cost a total of $81.2 billion dollars [U.S. Department of Commerce (2006)] and 1,836 people lost their lives [Louisiana Department of Health and Hospitals (2006)]. Moreover, Hurricane Katrina engulfed 45 bridges, and the traffic that resulted from re-routing significantly increased the congestion on highway I-10 [cf. DesRoches and Rix (2006)]. As another example, the 2008 Chinese winter storm caused $12 billion in damage to the nation's economy [BBC News (2008a)]. Due to the transport breakdowns, more than 180 million people were stranded on their way home during the Chinese holiday season. Moreover, millions of people suffered from power outages, plus there was substantial food price inflation because of the transportation delays in the delivery of raw materials and crops [BBC News (2008b)]. All these disasters remind us of how vulnerable our transportation networks are and how their absence and/or deterioration can significantly affect not only our daily lives but also our economies.

The impact of the degradation of transportation network infrastructure is now being increasingly documented. Indeed, the gradual deterioration of the U.S. transportation

Fragile Networks. By Anna Nagurney and Qiang Qiang

networks is now affecting the nation's economy. In particular, a recent American Society of Civil Engineers (ASCE) survey (2005) noted that over one quarter of the nation's 590,750 bridges were rated as structurally deficient or functionally obsolete. The degradation of transportation networks due to poor maintenance, natural disasters, deterioration over time, and unforeseen attacks now leads to estimates of $94 billion in terms of needed repairs for U.S. roads alone. Poor road conditions in the United States cost motorists $54 billion in repairs and operating costs annually.

Due to the deteriorating conditions of U.S. transportation networks, coupled with increasing number of vehicles, American commuters now spend 3.5 billion hours a year stuck in traffic, which translates to a cost of $63.2 billion a year to the economy [ASCE (2005)]. At the same time, and as noted in Chapter 1, a report from the U.S. Department of Transportation Federal Highway Administration (2006a) states that the country is experiencing a freight capacity crisis that threatens the strength and productivity of its economy. According to the American Road & Transportation Builders Association [see Jeanneret (2006)], nearly 75% of U.S. freight is carried on highways, and bottlenecks are causing truckers 243 million hours of delay each year, with an estimated associated cost of $8 billion. The U.S. government is facing a $1.6 trillion deficit over the next 5 years in terms of repairing and reconstructing the infrastructure, according to one estimate [Environment News Service (2008)]. Hence the construction of suitable transportation network robustness measures is of both theoretical and practical importance.

In addition, the increase in the number of vehicles over the years has well surpassed the development of transportation infrastructure, which adds another stress to the already ailing transportation networks. According to a recent report by the U.S. Department of Transportation Federal Highway Administration (2006b), the increase in congestion and associated congestion costs reflect the fact that over the last 20 years the vehicle miles of travel (VMT) on U.S. roads have nearly doubled while the length of roads in terms of lane miles has increased only about 4%. Figure 4.1 compares the growth in VMT to the growth in lane miles.

On the other hand, the increasing amount of congestion escalates environmental emissions, which, in turn, further affect the environment negatively, which expedites the deterioration of the critical road infrastructure, thus, creating a vicious cycle. Recently, a report conducted by the Committee on Climate Change and the U.S. Transportation Research Board (2006) studied the impact of climate change on the nation's transportation networks. The study showed that crucial buffer zones that once protected infrastructure are diminishing due to erosion and loss of wetlands with an estimated 60,000 miles of coastal highways already exposed to periodic storm flooding. Moreover, climate change not only affects the coastal areas but the Midwest of the United States, as well. For example, the study further showed that the intense precipitation that occurred in 1993 caused severe flooding that damaged hundreds of farmland, towns, and transportation routes along 500 miles of the Mississippi and Missouri river systems. The interaction between climate change and the effects on transportation networks can also be found in numerous other studies [cf. National Assessment Team (2001), U.S. Department of Transportation (2002), Smith and Levasseur (2002), Zimmerman (2003), Arkell and Darch (2006), and Schulz (2007)].

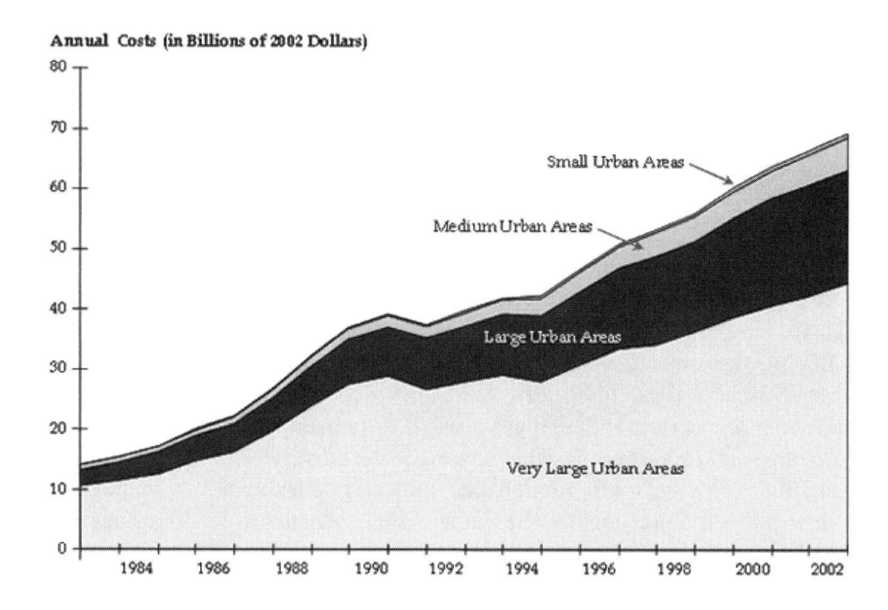

Figure 4.1: Vehicle miles of travel and roadway lane miles growth index, 1980 to 2002 [Source: U.S. Department of Transportation Federal Highway Administration (2006b)]

Emissions generated through transportation are some of the biggest contributors to climate change and global warming. According to a U.S. EPA (2006) report, the transportation sector in 2003 accounted for 27% of the total greenhouse gas emissions in the U.S., and the increase in this sector was the largest of any in the period 1990 - 2003. In addition, the energy use due to transportation is expected to increase by 48% between 2003 and 2025, even with modest improvements in the efficiency of vehicular engines. It is interesting that a study undertaken by a Norwegian research organization [Knudsen and Bang (2007)] claims that infrastructure capacity increases are directly linked to decreases in polluting emissions from motor vehicles. Using a traffic microsimulation, the study showed, for example, that upgrading narrow, winding roads or adding a lane to a congested motorway can yield decreases of up to 38% in CO_2 emissions, 67% in CO emissions, and 75% in NO_x emissions, without generating substantially more car trips. Therefore, it is important to investigate the relationship between the network capacity and the corresponding environmental effects. With a robust transportation network, we can expect less congestion and, therefore, fewer emissions, which, in turn, is expected to slow down global warming and to reduce the possibility of extreme weather. Thus, a positive cycle is established.

From this discussion, we can see that to better protect transportation networks, the network administrator needs to have a clear idea as to where the vulnerable nodes and links are in order to hedge against unexpected disruptions, be it natural disasters, terrorist attacks, or unintentional accidents. In addition, it is crucial to understand how a transportation network reacts to capacity degradation under alternative user travel behaviors. Furthermore, with the closely interwoven relationship between climate change and transportation network degradation, we believe that it is also important to discuss the environmental "robustness" of transportation networks in terms of the total emissions using empirically calibrated emission functions that are now available.

More and more researchers in transportation have realized the urgent need to study transportation network vulnerability and robustness. Nicholson and Du (1997) used system surplus as a performance measure to analyze degradable transportation networks with elastic demand. Jenelius, Petersen, and Mattsson (2006) proposed several link importance indicators and applied them to the road transportation network in northern Sweden. These indicators are distinct, depending on whether or not there exist disconnected origin and destination pairs in a network. Moreover, Murray-Tuite and Mahmassani (2004) proposed two scenario-based network indices, namely, the vulnerability index and the disruption index, to study the vulnerability of the network that faces an intentional attack. The vulnerability index is for an origin/destination pair under each scenario, and the set of links with high vulnerability index values is considered to be important for connecting that origin/destination pair. The disruption index for the network under a scenario is defined as the aggregation of these vulnerability indices across all the O/D pairs. The authors further used the disruption index as the input to a bilevel programming problem that they proposed to model the game between the attacker and the network administrator. Bell (2000) used a game theory approach to analyze transportation network performance between the users who try to minimize the expected travel cost and an "evil" entity who wishes to maximize the total travel cost on the network. Sakakibara, Kajitani, and Okada (2004) proposed a topological index and considered a transportation network to be robust if it is "dispersed" in terms of the number of links connected to each node. Scott et al. (2006), on the other hand, examined transportation network robustness by computing the increase in the total network cost evaluated at the U-O flow pattern after the removal of each link.

Moreover, Taylor, Sekhar, and D'Este (2006) applied three accessibility indices, namely, the change in generalized travel cost, the Hansen integral accessibility index, and the ARIA index, to study the vulnerability of the Australian road network. A node is considered to be vulnerable if the loss (or the substantial degradation) of a small number of links significantly diminishes the accessibility of the node, while a link is deemed to be critical if the loss (or substantial degradation) of the link significantly diminishes the accessibility of the network or of particular nodes. Dueñas-Osorio, Craig, and Goodno (2005) defined a disruption index for measuring the network vulnerability based on the average service reduction after disruptions occur.

In this chapter, we apply the network measures proposed in Chapter 3 to study the vulnerability and robustness of transportation networks. Furthermore, we also analyze the environmental impact of transportation network capacity degradation. This chapter is organized as follows. In Section 4.1, the network performance measure $\mathcal{E}$ [cf. (3.2)] is applied to study a larger network example, for which the importance of links is computed and the links are ranked. In Section 4.2, we apply the network

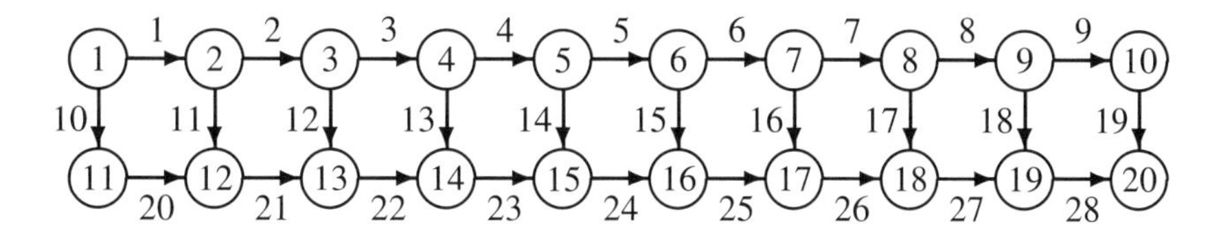

Figure 4.2: Transportation network example in Section 4.1

performance measure $\mathcal{E}$ to study link importance of the Sioux Falls (South Dakota) network based on the BPR link cost functions [cf. (3.7a)]. Furthermore, the network robustness measure and the relative total cost indices are calculated. In Section 4.3, to demonstrate the relevance and practical usefulness of the network measures introduced in Chapter 3, we calculate the network robustness measure and the relative total cost indices for a larger network, the Anaheim (California) network. In Section 4.4, based on carbon monoxide (CO) emission functions from the literature, we propose environmental impact assessment indices for transportation networks under different travel behaviors. In addition, we introduce a link importance indicator that allows one to determine the ranking of the links in terms of their impact on the environment in case they are removed or destroyed. We then apply the environmental impact assessment indices in Section 4.5 to study the Sioux Falls network and the Anaheim (California) network. We summarize this chapter in Section 4.6.

4.1 A LARGER FIXED DEMAND NETWORK

The transportation network example used for this section consisted of 20 nodes, 28 links, and 2 O/D pairs and is depicted in Figure 4.2.

A similar transportation network was used in Nagurney (1984) where it is referred to as Network 20 [see also Dhanda, Nagurney, and Ramanujam (1999)]. For simplicity, and easy reproducibility, we considered separable user link cost functions, which were adapted from Network 20 in Nagurney (1984) with the cross-terms removed.

The O/D pairs were $w_1 = (1, 20)$ and $w_2 = (1, 19)$, and the travel demands were $d_{w_1} = 100$, and $d_{w_2} = 10$. The link cost functions are given in Table 4.1.

The projection method (cf. Section 2.4.2) was used along with the embedded Dafermos and Sparrow (1969) equilibration algorithm (cf. Section 2.4.1) to compute the equilibrium solutions and to determine the network efficiency $\mathcal{E}$ according to (3.2) and the importance values $I(g)$ of the links according to (3.5) and their importance rankings.

The computed efficiency measure for this transportation network is $\mathcal{E} = .002518$. The computed importance values of the links and their rankings for this network are also reported in Table 4.1.

Table 4.1: Transportation network example in Section 4.1 - links, link cost functions, importance values, and importance rankings

Link a	Link Cost Function $c_a(f_a)$	Importance Value	Importance Ranking
1	$.00005f_1^4 + 5f_1 + 500$	.9086	3
2	$.00003f_2^4 + 4f_2 + 200$	.8984	4
3	$.00005f_3^4 + 3f_3 + 350$	.8791	6
4	$.00003f_4^4 + 6f_4 + 400$	.8672	7
5	$.00006f_5^4 + 6f_5 + 600$	.8430	9
6	$7f_6 + 500$	.8226	11
7	$.00008f_7^4 + 8f_7 + 400$	.7750	12
8	$.00004f_8^4 + 5f_8 + 650$	.5483	15
9	$.00001f_9^4 + 6f_9 + 700$	.0362	17
10	$4f_{10} + 800$	.6641	14
11	$.00007f_{11}^4 + 7f_{11} + 650$	.0000	22
12	$8f_{12} + 700$	.0006	20
13	$.00001f_{13}^4 + 7f_{13} + 600$	.0000	22
14	$8f_{14} + 500$	.0000	22
15	$.00003f_{15}^4 + 9f_{15} + 200$	.0000	22
16	$8f_{16} + 300$	.0001	21
17	$.00003f_{17}^4 + 7f_{17} + 450$	.0000	22
18	$5f_{18} + 300$	.0175	18
19	$8f_{19} + 600$	.0362	17
20	$.00003f_{20}^4 + 6f_{20} + 300$	.6641	14
21	$.00004f_{21}^4 + 4f_{21} + 400$	.7537	13
22	$.00002f_{22}^4 + 6f_{22} + 500$	.8333	10
23	$.00003f_{23}^4 + 9f_{23} + 350$	.8598	8
24	$.00002f_{24}^4 + 8f_{24} + 400$	.8939	5
25	$.00003f_{25}^4 + 9f_{25} + 450$	.4162	16
26	$.00006f_{26}^4 + 7f_{26} + 300$	.9203	2
27	$.00003f_{27}^4 + 8f_{27} + 500$	.9213	1
28	$.00003f_{28}^4 + 7f_{28} + 650$	.0155	19

Figure 4.3 displays the importance values and the importance rankings of the links for this transportation network example. From the results in the figure, it is clear that transportation planners and network security officials should pay most attention to links 27, 26, 1, and 2 because these are the top four links in terms of importance rankings. On the other hand, the elimination of links 11, 13, 14, 15, and 17 should have no impact on the network performance/efficiency. Of course, it is important to note that in this example we considered fixed travel demands associated with the O/D pairs of travel. Hence, such an approach would apply, for example, to the commuting

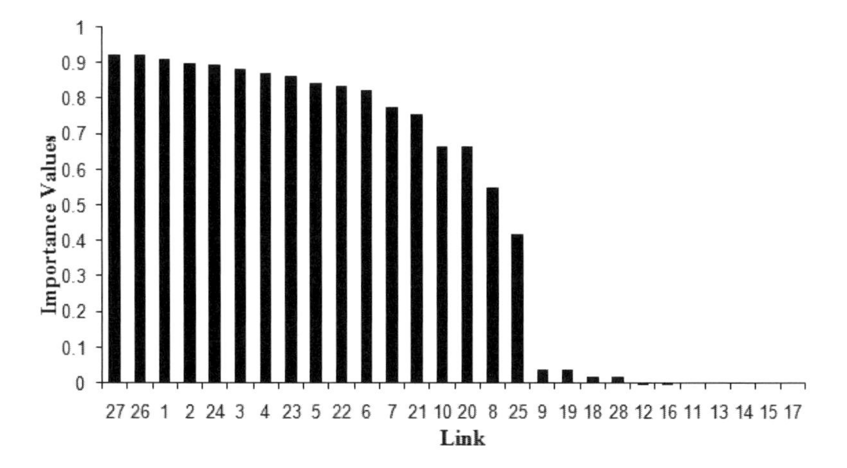

Figure 4.3: Link importance rankings of the network example in Section 4.1

periods, when one would expect the demand for travel would be the highest (and also consistent and reproducible over the commuting time periods under consideration).

4.2 THE SIOUX FALLS NETWORK

In this section, we apply the network measure $\mathcal{E}$ to the Sioux Falls network [cf. LeBlanc, Morlok, and Pierskalla (1975)]. The user link cost functions in this network are of the form of the BPR link travel cost functions [cf. (3.7a)]. The network topology of the Sioux Falls network is displayed in Figure 4.4.

4.2.1 Importance of Links in the Sioux Falls Network

There are 24 nodes, 76 links, and 528 O/D pairs of nodes in the Sioux Falls network. The relevant data and parameters in the BPR link cost functions are from LeBlanc, Morlok, and Pierskalla (1975) and the transportation network datasets maintained by Bar-Gera (http://www.bgu.ac.il/~bargera/tntp/). We used the projection method (cf. Section 2.4.2) with the embedded equilibration algorithm (cf. Section 2.4.1) and the column generation algorithm [cf. Leventhal, Nemhauser, and Trotter (1973)] to compute the equilibrium solutions. Then, based on these solutions, the network efficiency according to (3.2) and the importance values and the importance rankings of the links according to (3.5) were determined. The computational schemes were implemented in MATLAB (www.mathworks.com) on an IBM T61 computer.

We note that Bar-Gera (2002) proposed an origin-based algorithm to solve transportation network equilibrium problems. The author shows that the algorithm out-

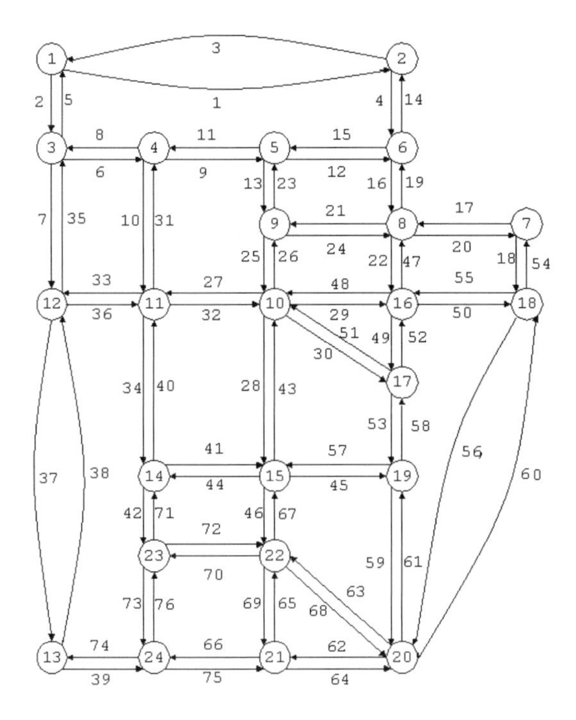

Figure 4.4: The Sioux Falls network

performs several existing algorithms. However, due to the scope of this book, we do not compare the efficiency of different algorithms here. Our solution approach serves our purpose sufficiently.

The computed network efficiency measure $\mathcal{E}$ for the Sioux Falls network is $\mathcal{E} = 47.6092$. The resulting link importance values $I(g)$ and rankings are depicted in Figure 4.5. From Figure 4.5, it can be seen that links 56, 60, 36, and 37 are the most important links, and hence special attention should be paid to protect these links accordingly, while the removal of links 10, 31, 4, and 14 would cause the least efficiency loss.

4.2.2 Robustness of the Sioux Falls Network

We can also study the robustness of the Sioux Falls network by using the network robustness measure defined by (3.6). The results for different capacity retention ratios are displayed in Figure 4.6. From the figure we can see that the robustness of the Sioux Falls network drops quickly when γ varies from .9 to .4, while it remains comparatively constant when γ is below .3.

Moreover, we can also study the robustness of a network when the network capacities are enhanced (cf. Remark 3.2). Figure 4.7 presents the network robustness

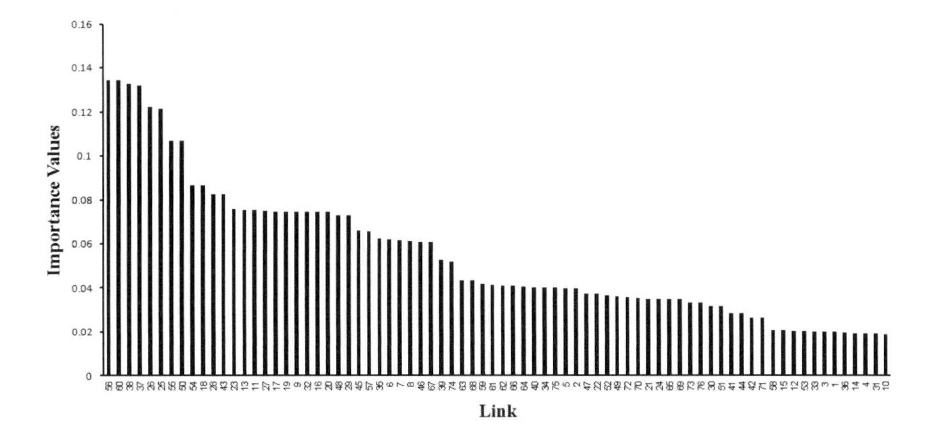

Figure 4.5: The Sioux Falls network link importance rankings

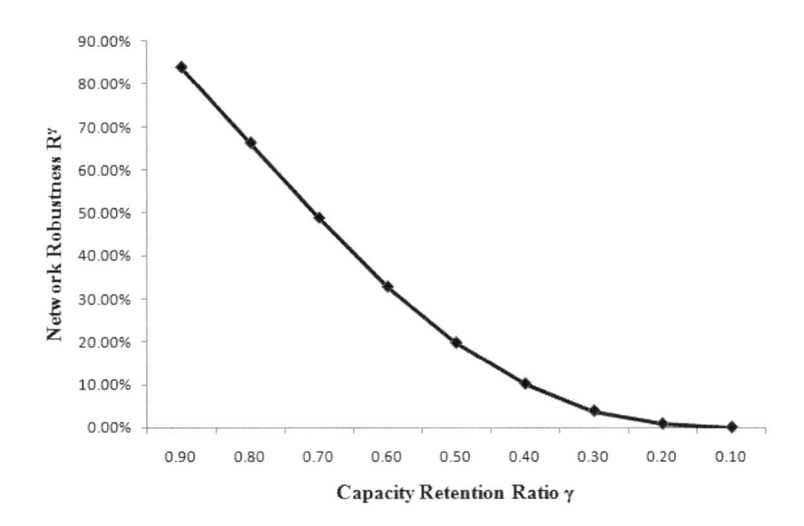

Figure 4.6: Robustness vs. capacity retention ratio for the Sioux Falls network

for the Sioux Falls network with capacity enhancements. The concave shape of the curve in Figure 4.7 suggests that the Sioux Falls network has a decreasing rate of robustness improvement under uniform capacity increases.

4.2.3 The Relative Total Cost Indices for the Sioux Falls Network

We now compute the relative total cost indices I^{γ}_{U-O} and I^{γ}_{S-O} according to (3.10a) and (3.10b) for the Sioux Falls network to study its network robustness under alter-

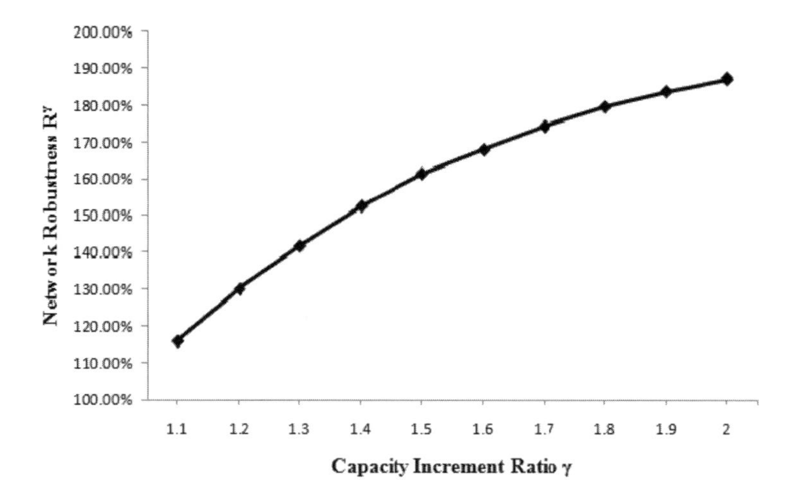

Figure 4.7: Robustness vs. capacity increment ratio for the Sioux Falls network

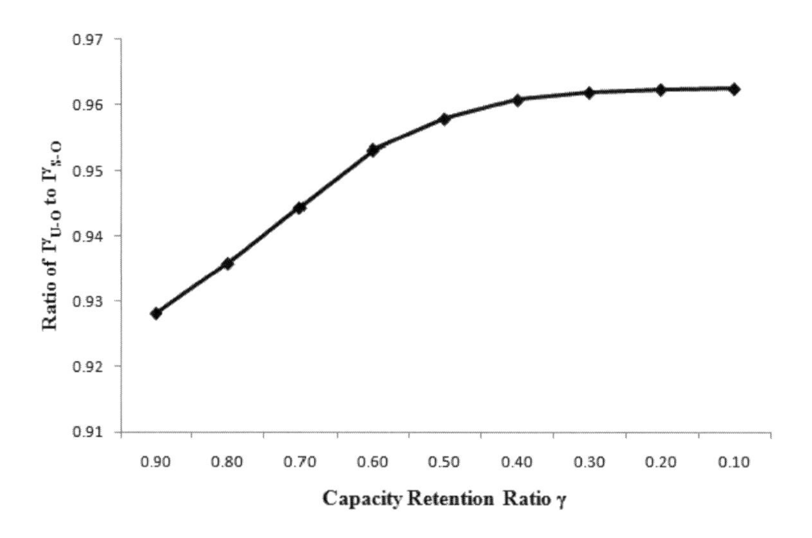

Figure 4.8: Ratio of I^{γ}_{U-O} to I^{γ}_{S-O} for the Sioux Falls network under the capacity retention ratio γ

native user behaviors. Figure 4.8 displays the ratio of I^{γ}_{U-O} to I^{γ}_{S-O} for the Sioux Falls network. It is interesting that from Figure 4.8 we can see that although the S-O flow pattern leads to the minimum total cost for the Sioux Falls network, the network is more robust under the U-O flow pattern.

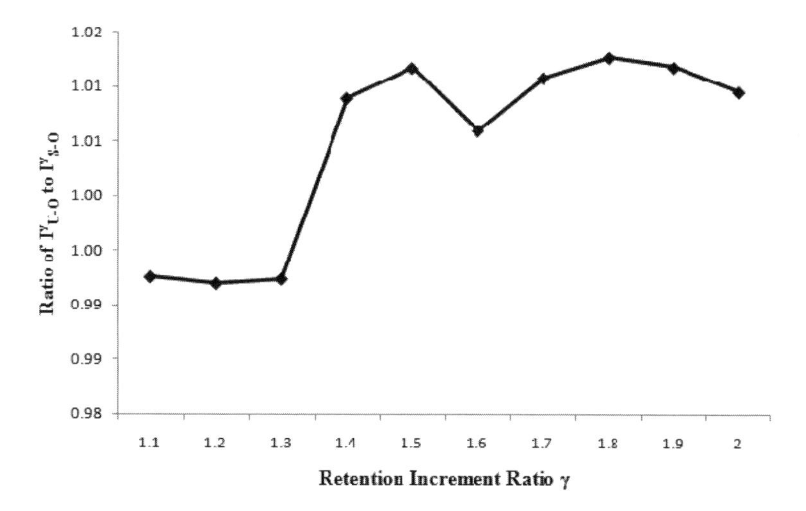

Figure 4.9: Ratio of I^{γ}_{U-O} to I^{γ}_{S-O} for the Sioux Falls network under capacity increment ratio γ

Figure 4.9 displays the ratio of I^{γ}_{U-O} to I^{γ}_{S-O} for the Sioux Falls network when the capacity is enhanced (cf. Remark 3.3). It can be seen from Figure 4.9 that the Sioux Falls network has a better relative total cost improvement under the U-O flow pattern when γ is above 1.4, whereas the relative total cost under the S-O flow pattern outperforms that under the U-O user behavior when γ is below 1.3.

4.3 THE ANAHEIM NETWORK

In this section, we apply the robustness measure [cf. (3.6)] to study the Anaheim network. Furthermore, the total relative cost indices [cf. (3.10a) and (3.10b)] are used to study the robustness of this transportation network under different user behaviors. Similar to the network studied in Section 4.2, each link of the Anaheim network has a link travel cost functional form that is of the BPR form [cf. (3.7a)].

The network topology is given in Figure 4.10.

There are 461 nodes, 914 links, and 1,406 O/D pairs in the Anaheim network. The relevant data and parameters in the BPR link cost functions are from the transportation network datasets maintained by Bar-Gera (http://www.bgu.ac.il/~bargera/tntp/). We used the projection method (cf. Section 2.4.2) with the embedded equilibration algorithm (cf. Section 2.4.1) and the column generation algorithm [cf. Leventhal, Nemhauser, and Trotter (1973)] to compute the equilibrium solutions. Then, based on these solutions, the network efficiency according to (3.2) and the importance values and the importance rankings of the links according to (3.5) were determined. The above computation schemes were implemented in MATLAB (www.mathworks.com)

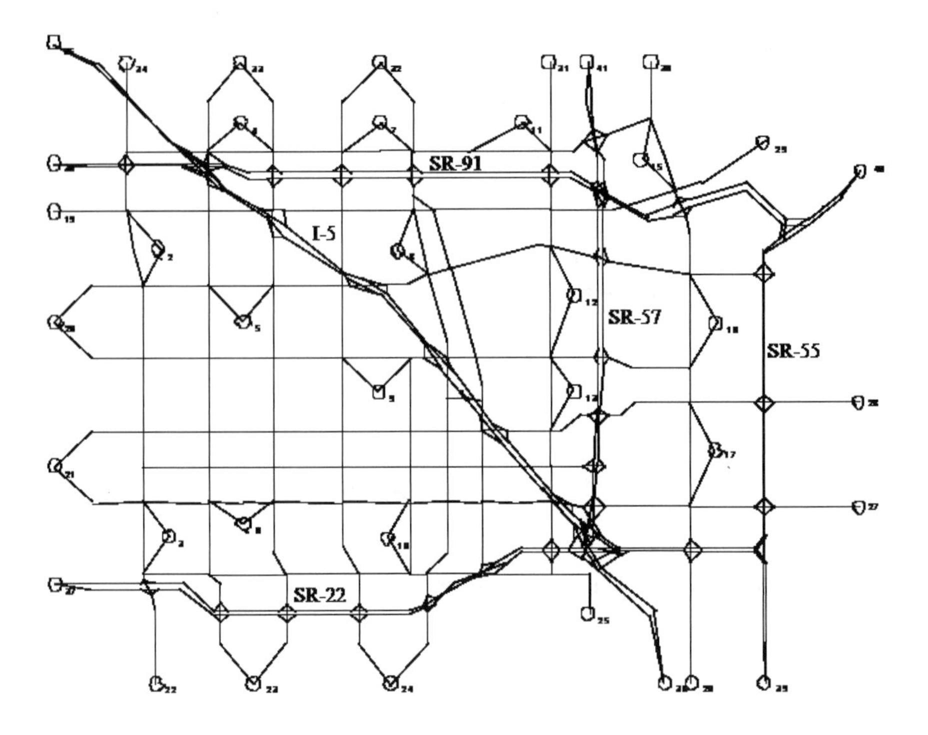

Figure 4.10: The Anaheim network (Source: http://www.bgu.ac.il/~bargera/tntp/)

on an IBM T61 computer. The computed network efficiency measure for the Anaheim network is $\mathcal{E} = 7.3651$.

4.3.1 Robustness of the Anaheim Network

We now study the robustness, defined by (3.6), of the Anaheim network. The results for different capacity retention ratios are displayed in Figure 4.11.

Figure 4.12 presents the robustness of the Anaheim network with capacity enhancement (cf. Remark 3.2). We see that the robustness of the Anaheim network keeps increasing smoothly except for when γ is between 1.2 and 1.3, where the robustness increment is "flat."

4.3.2 The Relative Total Cost Indices for the Anaheim Network

We now compute the relative total cost indices I^{γ}_{U-O} and I^{γ}_{S-O} according to (3.10a) and (3.10b) for the Anaheim network to study its robustness under alternative user behaviors. In Figure 4.13, we graph the ratio I^{γ}_{U-O} to I^{γ}_{S-O} for this network.

From Figure 4.13, we can see that the Anaheim network under the S-O solution is more robust in terms of the relative total cost increase when the capacity retention ratio

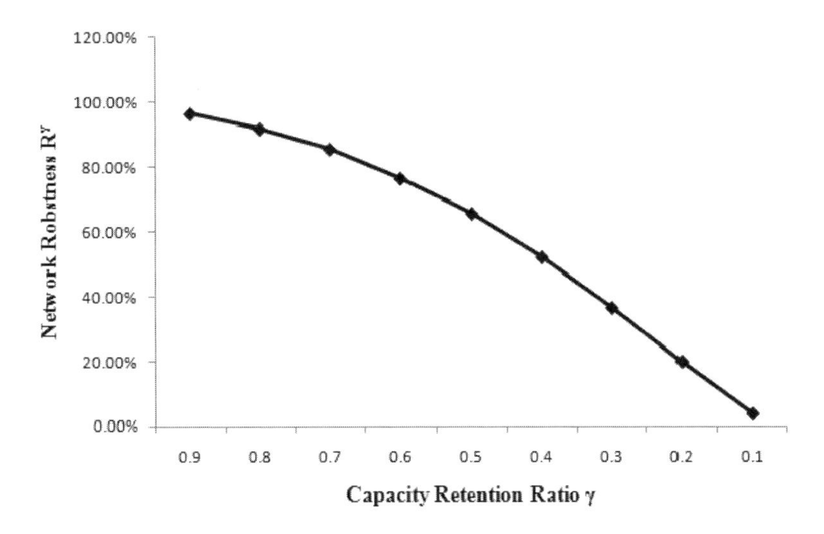

Figure 4.11: Robustness vs. capacity retention ratio for the Anaheim network

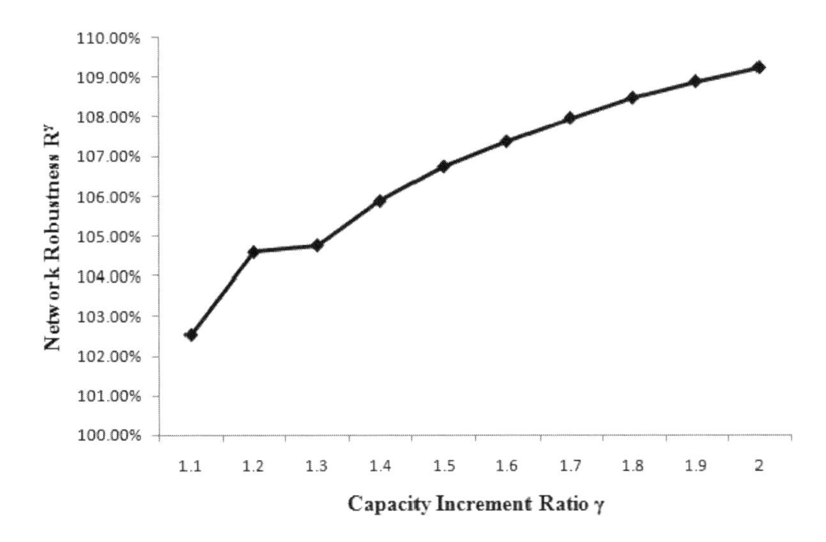

Figure 4.12: Robustness vs. capacity increment ratio for the Anaheim network

γ is above .3, whereas the U-O solution leads to lower relative total cost increases; therefore, the network is more robust when γ is below .3.

Figure 4.14 displays the ratio of I^{γ}_{U-O} to I^{γ}_{S-O} for the Anaheim network when the capacity is enhanced (cf. Remark 3.3). It is interesting that unlike the Sioux Falls network, when the capacity incremental ratio γ is below 1.2, the U-O solution leads

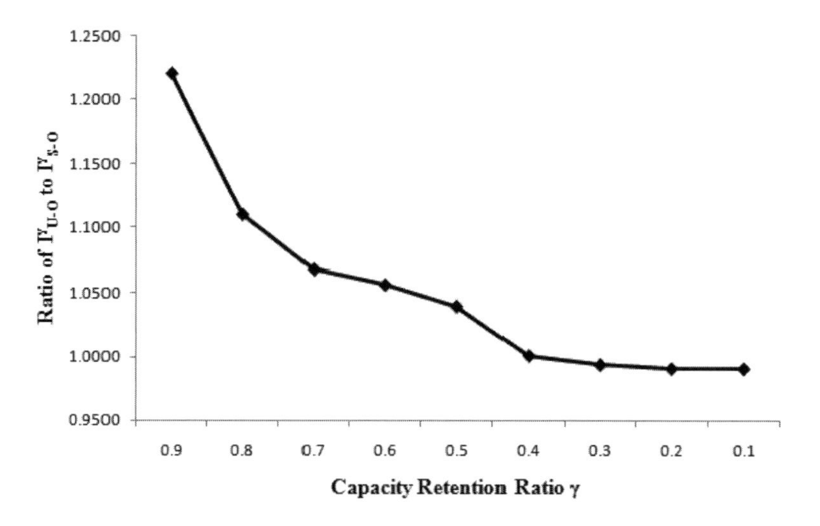

Figure 4.13: Ratio of I^{γ}_{U-O} to I^{γ}_{S-O} for the Anaheim network under capacity retention ratio γ

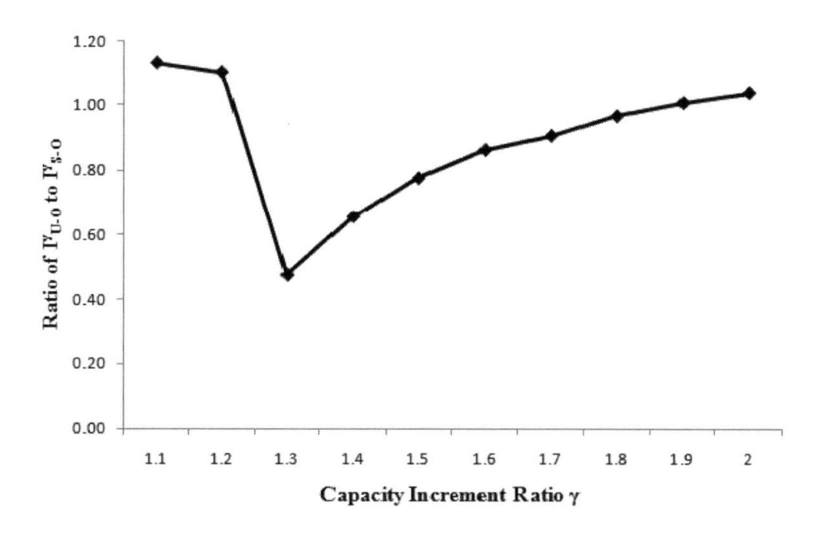

Figure 4.14: Ratio of I^{γ}_{U-O} to I^{γ}_{S-O} for the Anaheim network under capacity increment ratio γ

to a better relative total cost improvement in the Anaheim network, whereas the S-O solution results in a better relative total cost saving when γ is above 1.3.

4.4 THE ENVIRONMENTAL ROBUSTNESS AND LINK IMPORTANCE OF TRANSPORTATION NETWORKS

In this section, we turn to the rigorous assessment of the environmental robustness of transportation networks. We begin with a discussion of empirical environmental emission functions.

4.4.1 Environmental Emissions

Alexopoulos and Assimacopoulos (1993) argued that because carbon monoxide (CO) is emitted exclusively by vehicular traffic it is important as an indicator for the level of atmospheric pollution generated by such traffic. Moreover, it has been shown that CO is the most significant pollutant among all other types of vehicle emissions [cf. U.S. Department of Transportation Federal Highway Administration (2006c)]. Furthermore, it is noted that other pollutants that are related to congestion exhibit similar behavior [cf. Hizir (2006) and the California Air Resource Board (2005)]. Analogous arguments have also been made by, among others, Yin and Lawphongpanich (2006) [see also Rilett and Benedek (1994), Wallace et al. (1998), and Sugawara and Niemeier (2002)]. In particular, we note that Yin and Lawphongpanich (2006) utilized the following function to estimate vehicular CO emissions, which was adopted from the macroscopic relationship model of Wallace et al. (1998)

$$e_a(f_a) = .2038 \times c_a(f_a) \times e^{.7962 \times (\frac{l_a}{c_a(f_a)})}, \quad \forall a \in \mathcal{L}, \tag{4.1}$$

where l_a denotes the length of link a and c_a corresponds to the travel time (in minutes) to traverse link a. The length l_a is measured in kilometers for each link $a \in \mathcal{L}$ and the emissions are in grams per hour. The expression for total CO emissions on a link a, denoted by $\hat{e}_a(f_a)$, is then given by

$$\hat{e}_a(f_a) = e_a(f_a) \times f_a, \quad \forall a \in \mathcal{L}. \tag{4.2}$$

The total emissions of CO generated on a network is denoted by TE and is hence

$$\text{TE} = \sum_{a \in \mathcal{L}} \hat{e}_a(f_a). \tag{4.3}$$

In this section, we are interested in determining TE evaluated at the U-O solution and at the S-O solution for given networks as the capacity of the links on the networks degrades.

We point out that Hizir (2006) derived total emission functions in the case of multiple pollutants, notably, of carbon dioxide (CO_2) and of nitrous oxide (NO_x) that are similar to the above total emission functions, except with differing parameters. Moreover, the emission functions also included link capacities, as in the case of (4.1) with BPR functions. A similar approach was taken by Akcelik and Besley (2003) in the case of CO, CO_2, hydrocarbon (HC), and NO_x.

4.4.2 Environmental Impact Assessment Indices for Transportation Networks

We here propose indices based on the relative total emissions generated that assess the environmental impact of the degradation of a transportation network based on the two behavioral solution concepts, namely, under the user-optimizing flow pattern, denoted by TE_{U-O}, and the system-optimizing flow pattern, denoted by TE_{S-O}. In particular, TE_{U-O} denotes the total emissions on the network as given by expression (4.2), where the vector f is the solution to the user-optimizing (or transportation network equilibrium) conditions (cf. (2.19), (2.23), and (2.28) in Chapter 2). On the other hand, TE_{S-O} represents the total emissions generated on the network according to expression (4.2) but now evaluated at the flow pattern given by the solution to the S-O problem (cf. (2.38) in Chapter 2). We believe that the total emissions generated are an appropriate measure since this value represents the total emissions to society associated with travel on transportation networks. Moreover, as the links degrade and the practical capacity of links decreases, the total emissions are expected to increase; hence the relative total emissions of a transportation network reflect the environmental impact.

The environmental impact assessment index for a transportation network $\mathcal{G}$ with the vector of demands d, the vector of user link cost functions c, and the vector of link capacities u is defined as the relative total emission increase under a given uniform capacity retention ratio γ ($\gamma \in (0,1]$) so that the new capacities [cf. (3.7a)] are given by γu. Let c denote the vector of BPR user link cost functions and let d denote the vector of O/D pair travel demands. The mathematical definition of the environmental impact assessment index under the user-optimizing flow pattern, denoted by EI^{γ}_{U-O}, is then

$$\text{EI}^{\gamma}_{U-O} = \text{EI}_{U-O}(\mathcal{G}, c, d, \gamma, u) = \frac{\text{TE}^{\gamma}_{U-O} - \text{TE}_{U-O}}{\text{TE}_{U-O}}, \qquad (4.4a)$$

where TE_{U-O} and TE^{γ}_{U-O} are the total emissions generated under the user-optimizing flow pattern with the original capacities and the remaining capacities (i.e., γu), respectively.

The mathematical definition of the environmental impact assessment index under the system-optimizing flow pattern, denoted by EI^{γ}_{S-O}, is

$$\text{EI}^{\gamma}_{S-O} = \text{EI}_{S-O}(\mathcal{G}, c, d, \gamma, u) = \frac{\text{TE}^{\gamma}_{S-O} - \text{TE}_{S-O}}{\text{TE}_{S-O}}, \qquad (4.4b)$$

where TE_{S-O} and TE^{γ}_{S-O} are the total emissions generated at the system-optimizing flow pattern with the original capacities and the remaining capacities (i.e., γu), respectively.

For example, similar to the discussion after Definition 3.3, if $\gamma = .9$, the user link cost functions given by (3.7a) now have the link capacities given by $.9u_a$ for $a \in \mathcal{L}$; if $\gamma = .7$, the link capacities become $.7u_a$ for all links $a \in \mathcal{L}$, and so on. Such changes also affect the emission functions (4.1) and (4.2).

Note that (4.4a) and (4.4b) capture the impact of alternative behaviors on the environment as the transportation network is subject to link capacity degradations.

These indicators focus on uniform link capacity degradations where each link capacity denoted by u_a for links $a \in \mathcal{L}$ is degraded to γu_a for all $a \in \mathcal{L}$ where $\gamma \in (0, 1]$. One can also construct extensions of (4.4a) and (4.4b) to handle nonuniform deterioration in link capacities where γ is now replaced by $\tilde{\gamma}$, and where $\tilde{\gamma}$ is a vector of link retention ratios.

From these definitions, a transportation network under a given capacity retention/deterioration ratio γ (and under either S-O or U-O travel behavior) is considered to be *environmentally* robust if the index EI^{γ} is low. This means that the relative total emissions do not change much; hence the transportation network may be viewed as being more robust, from an environmental perspective, than if the relative total emissions value is large.

4.4.3 Environmental Link Importance Identification and Ranking

As we discussed earlier in this chapter, damage caused to roads and bridges is expected to result in the rerouting of traffic (under U-O or S-O behavior), which may result in congestion on the other part of the transportation network. Furthermore, such congestion may further contribute to greenhouse emissions. Therefore, a thorough understanding of the environmental impacts after the destruction of a link/node or a set of links and nodes is crucial for making strategic plans to protect the environment. Note that the importance measure introduced in Chapter 3 is focused on network performance and does not apply directly to environmental impact assessment. In addition, the importance measure does not allow for S-O behavior. Hence we now propose environmental link importance indicators that can be used to study the environmental link importance in a transportation network with alternative user behaviors. The formal definitions are given as

$$\mathrm{I}^l_{U-O} = \frac{\mathrm{TE}_{U-O}(\mathcal{G} - l) - \mathrm{TE}_{U-O}}{\mathrm{TE}_{U-O}} \tag{4.5a}$$

$$\mathrm{I}^l_{S-O} = \frac{\mathrm{TE}_{S-O}(\mathcal{G} - l) - \mathrm{TE}_{S-O}}{\mathrm{TE}_{S-O}}, \tag{4.5b}$$

where I^l_{U-O} denotes the importance indicator for link l assuming U-O behavior and I^l_{S-O} denotes the analogue under S-O behavior; $\mathrm{TE}_{U-O}(\mathcal{G} - l)$ denotes the total emissions generated under U-O behavior if link l is removed from the network, and $\mathrm{TE}_{S-O}(\mathcal{G} - l)$ denotes the same but under S-O behavior. Note that a link may be, in effect, removed from a transportation network due to extreme events such as a bridge collapsing, or a road becoming impassable. Based on the specific values of (4.5a) and (4.5b), the links for a given transportation network can then be ranked. Clearly, the most important links should be maintained and secured at a higher level because their removal will have the largest environmental impact.

Obviously, to make a measure such as (4.5a) or (4.5b) applicable and well-defined, it is essential that after the elimination of a given link l there is still a path or route available between each O/D pair.

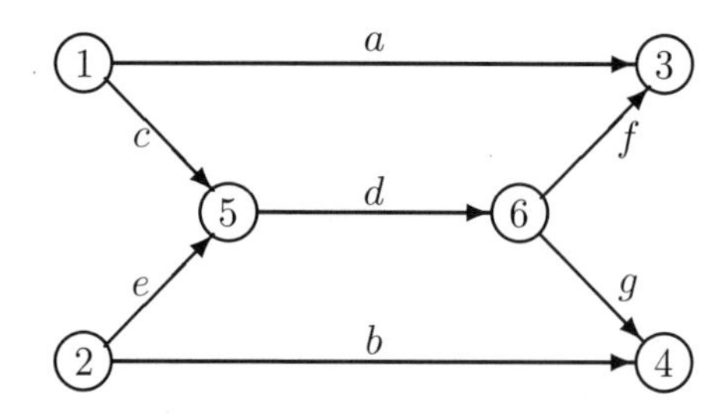

Figure 4.15: The network topology for the example in Section 4.4.4.1

4.4.4 Numerical Examples

We now explore the concepts described in Sections 4.4.2 and 4.4.3 in the context of concrete numerical examples. In Section 4.4.4.1, we consider a transportation network from the recent literature. In Section 4.4.4.2, we study the Sioux Falls network. We then use the environmental network measures to examine the Anaheim network in Section 4.4.4.3.

4.4.4.1 Application to a Network from the Literature The topology of the first transportation network that we studied in this section is depicted in Figure 4.15. This transportation network, but without the investigation of capacity degradation, was proposed by Yin and Lawphongpanich (2006). There are two O/D pairs in the network $w_1 = (1, 3)$ and $w_2 = (2, 4)$, with demands of $d_{w_1} = 3000$ vehicles per hour and $d_{w_2} = 3000$ vehicles per hour.

The user link cost functions, which here correspond to travel time in minutes, are of the BPR form given by (3.7a) and are as follows

$$c_a(f_a) = 8(1 + .15(f_a/2000)^4), \quad c_b(f_b) = 9(1 + .15(f_b/2000)^4),$$

$$c_c(f_c) = 2(1 + .15(f_c/2000)^4), \quad c_d(f_d) = 6(1 + .15(f_d/4000)^4),$$

$$c_e(f_e) = 3(1 + .15(f_e/2000)^4), \quad c_f(f_f) = 3(1 + .15(f_f/2500)^4),$$

$$c_g(f_g) = 4(1 + .15(f_g/2500)^4).$$

The lengths of the links, in kilometers, in turn, which are needed to compute the environmental emissions [cf. (4.1) - (4.3)] and constructed by Yin and Lawphongpanich (2006) are given by:

$$l_a = 8.0, \quad l_b = 9.0, \quad l_c = 2.0, \quad l_d = 6.0, \quad l_e = 3.0, \quad l_f = 3.0, \quad l_g = 4.0.$$

Table 4.2 presents the U-O link flow solutions as the capacity ratio γ changes in the range from 0. to 1. Table 4.3 reports the corresponding S-O link flow solutions. Table 4.4 displays the total CO emissions generated under these two distinct behavioral

Table 4.2: U-O solutions as capacity deteriorates – link flows for the network in Section 4.4.4.1

Link	$\gamma = 1.$	$\gamma = .9$	$\gamma = .8$	$\gamma = .7$	$\gamma = .6$	$\gamma = .5$	$\gamma = .4$	$\gamma = .3$	$\gamma = .2$	$\gamma = .1$
a	2514.99	2268.31	2038.30	1842.66	1699.60	1609.46	1559.85	1536.47	1527.82	1525.83
b	2624.17	2365.60	2120.00	1905.80	1742.40	1635.32	1574.73	1545.75	1534.96	1532.47
c	485.01	731.69	961.70	1157.34	1300.40	1390.54	1440.15	1463.53	1472.18	1474.17
d	860.84	1366.09	1841.70	2251.44	2558.00	2755.23	2865.42	2917.78	2937.21	2941.70
e	375.83	634.40	880.00	109.20	1257.60	1364.68	1425.27	1454.25	1465.04	1467.17
f	485.01	731.69	961.70	1157.34	1300.40	1390.54	1440.14	1463.53	1472.18	1474.17
g	375.83	634.40	880.00	1094.20	1257.60	1364.68	1425.27	1454.25	1465.04	1467.53

Table 4.3: S-O solutions as capacity deteriorates – link flows for the network in Section 4.4.4.1

Link	$\gamma = 1.$	$\gamma = .9$	$\gamma = .8$	$\gamma = .7$	$\gamma = .6$	$\gamma = .5$	$\gamma = .4$	$\gamma = .3$	$\gamma = .2$	$\gamma = .1$
a	1791.87	1701.73	1635.64	1589.96	1560.28	1542.34	1532.50	1527.85	1526.12	1525.72
b	1848.60	1744.93	1666.79	1611.65	1575.26	1553.05	1540.81	1535.00	1532.84	1532.34
c	1208.13	1298.27	1364.36	1410.04	1439.72	1457.67	1467.50	1472.15	1473.88	1474.28
d	2359.53	2553.34	2697.57	2798.40	2864.46	2904.62	2926.69	2937.16	2941.04	2941.94
e	1151.40	1255.07	1333.21	1388.35	1424.74	1446.95	1459.19	1465.00	1467.16	1467.66
f	1208.13	1298.27	1364.36	1410.04	1439.72	1457.67	1467.50	1472.15	1473.88	1474.28
g	1151.40	1255.07	1333.21	1388.35	1424.74	1446.95	1459.19	1465.00	1467.16	1467.66

Table 4.4: Total emissions generated (grams/hour) and environmental impact indicators for varying degradable capacities for the network in Section 4.4.4.1

γ	TE^{γ}_{U-O}	EI^{γ}_{U-O}	$TE^{\gamma}{}_{S-O}$	EI^{γ}_{S-O}
1.	26,744.62	.0000	27,140.19	.0000
.9	27,336.24	.0221	27,565.00	.0157
.8	27,982.55	.0463	28,045.61	.0334
.7	28,820.11	.0776	28,753.98	.0595
.6	30,291.05	.1326	30,162.84	.1114
.5	33,874.37	.2666	33,758.30	.2438
.4	45,033.94	.6839	44,970.11	.6570
.3	88,964.12	2.3364	88,943.63	2.2772
.2	355,639.84	12.2976	355,636.28	12.1037
.1	5,351,015.00	199.0782	5,351,016.50	196.1621

assumptions and under the same capacity retention ratios γ. In addition, Table 4.4 documents the two environmental impact assessment indices.

Figure 4.16 depicts the ratio of TE^{γ}_{U-O} to $TE^{\gamma}{}_{S-O}$. Figure 4.17 plots the environmental impact indicators under U-O and S-O behaviors.

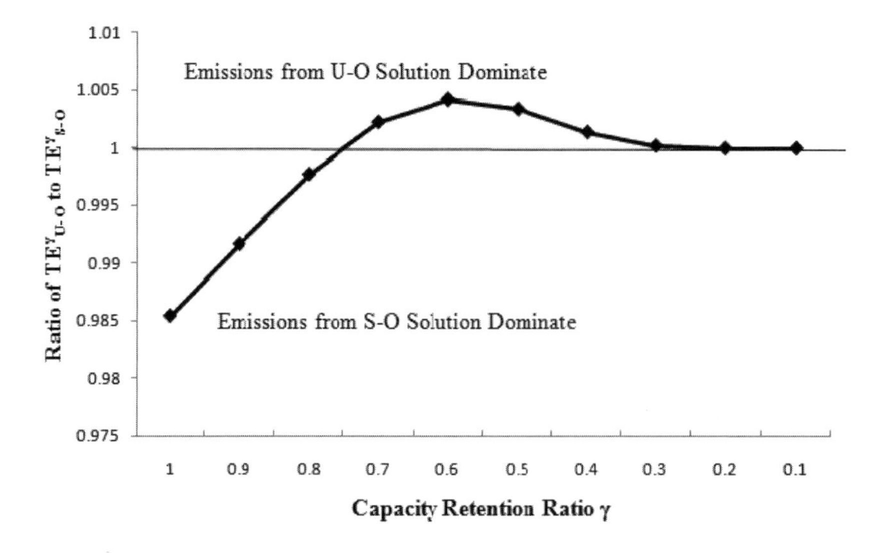

Figure 4.16: Ratio of TE^{γ}_{U-O} to $\text{TE}^{\gamma}{}_{S-O}$ for the network in Section 4.4.4.1

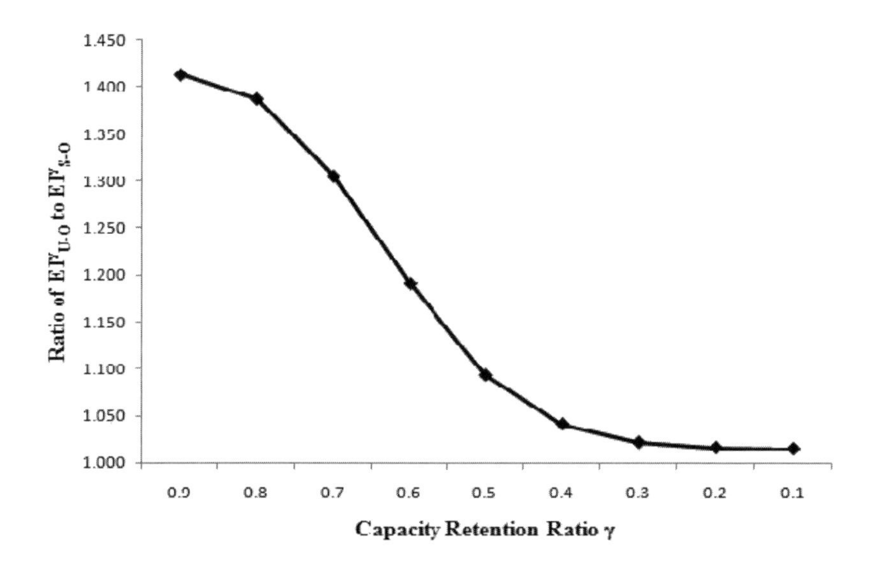

Figure 4.17: Ratio of EI^{γ}_{U-O} to EI^{γ}_{S-O} for the network in Section 4.4.4.1

As can be seen from Table 4.4 and Figure 4.16, the total emissions generated are lower under the U-O behavioral principle from $\gamma = 1$ [this case was noted also by Yin and Lawphongpanich (2006)] until $\gamma = .7$. For $\gamma = .7, .6$, and so on, through $\gamma = .1$

Table 4.5: Environmental link importance indicators under U-O behavior for the network in Section 4.4.4.1

Link l	$\mathrm{TE}_{U-O}(\mathcal{G}-l)$	I^l_{U-O}
a	30439.95	.14
b	31823.13	.19
c	27802.31	.04
d	28752.22	.07
e	27692.11	.03
f	27802.31	.04
g	27692.11	.03

Table 4.6: Environmental link importance indicators under S-O behavior for the network in Section 4.4.4.1

Link l	$\mathrm{TE}_{S-O}(\mathcal{G}-l)$	I^l_{S-O}
a	30493.30	.12
b	31856.33	.17
c	28070.21	.03
d	28752.22	.06
e	27909.01	.03
f	28070.21	.03
g	27909.01	.03

the total emissions generated under S-O behavior are lower than those generated under U-O behavior. However, once $\gamma = .3$, the difference is not appreciable. One can easily see from Table 4.4 and Figure 4.17, in turn, that under S-O behavior the transportation network may be viewed as being more robust from an environmental perspective in that, for a given value of γ that is less than 1, the value for S-O is lower than the value for U-O, indicating that the relative increase in emissions for this example is lower when the transportation network link capacities decrease in the case of S-O behavior.

We proceed to now determine the link importance indicators according to (4.4a) and (4.4b). The results are reported in Tables 4.5 and 4.6, respectively.

Figure 4.18 displays the importance indicator values and the rankings of the links from most important to least important under both U-O and S-O behaviors. It is clear from Figure 4.18 that the rankings of the links are identical for this numerical example when the travelers behave in either a U-O or in a S-O manner. In particular, link b is most important, followed by link a and then link d. Links c and f are equal

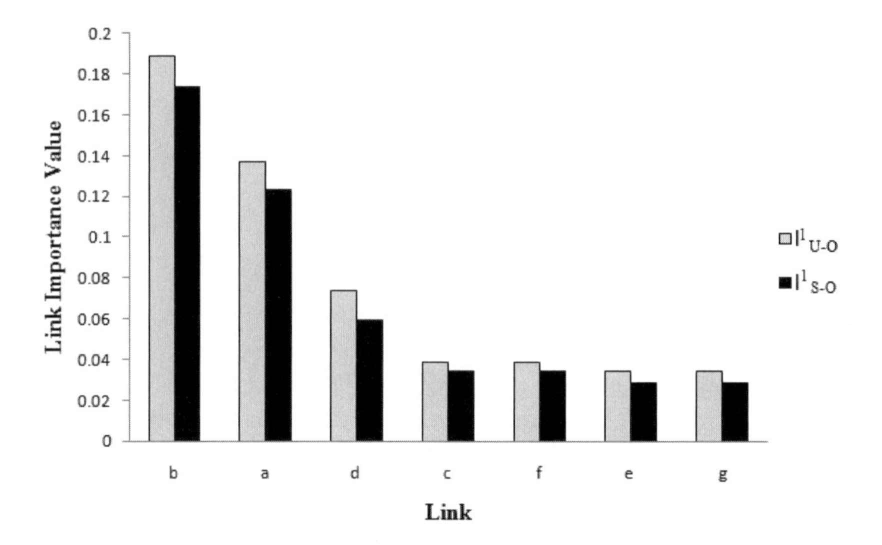

Figure 4.18: Environmental link importance values and rankings under U-O and S-O behavior for the network in Section 4.4.4.1

in importance, followed by links e and g, which also share the same importance indicator value. Hence, from an environmental planning, maintenance, and security perspective, the links should be "protected" accordingly.

From the data in Tables 4.2 and 4.3, we see that link b carries the largest amount of flow in both U-O and S-O solutions. Therefore, the removal of link b causes a significant amount of traffic flow to reroute, which causes more congestion on the other links. This resulting congestion is the most significant contributor to the total emissions. On the other hand, links e and g carry the least amount of flow in both U-O and S-O solutions. Hence the removal of them results in a minimal amount of rerouting of flows and causes the least additional congestion on the other links. As a consequence, these two links are not important from an environmental perspective.

4.4.4.2 The Environmental Robustness of the Sioux Falls Network We here determine the values of the environmental measures to ascertain the environmental robustness of the Sioux Falls network. The network topology is shown in Figure 4.4. In Table 4.7, we present the total CO emissions generated under the two distinct behavioral assumptions and under the same capacity retention ratios γ. In addition, Table 4.7 also documents the two environmental impact assessment indices.

Figure 4.19 depicts the ratio of $\mathrm{TE}^{\gamma}_{U-O}$ to $\mathrm{TE}^{\gamma}{}_{S-O}$. In Figure 4.20, we plot the ratio of $\mathrm{EI}^{\gamma}_{U-O}$ to $\mathrm{EI}^{\gamma}{}_{S-O}$.

As can be seen from Table 4.7 and Figure 4.19, the total emissions generated under the U-O behavioral principle are consistently lower than those under S-O behavior. Moreover, the difference gets larger with the increase in the capacity retention ratio γ. One can also easily see from Table 4.7 and Figure 4.20, in turn, that under U-O

Table 4.7: Total emissions generated (grams/hour) and environmental impact indicators for varying degradable capacities for the Sioux Falls network

γ	TE^{γ}_{U-O} $(\times 10^6)$	EI^{γ}_{U-O}	$\text{TE}^{\gamma}{}_{S-O}(\times 10^6)$	EI^{γ}_{S-O}
1.	2.2323	.0000	5.2578	.0000
.9	2.5964	.1636	7.2030	.3700
.8	3.2447	.4542	10.6592	1.0273
.7	4.4820	1.0087	17.2023	2.2717
.6	7.0677	2.1675	30.7401	4.8466
.5	13.1320	4.8854	62.3732	10.8631
.4	30.0623	12.4720	150.5202	27.6279
.3	92.0500	40.2540	473.1643	88.9920
.2	460.5401	205.3999	2390.7321	453.6959
.1	7348.6102	3292.4000	38235.1022	7271.0529

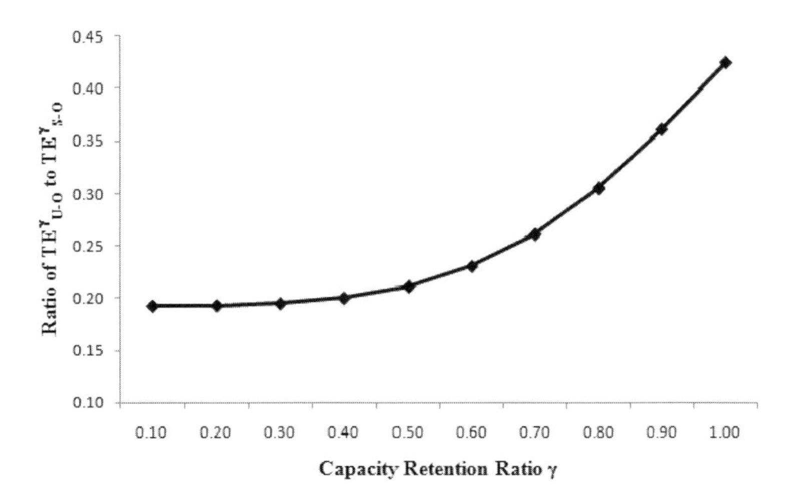

Figure 4.19: Ratio of TE^{γ}_{U-O} to TE^{γ}_{S-O} for the Sioux Falls network

behavior the transportation network may be viewed as being more robust from an environmental perspective in that, for a given value of γ less than 1, the value for U-O is lower than the value for S-O, indicating that the relative increase in emissions for the Sioux Falls network is lower when the transportation network link capacities decrease in the case of U-O behavior. This result is different from that of the example in Section 4.4.4.1 because the network topology and the cost and demand information are different. Therefore, we have to study networks individually and

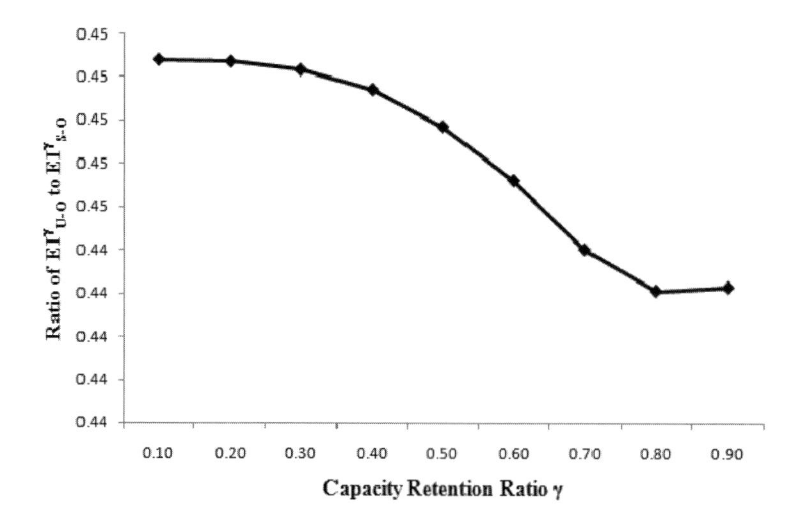

Figure 4.20: Ratio of $\mathrm{EI}^{\gamma}_{U-O}$ to $\mathrm{EI}^{\gamma}_{S-O}$ for the Sioux Falls network

Table 4.8: Total emissions generated (grams/hour) and environmental impact indicators for varying degradable capacities for the Anaheim network

γ	$\mathrm{TE}^{\gamma}_{U-O}\ (\times 10^6)$	$\mathrm{EI}^{\gamma}_{U-O}$	$\mathrm{TE}^{\gamma}{}_{S-O}(\times 10^6)$	$\mathrm{EI}^{\gamma}_{S-O}$
1.	6.8695	.0000	6.8510	.0000
.9	6.9382	.0100	6.8960	.0066
.8	7.0651	.0285	7.0254	.0255
.7	7.3300	.0664	7.2955	.0649
.6	7.9086	.1513	7.8901	.1517
.5	9.3661	.3634	9.3568	.3658
.4	13.6260	.9836	13.6140	.9872
.3	29.4370	3.2852	29.4050	3.2921
.2	124.3400	17.1003	124.3000	17.1433
.1	1903.3000	276.0653	1902.8000	276.7405

with the appropriate associated travel behavior to evaluate the environmental impact in the case of network link degradation.

4.4.4.3 The Environmental Robustness of the Anaheim Network We now compute the environmental impact indices for the Anaheim network. In Table 4.8, the total CO emissions under the two alternative travel behaviors are listed with capacity retention ratios γ. In addition, Table 4.8 also documents the two environmental impact assessment indices.

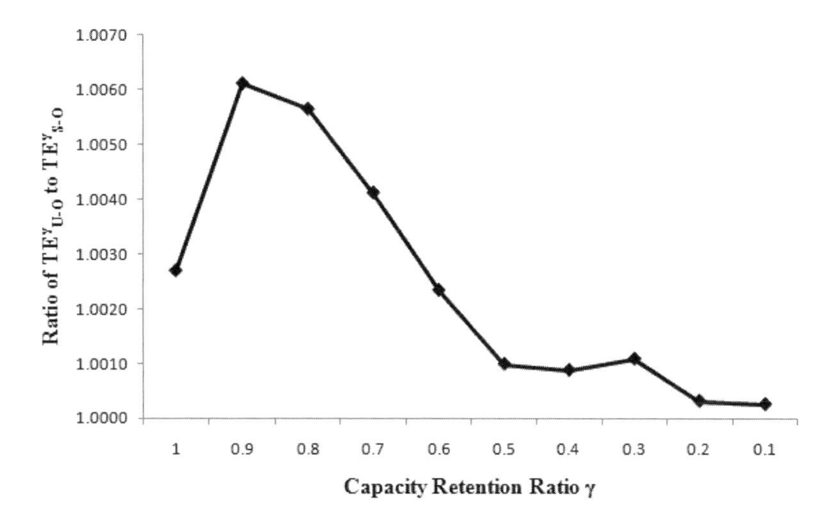

Figure 4.21: Ratio of $\mathrm{TE}^{\gamma}_{U-O}$ to $\mathrm{TE}^{\gamma}_{S-O}$ for the Anaheim network

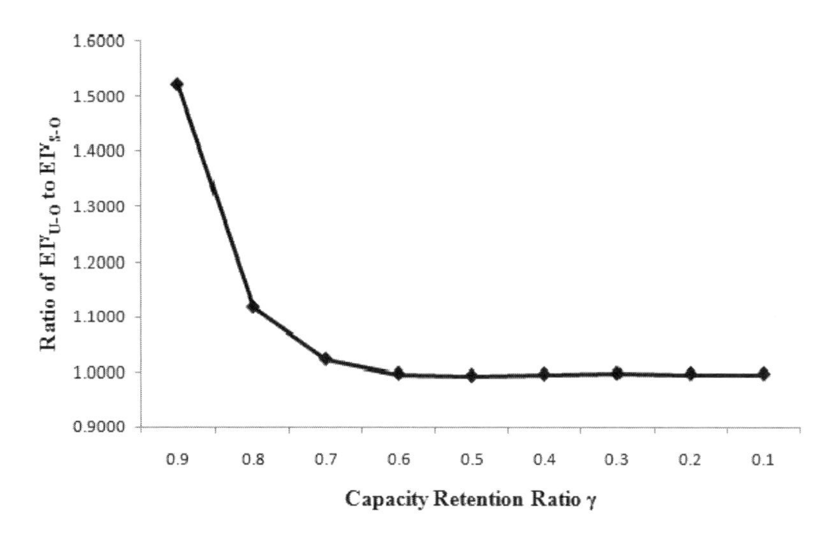

Figure 4.22: Ratio of $\mathrm{EI}^{\gamma}_{U-O}$ to $\mathrm{EI}^{\gamma}_{S-O}$ for the Anaheim network

Figure 4.21 depicts the ratio of $\mathrm{TE}^{\gamma}_{U-O}$ to $\mathrm{TE}^{\gamma}{}_{S-O}$. Figure 4.22 plots the ratio of $\mathrm{EI}^{\gamma}_{U-O}$ to $\mathrm{EI}^{\gamma}{}_{S-O}$.

We can see from Figure 4.21 that the total emissions under the U-O flow pattern are always greater than those under the S-O flow pattern, and when γ is equal to .9 the ratio reaches a maximum.

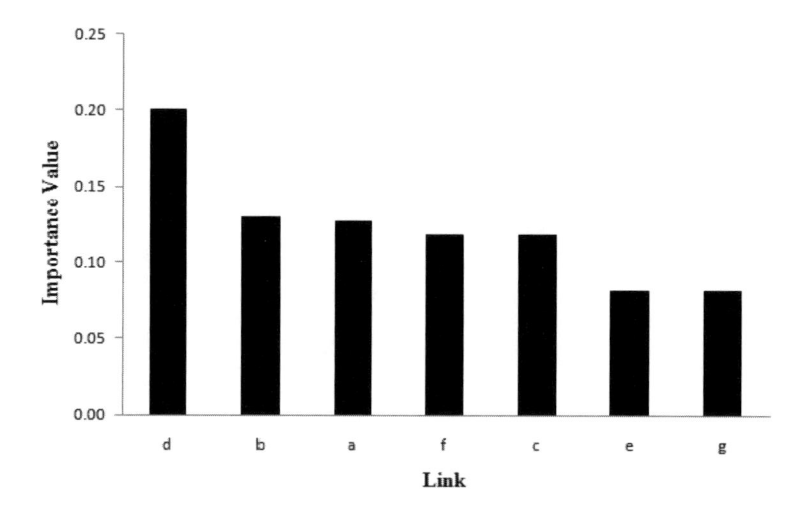

Figure 4.23: Importance and ranking of links for the network in Section 4.4.4.1 using (3.5)

From Figure 4.22, we can see that the Anaheim network is considered environmentally robust under S-O user travel behavior when γ is above .7, while it is slightly better off under U-O user travel behavior when γ is under .6.

4.5 SUMMARY AND CONCLUSIONS

As argued earlier in this chapter, transportation networks are one of the most important critical infrastructure networks because they provide life-line support to societies and economies. To assess the functionality of transportation networks, we have to consider flows with different user behaviors. In this chapter, we applied the network measures introduced in Chapter 3 to study some real-world large transportation network examples, including the Sioux Falls network and the Anaheim network. In particular, we showed that, with the proposed network measures, one can analyze transportation network performance and robustness under different user behaviors.

Furthermore, congestion on transportation networks is a major contributor to greenhouse emissions, which result in more and more disastrous weather that causes further damage to transportation networks. To study the environmental impact of transportation deterioration, we proposed environmental impact indices to measure how environmentally robust a transportation network is after capacity degradation under different user behaviors. Using similar concepts, we also studied the environmental importance of transportation links after an outright removal of a link. This chapter clearly demonstrates that the proposed network measures can be used to study vulnerability and robustness from the perspectives of network performance, total cost, and/or environmental emissions, for large transportation networks.

Moreover, by using the network efficiency/performance measure (3.2) proposed in Chapter 3, and those described in this chapter [cf. (4.5a) and (4.5b)], one can study the importance of network components from different perspectives. For example, the network performance measure $\mathcal{E}$ (cf. (3.2)) and its resulting link/node importance measure $I(g)$ [cf. (3.5)] should be used when one is interested in analyzing transportation network vulnerability from the network efficiency/performance point of view. On the other hand, if one is concerned with the environmental impact of the removal of a link under different user behaviors, the environmental importance indicators [cf. (4.5a) and (4.5b)] are appropriate to use. For completeness, we also display in Figure 4.23 the link importance and rankings based on (3.5) for the network example in Section 4.4.4.1.

From Figure 4.23, one can see that from a network performance perspective link d is the most important link because it is shared by both O/D pairs and the removal of it will cause rerouting, which will significantly increase the costs on the remaining paths. However, by comparing Figures 4.18 and 4.23, we know that links g and e (and then links c and f) remain the least important under the network efficiency link importance measure as well as either (4.5a) or (4.5b).

Nevertheless, we need to point out that these network importance measures are based on different assumptions and, therefore, have different possible applications. For example, as discussed in Section 3.2.2, the importance measure based on (3.2) can handle a network with disconnected O/D pairs under U-O flow patterns, whereas the definitions of the environmental importance indicators assume that every O/D pair remains connected after the removal of a link for both U-O and S-O solutions. Moreover, the network importance measure (3.5) can also be used to study the importance of nodes, while the environmental importance indicators focus on the importance of links. Obviously, if the removal of a node will not cause any disconnected O/D pair, (4.5a) and (4.5b) can also be used to study the environmental importance of a node under different user behaviors.

4.6 SOURCES AND NOTES

Section 4.1 is based on Section 5 in Nagurney and Qiang (2007b). The relative total cost index results for the Sioux Falls network in Section 4.2 are from Nagurney and Qiang (2009). However, in this chapter, we extended the results by also computing the robustness measure results for the Sioux Falls network in Section 4.2.2. In addition, new capacity enhancement results are presented in this chapter. Furthermore, we applied the network robustness measures to study a larger network, the Anaheim network, in Section 4.3. Finally, Section 4.4 is based on the paper by Nagurney, Qiang, and Nagurney (2008). In this chapter, however, we added new results for the large-scale Anaheim network.

We used $\epsilon = 0.1$ as the convergence tolerance for both the equilibration algorithm and the projection method (cf. Sections 2.4.1 and 2.4.2) for the Sioux Falls and the Anaheim network examples in Sections 4.2, 4.3 and 4.4. Note that the user link cost functions for these large-scale networks are separable. It is worth stating, to be fully explicit, that, in the computation of the network efficiency measure $\mathcal{E}$ [cf. (3.2)] and the importance of a network component $I(g)$ [see (3.5)], in the case of nonseparable user link cost functions (and, similarly, nonseparable disutility functions in the case of elastic demands), the removal of a link (O/D node), in effect, is handled as resulting in zero flow in the terms of the corresponding link cost functions (the disutility functions), because the link (the node) no longer exists. Furthermore, we can also directly apply the network efficiency measure $\mathcal{E}$ and the resulting network component importance measure $I(g)$ to study multimodal/multiclass transportation networks after the network is transformed into a single-modal/single-class network, as discussed in Section 2.5.

Clearly, different modes of transportation from public modes such as busses and trains to private ones from cars to bicycles affect congestion and environmental emissions in distinct ways. The quantitative tools constructed in this chapter can also be applied with straightforward adaptation to determine the impact on network efficiency and/or on the environment, of the removal/deterioration of transportation modes or the investment in modal capacity. Finally, given the importance of climate change and its impact on infrastructure, the tools in this chapter are of immediate relevance to practice [see, e.g., Schulz (2007)].

CHAPTER 5

SUPPLY CHAIN NETWORKS WITH DISRUPTION RISKS

5.1 INTRODUCTION

Supply chain networks are the underpinning skeletons of the business world. These networks, more and more, are global in nature, with products consisting of parts manufactured in different regions of the world, assembled in yet other locations, and then shipped across continents and oceans to retailers and consumers. Such complex networks consist of manufacturers (and their suppliers), shippers, and carriers using various modes of transportation, and distribution centers (and warehouses), where the products are stored and, ultimately, from which they are sent from to the customers. Supply chains involve many decision-makers interacting with one another, sometimes competing and at other times necessarily cooperating.

Indeed, as a result of globalization, more and more companies and customers rely on overseas suppliers and products. For example, according to the U.S. Customs and Border Protection (2007), the United States currently imports approximately $2 trillion worth of products annually from more than 150 countries. Furthermore, experts project that this amount will triple by 2015. Figure 5.1 shows the U.S. import value by fiscal year, which clearly demonstrates this upward trend.

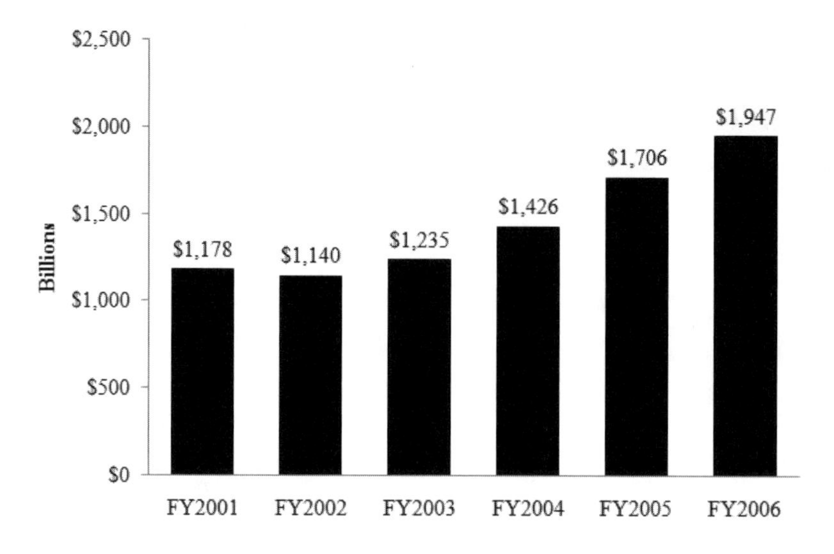

Figure 5.1: U.S. import value by fiscal year [Source: U.S. Customs and Border Protection (2007)]

Supply chain networks depend on infrastructure networks for their effective and efficient operations from manufacturing and logistical networks, to transportation networks, electric power networks, financial networks, and telecommunication networks, most notably, the Internet.

No supply chain, logistics system, or infrastructure system is immune to disruptions and as long as there have been supply chains there have been disruptions.

However, in the past decade there have been vivid high-profile examples of supply chain disruptions and their effects. As a now classical example, in March 2000, lightning struck a Philips semiconductor plant in Albuquerque, New Mexico, creating a 10-minute fire that resulted in smoke and water damage to millions of computer chips and the subsequent delay of delivery to its two largest customers: Finland's Nokia and Sweden's Ericsson. Ericsson used the Philips plant as its sole source and reported a $400 million loss because it did not receive the chip deliveries in a timely manner; Nokia moved quickly to tie up spare capacity at other Philips plants and refitted some of its phones so that it could use chips from other U.S. and from Japanese suppliers. Nokia managed to arrange alternative supplies and, therefore, mitigated the effect of the disruption. Ericsson learned a painful lesson from this disaster [cf. Handfield (2007)].

Another major supply chain disruption in the United States was due to the West Coast port lockout in 2002, which resulted in a 10 day shutdown of ports in early October, typically, the busiest month. About 42% of the U.S. trade products and 52% of imported apparel go through these ports, which include Los Angeles. The Pacific Maritime Association (PMA) locked out members of the International Longshore and

Warehouse Union (ILWU), forcing cargo ships to sit idly in ports from Seattle to San Diego, holding up the loading and unloading of potentially billions of dollars of cargo. Estimated losses were $1 billion per day. Wal-Mart and Costco planned months in advance and Wal-Mart anticipated the lockout and had extra inventory shipped to Hawaii and Alaska. Port operations and schedules did not return to normal until 6 months after the strike ended [cf. Lazaro (2002)].

Owing to Hurricane Katrina, which hit in 2005, 10 - 15% of the total U.S. gasoline production was halted, raising the oil price both in the United States and overseas [see, e.g., Canadian Competition Bureau (2006)]. In 2008, we saw how Hurricane Gustav affected gas availability (and prices) thousands of miles away from the points of impact.

The world price of coffee, in turn, rose 22% after Hurricane Mitch in 1998 struck the Central American republics of Nicaragua, Guatemala, and Honduras, which affected supply chains worldwide [Fairtrade Foundation (2002)].

The financial and economic crisis of 2008 is expected to result in failures not only of financial and other service providers but also of manufacturers, including, possibly, automobile manufacturers. Potential cascading failures may affect suppliers upstream in their supply chains as well as retailers downstream.

As summarized by Sheffi (2005), one of the main characteristics of disruptions in supply networks is "the seemingly unrelated consequences and vulnerabilities stemming from global connectivity."

In addition, to demonstrate the dependence of supply chain networks on other infrastructure networks, including the Internet, we highlight the following: According to Kembel (2000) the cost of downtime (in terms of lost revenue) for several online companies if their computers and/or telecomm networks are down (for one hour of downtime) ranges from $225,000 for eBay; $180,000 for Amazon, and $6,450,000 for a brokerage firm. These costs do not even include employee costs or loss of goodwill [see also Snyder and Shen (2006)].

It is interesting that, in recent decades, the focus has been on lean supply chains and, although such supply chains may work well when the environment is predictable and steady, they may be very sensitive to disruptions because they lack redundancy and slack in their systems. Furthermore, firms today may be much less vertically integrated (while, clearly, more global). Indeed, decades ago, the Ford Motor Company and other automobile manufacturers and even IBM produced their products essentially in their entirety.

Lynn (2006) argued that globalization has led to extremely fragile supply chains. Suppliers today may be in parts of the world that are unstable and subject to natural disasters, political instability, and strife. In fact, Craighead et al. (2007) emphasized that supply chain disruptions and the associated operational and financial risks are the most pressing issues faced by firms in today's competitive global environment.

The rigorous modeling and analysis of supply chain networks, in the presence of possible disruptions, is imperative because, as noted, disruptions may have lasting major financial consequences. Hendricks and Singhal (2005) analyzed 800 instances of supply chain disruptions experienced by firms whose stocks are publicly traded. They found that the companies that suffered supply chain disruptions experienced share price returns 33 - 40 % lower than the industry and the general market benchmarks. Furthermore, share price volatility was 13.5 % higher in these companies in the year after a disruption than in the year before the event. Based on their findings, it is evident that only well-prepared companies can effectively cope with supply chain disruptions. Wagner and Bode (2007), in turn, designed a survey to study empirically the responses from executives of firms in Germany regarding their opinions about the factors that affect supply chain vulnerability. The authors found that demand-side risks are related to customer dependence whereas supply-side risks are associated with supplier dependence, single sourcing, and global sourcing.

Supply chain disruptions and the associated risks are now major topics in both theoretical and applied research as well as in practice. Risk in the context of supply chains may be associated with the production/procurement processes, the transportation/shipment of the goods, and/or the demand markets. It is notable that the focus of research has been on demand-side risk, which is related to fluctuations in the demand for products, as opposed to the supply-side risk, which deals with uncertain conditions that affect the production and transportation processes of the supply chain. The goal of supply chain risk management is to alleviate the consequences of disruptions and risks or, simply put, to increase the *robustness* of a supply chain. However, there are very few quantitative models for measuring supply chain robustness. For example, Bundschuh, Klajan, and Thurston (2003) discussed the design of a supply chain from both reliability and robustness perspectives. The authors built a mixed integer programming supply chain model with constraints for reliability and robustness. The robustness constraint was formulated in an implicit form by requiring the suppliers' sourcing limit to exceed a certain level. In this way, the model builds redundancy into a supply chain.

Snyder and Daskin (2005) examined supply chain disruptions in the context of facility location. The objective of their model was to select locations for warehouses and other facilities that minimize the transportation costs to customers and, at the same time, account for possible closures of facilities that would result in re-routing of the product. However, Snyder and Shen (2006) note: "Although these are multi-location models, they focus primarily on the local effects of disruptions."

Indeed, to-date, most supply disruption studies have focused on a local point of view, in the form of a single-supplier problem [see, e.g., Gupta (1996) and Parlar (1997)] or a two-supplier problem [see, e.g., Parlar and Perry (1996)]. Very few papers have examined supply chain risk management in an environment with multiple decision-makers and in the case of uncertain demands [cf. Tomlin (2006)].

It is imperative to study supply chain risk management from a *holistic, system-wide* perspective and to capture the interactions among the multiple decision-makers in the various supply chain network tiers. Such a perspective has also been argued by Wu, Blackhurst, and Chidambaram (2006), who focused on inbound supply risk analysis. Toward that end, in this chapter, we take an entirely different approach, and we consider supply chain robustness in the context of multitiered supply chain networks with multiple decision-makers under equilibrium conditions.

However, to study supply chain robustness, an informative and effective performance measure is first required. Beamon (1998, 1999) reviewed the supply chain literature and suggested directions for research on supply chain performance measures, which should include criteria on efficient resource allocation, output maximization, and flexible adaptation to the environmental changes [see also, Lee and Whang (1999), Lambert and Pohlen (2001), and Lai, Ngai, and Cheng (2002)]. Different supply performance measures can be devised based on the specific nature of the problem. In any event, the discussion here is not meant to cover all the existing supply chain performance measures. It would be a challenging task to devise a supply chain performance measure that would cover all aspects of supply chains. Such a discussion, we expect, will be an ongoing research topic for years to come.

In this chapter, we study supply chain robustness based on the unified network performance measure for decentralized decision-making described in Section 3.2. Recall that the unified measure captures the network flows, the costs, and the decision-makers' behavior under network equilibrium conditions. Moreover, the model in this chapter extends the supply chain model of Nagurney, Dong, and Zhang (2002), which was the first to capture competition across tiers of decision-makers and cooperation between successive tiers in a general network equilibrium framework. Here we consider also random demands [cf. Nagurney et al. (2005)] and several other extensions. Specifically, to study supply chain robustness, the model contains the following novel features:

1. We associate each process in a supply chain with random cost parameters to represent the impact of disruptions to the supply chain.

2. We extend the aforementioned supply chain models to capture the attitude of the manufacturers and the retailers toward disruption risks.

3. We propose a weighted performance measure to evaluate different supply chain disruptions.

4. Different transportation modes are explicitly incorporated into the model [see also, e.g., Dong, Zhang, and Nagurney (2002) and Dong et al. (2005)].

In a multimodal transportation supply chain, alternative transportation modes can be used in the case of a failure of a specific transportation mode. Indeed, authors have emphasized that redundancy needs to be considered in the design of supply chains to prevent supply chain disruptions. For example, Wilson (2007) used a system dynamic simulation to study the relationship between transportation disruptions and supply chain performance and found that the existence of transportation alternatives significantly improved supply chain performance in the case of transportation disruptions.

In our modeling framework, we assume that the probability distributions of the disruption-related cost parameters are known. This assumption is not unreasonable given today's advanced information technology and increasing awareness of the risks among managers. A great deal of disruption-related information can be obtained from a careful examination and abstraction of the relevant data sources. Specifically, as indicated by Sheffi (2005), "as investigation boards and legal proceedings have revealed, in many cases relevant data are on the record but not funneled into a useful place or not analyzed to bring out the information in the data."

The organization of this chapter is as follows. In Section 5.2, we present the model of a supply chain network faced with (possible) disruptions and in the case of random demands and multiple transportation modes. In Section 5.3, we propose a weighted supply chain performance measure with consideration of robustness. In Section 5.4, we present the realization of the modified projection method for the supply chain network equilibrium problem with uncertainty and random demands. In Section 5.5, we provide numerical examples to illustrate the model and concepts. The results are summarized in Section 5.6.

5.2 THE SUPPLY CHAIN MODEL WITH DISRUPTION RISKS AND RANDOM DEMANDS

The supply chain model consists of m manufacturers, with a typical manufacturer denoted by i, n retailers with a typical retailer denoted by j, and o demand markets with a typical demand market denoted by k. Furthermore, we assume that there are g transportation modes from manufacturers to retailers, with a typical mode denoted by u, and that there are h transportation modes between retailers and demand markets, with a typical mode denoted by v. Typical transportation modes may include trucking, rail, air, and sea. By allowing multiple modes of transportation between successive tiers of the supply chain we also generalize the earlier models of Dong, Zhang, and Nagurney (2002) and Dong et al. (2005).

Manufacturers are assumed to produce a homogeneous product, which can be purchased by retailers, who, in turn, make the product available to demand markets. Each process in the supply chain is associated with some random parameters that affect the cost functions.

The topology of the supply chain network is depicted in Figure 5.2.

5.2.1 The Behavior of the Manufacturers

We assume a homogeneous product economy, meaning that all manufacturers produce the same product which is then shipped to the retailers, who, in turn, sell the product to the demand markets.

Let q_i denote the production output of manufacturer i; $i = 1, \ldots, m$. Let q_{ij}^u denote the product shipment between manufacturer i and retailer j using mode of transportation u. We group the production outputs and these product shipments into the respective vectors q and Q^1.

Because the total amount of the product shipped from a manufacturer via different transportation modes has to be equal to the amount of the production of each manufacturer, we have the following relationship between the production of manufacturer i and its shipments to the retailers

$$q_i = \sum_{j=1}^{n} \sum_{u=1}^{g} q_{ij}^u, \quad i = 1, \ldots, m. \tag{5.1}$$

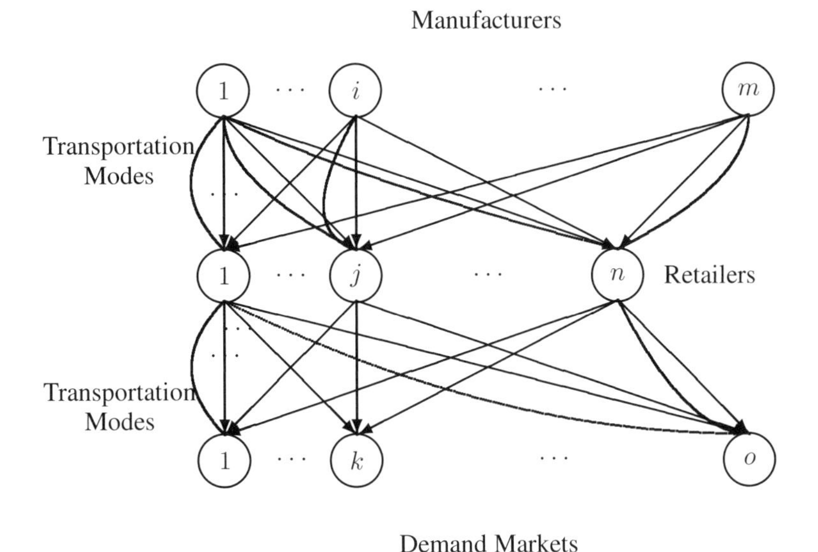

Figure 5.2: The multitiered and multimodal structure of the supply chain

We assume that disruptions will affect the production processes of manufacturers, the impact of which is reflected in the production cost functions. For each manufacturer i, there is a random parameter α_i that reflects the effect of disruptions to its production cost function with the corresponding cumulative distribution function given by $\mathcal{F}_i(\alpha_i)$. We let $f_i(q, \alpha_i) \equiv f_i(Q^1, \alpha_i)$ denote the production cost of manufacturer i with random parameter α_i for $i = 1, \ldots, m$.

The expected production cost function for firm i, denoted by $\hat{F}_i(Q^1)$, is given by

$$\hat{F}_i(Q^1) \equiv \int_{\alpha_i} f_i(Q^1, \alpha_i) d\mathcal{F}_i(\alpha_i), \quad i = 1, \ldots, m. \tag{5.2}$$

We denote the variance of the production cost function as $VF_i(Q^1)$, where $i = 1, \ldots, m$.

As noted earlier, we assume that each manufacturer has g types of transportation modes available to ship the product to the retailers, the cost of which is also subject to disruption impacts. The term β_{ij}^u denotes the random parameter associated with the transportation cost of shipment between i and j via mode u and the corresponding cumulative distribution function is denoted by $\mathcal{F}_{ij}^u(\beta_{ij}^u)$, for all $i = 1, \ldots, m$; $j = 1, \ldots, n$; and $u = 1, \ldots, g$. The transportation/transaction cost between manufacturer i and retailer j via transportation mode u with the random parameter β_{ij}^u is denoted by $c_{ij}^u(q_{ij}^u, \beta_{ij}^u)$ for all i, j, u.

The expected transportation cost function, denoted by $\hat{C}^u_{ij}(q^u_{ij})$, is then given by

$$\hat{C}^u_{ij}(q^u_{ij}) \equiv \int_{\beta^u_{ij}} c^u_{ij}(q^u_{ij}, \beta^u_{ij}) d\mathcal{F}^u_{ij}(\beta^u_{ij}), \ i = 1, \ldots, m; \ j = 1, \ldots, n; \ u = 1, \ldots, g. \tag{5.3}$$

We denote the variance of the transportation cost function as $VC^u_{ij}(q^u_{ij})$ where $i = 1, \ldots, m; \ j = 1, \ldots, n; \ u = 1, \ldots, g$.

It is well-known that variance may be used to measure risk [see, e.g., Silberberg and Suen (2000)]; Tomlin (2006) used such an approach to study risks in applications to supply chains. Therefore, we assign a nonnegative weight θ_i to the variance of the cost functions for each manufacturer i to reflect its attitude towards disruption risks. The larger the weight is, the larger the penalty a manufacturer imposes on the risk and, therefore, the more risk-averse the manufacturer is.

Let ρ^{u*}_{1ij} denote the price charged for the product by manufacturer i to retailer j when the product is shipped via transportation mode u. Hence manufacturers can price according to their locations as well as according to the transportation modes utilized. Each manufacturer faces two objectives: to maximize his expected profit and to minimize the disruption risks adjusted by his risk attitude. Therefore, the objective function for manufacturer i; $i = 1, \ldots, m$, can be expressed as follows

$$\text{Maximize} \quad \sum_{j=1}^{n} \sum_{u=1}^{g} \rho^{u*}_{1ij} q^u_{ij} - \hat{F}_i(Q^1) - \sum_{j=1}^{n} \sum_{u=1}^{g} \hat{C}^u_{ij}(q^u_{ij})$$

$$-\theta_i [VF_i(Q^1) + \sum_{j=1}^{n} \sum_{u=1}^{g} VC^u_{ij}(q^u_{ij})] \tag{5.4}$$

subject to:

$$q^u_{ij} \geq 0, \ \text{for all } i, j, \text{ and } u.$$

The first term in (5.4) represents the revenue. The second term is the expected disruption-related production cost. The third term is the expected disruption-related transportation cost. The fourth term is the cost of disruption risks adjusted by each manufacturer's attitude.

We assume that, for each manufacturer, the production cost function and the transaction cost functions without disruptions are continuously differentiable and convex. It is easy to verify that $\hat{F}_i(Q^1)$, $VF_i(Q^1)$, $\hat{C}^u_{ij}(q^u_{ij})$, and $VC^u_{ij}(q^u_{ij})$ are also continuously differentiable and convex. Furthermore, we assume that manufacturers compete in a noncooperative fashion in the sense of Nash (1950, 1951). Hence the optimality conditions for all manufacturers *simultaneously* can be expressed as the following variational inequality [see also Bazaraa, Sherali, and Shetty (1993) and Nagurney, Dong, and Zhang (2002)]: Determine $Q^{1*} \in R^{mng}_+$ satisfying

$$\sum_{i=1}^{m} \sum_{j=1}^{n} \sum_{u=1}^{g} [\frac{\partial \hat{F}_i(Q^{1*})}{\partial q^u_{ij}} + \frac{\partial \hat{C}^u_{ij}(q^{u*}_{ij})}{\partial q^u_{ij}} + \theta_i (\frac{\partial VF_i(Q^{1*})}{\partial q^{u*}_{ij}} + \frac{\partial VC^u_{ij}(q^{u*}_{ij})}{\partial q^{u*}_{ij}}) - \rho^{u*}_{1ij}]$$

$$\times [q^u_{ij} - q^{u*}_{ij}] \geq 0, \quad \forall Q^1 \in R^{mng}_+. \tag{5.5}$$

5.2.2 The Behavior of the Retailers

The retailers, in turn, are involved in transactions both with the manufacturers and the demand markets because they must obtain the product to deliver to the consumers at the demand markets.

Let ρ_{2jk}^{v*} denote the price charged for the product by retailer j to demand market k when the product is shipped via transportation mode v. Hence retailers can price according to their locations as well as according to the transportation modes utilized. This price is determined endogenously in the model along with the prices associated with the manufacturers, that is, the ρ_{1ij}^{u*}, for all i, j and u. We assume that certain disruptions will affect the retailers' handling processes (e.g., the storage and display processes). An additional random risk/disruption-related random parameter η_j is associated with the handling cost of retailer j. Recall that we also assume that there are h types of transportation modes available to each retailer for shipping the product to the demand markets.

Let η_j denote the random parameter associated with the handling cost of retailer j, with the corresponding cumulative distribution function denoted by $\mathcal{F}_j(\eta_j)$ for all retailers j; $j = 1, \ldots, n$. Let q_{jk}^v denote the product shipment between retailer j and demand market k shipped via mode k. We group all such shipments into the vector Q^2. The handling cost, in turn, associated with retailer j is denoted by c_j, where $c_j = c_j(Q^1, Q^2, \eta_j)$ for all retailers j; $j = 1, \ldots, n$.

The expected handling cost, denoted by $\hat{C}_j^1(Q^1, Q^2)$, for retailer j; $j = 1, \ldots, n$, is then given by

$$\hat{C}_j^1(Q^1, Q^2) \equiv \int_{\eta_j} c_j(Q^1, Q^2, \eta_j) d\mathcal{F}_j(\eta_j), \; j = 1, \ldots, n. \tag{5.6}$$

We denote the variance of the handling cost function as $VC_j^1(Q^1, Q^2)$ where $j = 1, \ldots, n$.

Furthermore, similar to the case for the manufacturers, we associate a nonnegative weight ϖ_j to the variance of each retailer's handling cost according to its attitude toward risk. Each retailer faces two objectives: to maximize its expected profit and to minimize the disruption risks adjusted by its risk attitude. Therefore, the objective function for retailer j; $j = 1, \ldots, n$, can be expressed as follows

$$\text{Maximize} \; \sum_{k=1}^{o} \sum_{v=1}^{h} \rho_{2jk}^{v*} q_{jk}^v - \hat{C}_j^1(Q^1, Q^2) - \sum_{i=1}^{m} \sum_{u=1}^{g} \rho_{1ij}^{u*} q_{ij}^u - \varpi_j VC_j^1(Q^1, Q^2) \tag{5.7}$$

subject to:

$$\sum_{k=1}^{o} \sum_{v=1}^{h} q_{jk}^v \leq \sum_{i=1}^{m} \sum_{u=1}^{g} q_{ij}^u \tag{5.8}$$

and the nonnegativity constraints: $q_{ij}^u \geq 0$ for all i, j, and u; $q_{jk}^v \geq 0$ for all j, k, and v.

Objective function (5.7) expresses that the difference between the revenues minus the expected handling cost and the payout to the manufacturers, and the weighted disruption risk is to be maximized. Constraint (5.8) states that retailers cannot purchase more product from a retailer than is available in stock.

The term γ_j denotes the Lagrange multiplier associated with constraint (5.8) for retailer j. Furthermore, we assume that, for each retailer, the handling cost without disruptions is continuously differentiable and convex. It is easy to verify that $\hat{C}_j^1(Q^1, Q^2)$ and $VC_j^1(Q^1, Q^2)$ are also continuously differentiable and convex. We assume that retailers compete with one another in a noncooperative manner, seeking to determine their optimal shipments from the manufacturers and to the demand markets. The optimality conditions for all retailers simultaneously coincide with the solution of the following variational inequality: Determine $(Q^{1*}, Q^{2*}, \gamma^*) \in R_+^{mng+noh+n}$ satisfying

$$\sum_{i=1}^{m}\sum_{j=1}^{n}\sum_{u=1}^{g}[\frac{\partial \hat{C}_j^1(Q^{1*}, Q^{2*})}{\partial q_{ij}^u} + \rho_{1ij}^{u*} + \varpi_j \frac{\partial VC_j^1(Q^{1*}, Q^{2*})}{\partial q_{ij}^u} - \gamma_j^*] \times [q_{ij}^u - q_{ij}^{u*}]$$

$$+\sum_{j=1}^{n}\sum_{k=1}^{o}\sum_{v=1}^{h}[-\rho_{2jk}^{v*} + \gamma_j^* + \frac{\partial \hat{C}_j^1(Q^{1*}, Q^{2*})}{\partial q_{jk}^v} + \varpi_j \frac{\partial VC_j^1(Q^{1*}, Q^{2*})}{\partial q_{jk}^v}]$$

$$\times [q_{jk}^v - q_{jk}^{v*}]$$

$$+\sum_{j=1}^{n}[\sum_{i=1}^{m}\sum_{u=1}^{g} q_{ij}^{u*} - \sum_{k=1}^{o}\sum_{v=1}^{h} q_{jk}^{v*}] \times [\gamma_j - \gamma_j^*] \geq 0, \ \forall (Q^1, Q^2, \gamma) \in R_+^{mng+noh+n}. \tag{5.9}$$

5.2.3 The Market Equilibrium Conditions

We now turn to a discussion of the market equilibrium conditions. Subsequently, we construct the equilibrium condition for the entire supply chain network.

Let $c_{jk}^v(Q^2)$ denote the unit transaction/transportation cost associated with shipping the product between retailer j and demand market k via transportation mode v for all j, k, v. Also, let ρ_{3k} denote the demand price at demand market k for $k = 1, \ldots, o$. We group the demand market prices into the vector ρ_3. The random demand at demand market k is denoted by $d_k(\rho_3)$, with the expected value $\hat{d}_k(\rho_3)$ for all retailers k; $k = 1, \ldots, o$.

The equilibrium conditions associated with the product shipments that take place between the retailers and the consumers are the *stochastic economic equilibrium* conditions, which, mathematically, take on the following form: For any retailer with associated demand market k; $k = 1, \ldots, o$

$$\hat{d}_k(\rho_3^*) \begin{cases} \leq \sum_{j=1}^{o}\sum_{v=1}^{h} q_{jk}^{v*}, & \text{if} \quad \rho_{3k}^* = 0, \\ = \sum_{j=1}^{o}\sum_{v=1}^{h} q_{jk}^{v*}, & \text{if} \quad \rho_{3k}^* > 0, \end{cases} \tag{5.10a}$$

$$\rho_{2jk}^{v*} + c_{jk}^v(Q^{2*}) \begin{cases} \geq \rho_{3k}^*, & \text{if} \quad q_{jk}^{v*} = 0, \\ = \rho_{3k}^*, & \text{if} \quad q_{jk}^{v*} > 0. \end{cases} \tag{5.10b}$$

Conditions $(5.10a)$ state if the expected demand price at demand market k is positive, then the quantities purchased by consumers at the demand market from the retailers in the aggregate is equal to the demand at demand market k. Conditions $(5.10b)$ state, in turn, that, in equilibrium, if the consumers at demand market k purchase the product from retailer j via transportation mode v, then the price charged by the retailer for the product plus the unit transaction cost is equal to the price that the consumers are willing to pay for the product. If the price plus the unit transaction cost exceeds the price the consumers are willing to pay at the demand market, then there will be no transaction between the retailer and demand market via that transportation mode.

Equilibrium conditions $(5.10a)$ and $(5.10b)$ are equivalent to the following variational inequality problem, after summing over all demand markets: Determine $(Q^{2*}, \rho_3^*) \in R_+^{noh+o}$ satisfying

$$\sum_{k=1}^{o}(\sum_{j=1}^{n}\sum_{v=1}^{h} q_{jk}^{v*} - \hat{d}_k(\rho_3^*)) \times [\rho_{3k} - \rho_{3k}^*]$$

$$+\sum_{k=1}^{o}\sum_{j=1}^{n}\sum_{v=1}^{h}(\rho_{2jk}^{v*} + c_{jk}^{v}(Q^{2*}) - \rho_{3k}^*) \times [q_{jk}^{v} - q_{jk}^{v*}] \geq 0, \ \forall \rho_3 \in R_+^o, \forall Q^2 \in R_+^{noh}. \tag{5.11}$$

Remark 5.1

In this chapter, we are interested in the cases in which the expected demands are positive, that is, $\hat{d}_k(\rho_3) > 0, \forall \rho_3 \in R_+^o$, for $k = 1, \ldots, o$. Furthermore, we assume that the unit transaction costs: $c_{jk}^v(Q^2) > 0, \forall j, k, \forall Q^2 \neq 0$.

Under these assumptions, we have

$$\rho_{3k}^* > 0 \quad \text{and} \quad \hat{d}_k(\rho_3^*) = \sum_{j=1}^{n}\sum_{k=1}^{o}\sum_{v=1}^{h} q_{jk}^{v*}, \quad k = 1, \ldots, o.$$

This can be shown by contradiction. If there exists a $\bar{k}$, where $\rho_{3\bar{k}}^* = 0$, then, according to $(5.10a)$, we have $\sum_{j=1}^{n}\sum_{k=1}^{o}\sum_{v=1}^{h} q_{jk}^{v*} \geq \hat{d}_{\bar{k}}(\rho_3^*) > 0$. Hence there exists at least a $(j,\bar{k})$ pair such that $q_{j\bar{k}}^{v*} > 0$, which means that $c_{j\bar{k}}^{v}(Q^{2*}) > 0$ by assumption. From conditions $(5.10b)$, we have $\rho_{2j\bar{k}}^{v*} + c_{j\bar{k}}^{v}(Q^{2*}) = \rho_{3\bar{k}} > 0$, which leads to a contradiction.

5.2.4 The Equilibrium Conditions of the Supply Chain

In equilibrium, the optimality conditions for all manufacturers, as expressed by (5.4); the optimality conditions for all retailers, as expressed by (5.9); and the equilibrium conditions for all the demand markets, as expressed by (5.11) must hold simultaneously [see also Nagurney et al. (2005)]. Hence the product shipments of the manufacturers with the retailers must be equal to the product shipments that retailers accept from the manufacturers. We now formally state the equilibrium conditions for the entire supply chain network.

Definition 5.1: Supply Chain Network Equilibrium under Uncertainty and Random Demands
The equilibrium state of the supply chain network with disruption risks and random demands is one where the flows of the product between the tiers of the decision-makers coincide and the flows and prices satisfy the sum of conditions (5.4), (5.9), and (5.11).

The summation of inequalities (5.4), (5.9), and (5.11), after algebraic simplification, yields the following result [see also Nagurney (1999, 2006a)].

Theorem 5.1: Variational Inequality Formulation
A product shipment and price pattern $(Q^{1*}, Q^{2*}, \gamma^*, \rho_3^*) \in R_+^{mng+noh+n+o}$ *is an equilibrium pattern of the supply chain model according to Definition 5.1, if and only if it satisfies the variational inequality problem*

$$\sum_{i=1}^{m}\sum_{j=1}^{n}\sum_{u=1}^{g}[\frac{\partial \hat{F}_i(Q^{1*})}{\partial q_{ij}^u} + \frac{\partial \hat{C}_{ij}^u(q_{ij}^{u*})}{\partial q_{ij}^u} + \theta_i(\frac{\partial VF_i(Q^{1*})}{\partial q_{ij}^u} + \frac{\partial VC_{ij}^u(q_{ij}^{u*})}{\partial q_{ij}^u})$$

$$+\frac{\partial \hat{C}_j^1(Q^{1*}, Q^{2*})}{\partial q_{ij}^u} + \varpi_j \frac{\partial VC_j^1(Q^{1*}, Q^{2*})}{\partial q_{ij}^u} - \gamma_j^*] \times [q_{ij}^u - q_{ij}^{u*}]$$

$$+\sum_{j=1}^{n}\sum_{k=1}^{o}\sum_{v=1}^{g}[\frac{\partial \hat{C}_j^1(Q^{1*}, Q^{2*})}{\partial q_{jk}^v} + \varpi_j \frac{\partial VC_j^1(Q^{1*}, Q^{2*})}{\partial q_{jk}^v}$$

$$+\gamma_j^* + c_{jk}^v(Q^{2*}) - \rho_{3k}^*] \times [q_{jk}^v - q_{jk}^{v*}]$$

$$+\sum_{j=1}^{n}[\sum_{i=1}^{m}\sum_{u=1}^{g} q_{ij}^{u*} - \sum_{k=1}^{o}\sum_{v=1}^{h} q_{jk}^{v*}] \times [\gamma_j - \gamma_j^*]$$

$$+\sum_{k=1}^{o}[\sum_{j=1}^{n}\sum_{v=1}^{h} q_{jk}^{v*} - \hat{d}_k(\rho_3^*)] \times [\rho_{3k} - \rho_{3k}^*] \geq 0, \ \forall (Q^1, Q^2, \gamma, \rho_3) \in R_+^{mng+noh+n+o}. \tag{5.12}$$

For easy reference in the subsequent sections, variational inequality problem (5.12) can be rewritten in standard variational inequality form [cf. (2.1)] as follows: Determine $X^* \in \mathcal{K}$

$$\langle F(X^*)^T, X - X^* \rangle \geq 0, \ \ \forall X \in \mathcal{K} \equiv R_+^{mng+noh+n+o}, \tag{5.13}$$

where the vector $X \equiv (Q^1, Q^2, \gamma, \rho_3)$, and the vector $F(X)$ is defined as $F(X) \equiv (F_{iju}, F_{jkv}, F_j, F_k)_{i=1,\ldots,m; j=1,\ldots,n; k=1,\ldots,o; u=1,\ldots,g; v=1,\ldots,h}$, with the specific components of F given by the functional terms preceding the multiplication signs in (5.12). Recall that $< \cdot, \cdot >$ denotes the inner product in N-dimensional Euclidean space, where $N = mng + noh + n + o$.

Note that the equilibrium values of the variables in the model [which can be determined from the solution of either variational inequality (5.12) or (5.13)] are the equilibrium product shipments between manufacturers and the retailers given by Q^{1*}, the equilibrium product shipments transacted between the retailers and the demand markets given by Q^{2*}, and the equilibrium prices ρ_3^* and γ^*. We now discuss how to recover the prices ρ_1^* associated with the top tier of nodes of the supply chain network and the prices ρ_2^* associated with the middle tier.

First, note that from (5.5) we have if $q_{ij}^{u*} > 0$, then the price $\rho_{1ij}^{u*} = \frac{\partial \hat{F}_i(Q^{1*})}{\partial q_{ij}^u} + \frac{\partial \hat{C}_{ij}^u(q_{ij}^{u*})}{\partial q_{ij}^u} + \theta_i(\frac{\partial VF_i(Q^{1*})}{\partial q_{ij}^{u*}} + \frac{\partial VC_{ij}^u(q_{ij}^{u*})}{\partial q_{ij}^{u*}})$. On the other hand, from (5.9), it follows that if $q_{jk}^{v*} > 0$, the price $\rho_{2j}^* = \gamma_j^* + \frac{\partial \hat{C}_j^1(Q^{1*},Q^{2*})}{\partial q_{jk}^v} + \varpi_j \frac{\partial VC_j^1(Q^{1*},Q^{2*})}{\partial q_{jk}^v}$. These expressions can be utilized to obtain all such prices for all decision-making manufacturers and retailers using the different modes of transportation.

5.3 A WEIGHTED SUPPLY CHAIN PERFORMANCE MEASURE

In this section, we first propose a supply chain network performance measure. Then we provide the definition of supply chain network robustness and follow with the definition of a weighted supply chain performance measure.

5.3.1 A Supply Chain Network Performance Measure

Based on the measure (3.2), we propose the following definition of a supply chain network performance measure.

Definition 5.2: The Supply Chain Network Performance Measure
The supply chain network performance measure, $\mathcal{E}^{SCN}$, for a given supply chain network with topology $\mathcal{G}$ as in Figure 5.2 and expected demands: $\hat{d}_k$; $k = 1, 2, \ldots, o$, is defined as follows

$$\mathcal{E}^{SCN} \equiv \frac{\sum_{k=1}^{o} \frac{\hat{d}_k}{\rho_{3k}}}{o}, \tag{5.14}$$

where o is the number of demand markets in the supply chain network, and $\hat{d}_k$ and ρ_{3k} denote, respectively, the expected equilibrium demand and the equilibrium price at demand market k.

Note that the equilibrium price is equal to the unit production and transaction costs plus the weighted marginal risks for producing and transacting one unit from the manufacturers to the demand markets [see also Nagurney (2006b)]. According to the performance measure, a supply chain network performs well in network equilibrium if, on the average, and across all demand markets, a large demand can be satisfied at a low price. In this chapter, we apply the above performance measure to assess the robustness of particular supply chain networks. From the discussion in Section 5.2.3, we have $\rho_{3k} > 0$; $k = 1, \ldots, o$. Therefore, the above definition is well-defined.

Furthermore, since each individual may have different opinions as to the risks, we need a basis against which to compare supply chain performance under different risk attitudes and to understand how risk attitudes affect the performance of a supply chain. Hence we define $\mathcal{E}^0_{SCN}$ as the supply chain performance measure, where the $\hat{d}_k$ and the ρ_{3k}; $k = 1, \ldots, o$, are obtained by assuming that the weights that reflect the manufacturers' and the retailers' attitudes toward the disruption risks are zero. This definition excludes individuals' subjective differences in a supply chain and, with this definition, we are ready to study supply chain network robustness.

5.3.2 Supply Chain Robustness Measurement

Robustness has a broad meaning and is often couched in different settings. Generally, and as emphasized in Chapters 3 and 4, robustness means that the system performs well when exposed to uncertain future conditions and perturbations [cf. Bundschuh, Klabjan, and Thurston (2003), Snyder (2003), and Holmgren (2007)].

Therefore, we propose the following rationale to assess the robustness of a supply chain: Assume that all the random parameters take on a given threshold probability value, for example, 95%. Moreover, assume that all the cumulative distribution functions for random parameters have inverse functions. Hence we have $\alpha_i = \mathcal{F}_i^{-1}(.95)$, for $i = 1, \ldots, m$; $\beta^u_{ij} = \mathcal{F}^{u^{-1}}_{ij}(.95)$, for $i = 1, \ldots, m$; $j = 1, \ldots, n$, and so on. With the disruption-related parameters assumed as given, we can calculate the supply chain performance measure according to Definition 5.2 as in (5.14). Let $\mathcal{E}_w$ denote the supply chain performance measure with random parameters fixed at a certain level as described earlier. For example, when $w = .95$, $\mathcal{E}_w$ is the supply chain performance with all the random risk parameters fixed at the value of a 95% probability level. Then, the supply chain network robustness measure, $\mathcal{R}^{SCN}$, is given by

$$\mathcal{R}^{SCN} = \mathcal{E}^0_{SCN} - \mathcal{E}_w, \tag{5.15}$$

where $\mathcal{E}^0_{SCN}$ gauges the supply chain performance based on the model introduced in Section 5.2, but with weights related to risks being zero.

Hence $\mathcal{E}^0_{SCN}$ measures and quantifies the base supply chain performance, whereas $\mathcal{E}_w$ assesses the supply chain performance measure at some prespecified uncertainty level. If their difference is small, a supply chain maintains its functionality well and we consider the supply chain to be robust at the threshold disruption level. Hence the lower the value of $\mathcal{R}^{SCN}$, the more robust a supply chain is.

Note that the above robustness definition has implications for network resilience as well. *Resilience* is a general and conceptual term that is challenging to formally quantify. McCarthy (2007) defined resilience "as the ability of a system to recover from adversity, either back to its original state or an adjusted state based on new requirements." For a comprehensive discussion of resilience within the context of critical infrastructure protection, see Garbin and Shortle (2007), Perelman (2007), and Scalingi (2007). Because our supply chain measure is based on the network equilibrium model, a network that is qualified as being robust according to our measure is also resilient, provided that its performance after experiencing the disruption(s) is close to the "original value." Interestingly, this idea is in agreement with Hansson and Helgesson (2003), who proposed that robustness can be treated as a special case of resilience.

5.3.2.1 A Weighted Supply Chain Performance Measure

Note that different supply chains may have different requirements regarding the performance and robustness concepts introduced in the previous sections. For example, in the case of a supply chain of a toy product, one may focus on how to satisfy demand in the most cost efficient way and not care too much about supply chain robustness. A medical/healthcare supply chain, on the other hand, may have a requirement that the supply chain be highly robust when faced with uncertain conditions. Consequently, to be able to examine and to evaluate different application-based supply chains from both perspectives, we now define a *weighted* supply chain performance measure, denoted by $\hat{\mathcal{E}}^{SCN}$ as follows

$$\hat{\mathcal{E}}^{SCN} = (1-\omega)\mathcal{E}^0_{SCN} + \omega(-\mathcal{R}^{SCN}), \tag{5.16}$$

where $\omega \in [0,1]$ is the weight that is placed on the supply chain robustness.

When $\omega = 1$, the performance of a supply chain hinges only on the robustness measure, which may be the case for a medical/healthcare supply chain, as noted. In contrast, when $\omega = 0$, the performance of the supply chain depends solely on how well it can satisfy demands at low prices. The supply chain of a toy product falls into this category.

5.4 THE ALGORITHM

For completeness, we now provide the explicit form that the steps of the modified projection method (see (2.52) and (2.53) in Chapter 2) take for the solution of the variational inequality (5.12) governing the supply chain network equilibrium problem under uncertainty and random demands.

Step 0: Initialization
Set $(Q^{10}, Q^{20}, \gamma^0, \rho_3^0) \in R_+^{mng+noh+n+k}$. Let $\mathcal{T} = 1$ and set α such that $0 < \alpha \leq \frac{1}{L}$ where L is the Lipschitz constant for the problem [cf. (2.13)].

Step 1: Computation
Compute $(\bar{Q}^{1\mathcal{T}}, \bar{Q}^{2\mathcal{T}}, \bar{\gamma}^{\mathcal{T}}, \bar{\rho}_3^{\mathcal{T}}) \in R_+^{mng+noh+n+k}$ by solving the VI subproblem

$$\sum_{i=1}^{m}\sum_{j=1}^{n}\sum_{u=1}^{g}[\bar{q}_{ij}^{u\mathcal{T}} + \alpha(\frac{\partial \hat{F}_i(Q^{1\mathcal{T}-1})}{\partial q_{ij}^u} + \frac{\partial \hat{C}_{ij}^u(q_{ij}^{u\mathcal{T}-1})}{\partial q_{ij}^u} + \theta_i(\frac{\partial VF_i(Q^{1\mathcal{T}-1})}{\partial q_{ij}^u}$$

$$+\frac{\partial VC_{ij}^u(q_{ij}^{u\mathcal{T}-1})}{\partial q_{ij}^u}) + \frac{\partial \hat{C}_j^1(Q^{1\mathcal{T}-1}, Q^{2\mathcal{T}-1})}{\partial q_{ij}^u} + \varpi_j \frac{\partial VC_j^1(Q^{1\mathcal{T}-1}, Q^{2\mathcal{T}-1})}{\partial q_{ij}^u}$$

$$-\gamma_j^{\mathcal{T}-1}) - q_{ij}^{u\mathcal{T}-1}] \times [q_{ij}^u - \bar{q}_{ij}^{u\mathcal{T}}]$$

$$+\sum_{j=1}^{n}\sum_{k=1}^{o}\sum_{v=1}^{h}[\bar{q}_{jk}^{v\mathcal{T}}+\alpha(\frac{\partial\hat{C}_j^1(Q^{1\mathcal{T}-1},Q^{2\mathcal{T}-1})}{\partial q_{jk}^v}+\varpi_j\frac{\partial VC_j^1(Q^{1\mathcal{T}-1},Q^{2\mathcal{T}-1})}{\partial q_{jk}^v}$$

$$+\gamma_j^{\mathcal{T}-1}+c_{jk}^v(Q^{2\mathcal{T}-1})-\rho_{3k}^{\mathcal{T}-1})-q_{jk}^{v\mathcal{T}-1}]\times[q_{jk}^v-\bar{q}_{jk}^{v\mathcal{T}}]$$

$$+\sum_{j=1}^{n}[\bar{\gamma}_j^{\mathcal{T}}+\alpha(\sum_{i=1}^{m}\sum_{u=1}^{g}q_{ij}^{u\mathcal{T}-1}-\sum_{k=1}^{o}\sum_{v=1}^{h}q_{jk}^{v\mathcal{T}-1})-\gamma_j^{\mathcal{T}-1}]\times[\gamma_j-\bar{\gamma}_j^{\mathcal{T}}]$$

$$+\sum_{k=1}^{o}[\sum_{j=1}^{n}\sum_{v=1}^{h}\bar{\rho}_{3k}^{\mathcal{T}}+\alpha(q_{jk}^{v\mathcal{T}-1}-\hat{d}_k(\rho_3^{\mathcal{T}-1}))-\rho_{3k}^{\mathcal{T}-1}]\times[\rho_{3k}-\bar{\rho}_{3k}^{\mathcal{T}}]\geq 0,$$

$$\forall(Q^1,Q^2,\gamma,\rho_3)\in R_+^{mng+noh+n+k}. \tag{5.17}$$

Step 2: Adaptation
Compute $(Q^{1\mathcal{T}},Q^{2\mathcal{T}},\gamma^{\mathcal{T}},\rho_3^{\mathcal{T}})\in R_+^{mng+noh+n+k}$ by solving the VI subproblem

$$\sum_{i=1}^{m}\sum_{j=1}^{n}\sum_{u=1}^{g}[q_{ij}^{u\mathcal{T}}+\alpha(\frac{\partial\hat{F}_i(\bar{Q}^{1\mathcal{T}})}{\partial q_{ij}^u}+\frac{\partial\hat{C}_{ij}^u(\bar{q}_{ij}^{u\mathcal{T}})}{\partial q_{ij}^u}+\theta_i(\frac{\partial VF_i(\bar{Q}^{1\mathcal{T}})}{\partial q_{ij}^u}$$

$$+\frac{\partial VC_{ij}^u(\bar{q}_{ij}^{u\mathcal{T}})}{\partial q_{ij}^u})+\frac{\partial\hat{C}_j^1(\bar{Q}^{1\mathcal{T}},\bar{Q}^{2\mathcal{T}})}{\partial q_{ij}^u}+\varpi_j\frac{\partial VC_j^1(\bar{Q}^{1\mathcal{T}},\bar{Q}^{2\mathcal{T}})}{\partial q_{ij}^u}$$

$$-\bar{\gamma}_j^{\mathcal{T}})-q_{ij}^{u\mathcal{T}-1}]\times[q_{ij}^u-q_{ij}^{u\mathcal{T}}]$$

$$+\sum_{j=1}^{n}\sum_{k=1}^{o}\sum_{v=1}^{h}[q_{jk}^{v\mathcal{T}}+\alpha(\frac{\partial\hat{C}_j^1(\bar{Q}^{1\mathcal{T}},\bar{Q}^{2\mathcal{T}})}{\partial q_{jk}^v}+\varpi_j\frac{\partial VC_j^1(\bar{Q}^{1\mathcal{T}},\bar{Q}^{2\mathcal{T}})}{\partial q_{jk}^v}$$

$$+\bar{\gamma}_j^{\mathcal{T}}+c_{jk}^v(\bar{Q}^{2\mathcal{T}})-\bar{\rho}_{3k}^{\mathcal{T}})-q_{jk}^{v\mathcal{T}-1}]\times[q_{jk}^v-q_{jk}^{v\mathcal{T}}]$$

$$+\sum_{j=1}^{n}[\gamma_j^{\mathcal{T}}+\alpha(\sum_{i=1}^{m}\sum_{u=1}^{g}\bar{q}_{ij}^{u\mathcal{T}}-\sum_{k=1}^{o}\sum_{v=1}^{h}\bar{q}_{jk}^{v\mathcal{T}})-\gamma_j^{\mathcal{T}-1}]\times[\gamma_j-\gamma_j^{\mathcal{T}}]$$

$$+\sum_{k=1}^{o}[\sum_{j=1}^{n}\sum_{v=1}^{h}\rho_{3k}^{\mathcal{T}}+\alpha(\bar{q}_{jk}^{v\mathcal{T}}-\hat{d}_k(\bar{\rho}_3^{\mathcal{T}}))-\rho_{3k}^{\mathcal{T}-1}]\times[\rho_{3k}-\rho_{3k}^{\mathcal{T}}]\geq 0,$$

$$\forall(Q^1,Q^2,\gamma,\rho_3)\in R_+^{mng+noh+n+k}. \tag{5.18}$$

Step 3: Convergence Verification
If max $|q_{ij}^{u\mathcal{T}}-q_{ij}^{u\mathcal{T}-1}|\leq\epsilon$, $|q_{jk}^{v\mathcal{T}}-q_{jk}^{v\mathcal{T}-1}|\leq\epsilon$, $|\gamma_j^{\mathcal{T}}-\gamma_j^{\mathcal{T}-1}|\leq\epsilon$, $|\rho_{3k}^{\mathcal{T}}-\rho_{3k}^{\mathcal{T}-1}|\leq\epsilon$ for all $i=1,\ldots,m$; $j=1,\ldots,n$; $u=1,\ldots,g$; $k=1,\ldots,k$; ; $v=1,\ldots,h$, then stop; else, set $\mathcal{T}=\mathcal{T}+1$, and return to Step 1.

As discussed earlier, due the simplicity of the underlying feasible set, these iterative steps yield closed form expressions for the product flows and prices. In particular, Step 1 [cf. (5.17)], at each iteration $\mathcal{T}$, yields the following expressions.

The product flows are computed explicitly according to

$$\bar{q}_{ij}^{u\mathcal{T}} = \max\{0,\, q_{ij}^{u\mathcal{T}-1} - \alpha(\frac{\partial \hat{F}_i(Q^{1\mathcal{T}-1})}{\partial q_{ij}^u} + \frac{\partial \hat{C}_{ij}^u(q_{ij}^{u\mathcal{T}-1})}{\partial q_{ij}^u} + \theta_i(\frac{\partial VF_i(Q^{1\mathcal{T}-1})}{\partial q_{ij}^u}$$

$$+\frac{\partial VC_{ij}^u(q_{ij}^{u\mathcal{T}-1})}{\partial q_{ij}^u}) + \frac{\partial \hat{C}_j^1(Q^{1\mathcal{T}-1}, Q^{2\mathcal{T}-1})}{\partial q_{ij}^u} + \varpi_j \frac{\partial VC_j^1(Q^{1\mathcal{T}-1}, Q^{2\mathcal{T}-1})}{\partial q_{ij}^u}$$

$$-\gamma_j^{\mathcal{T}-1})\},\ \forall i, j, u; \tag{5.19}$$

$$\bar{q}_{jk}^{v\mathcal{T}} = \max\{0,\, q_{jk}^{v\mathcal{T}-1} - \alpha(\frac{\partial \hat{C}_j^1(Q^{1\mathcal{T}-1}, Q^{2\mathcal{T}-1})}{\partial q_{jk}^v} + \varpi_j \frac{\partial VC_j^1(Q^{1\mathcal{T}-1}, Q^{2\mathcal{T}-1})}{\partial q_{jk}^v})$$

$$+\gamma_j^{\mathcal{T}-1} + c_{jk}^v(Q^{2\mathcal{T}-1}) - \rho_{3k}^{\mathcal{T}-1}\},\ \forall j, k, v. \tag{5.20}$$

The prices, in turn, are computed explicitly according to

$$\bar{\gamma}_j^{\mathcal{T}} = \max\{0, \gamma_j^{\mathcal{T}-1} - \alpha(\sum_{i=1}^{m}\sum_{u=1}^{g} q_{ij}^{u\mathcal{T}-1} - \sum_{k=1}^{o}\sum_{v=1}^{h} q_{jk}^{v\mathcal{T}-1})\},\ \forall j; \tag{5.21}$$

$$\bar{\rho}_{3k}^{\mathcal{T}} = \max\{0, \rho_{3k}^{\mathcal{T}-1} - \alpha(q_{jk}^{v\mathcal{T}-1} - \hat{d}_k(\rho_3{}^{\mathcal{T}-1}))\},\ \forall k. \tag{5.22}$$

Similarly, Step 2 [cf. (5.18)] yields the following closed form expressions at each iteration $\mathcal{T}$.

The product flows are computed explicitly according to

$$q_{ij}^{u\mathcal{T}} = \max\{0, q_{ij}^{u\mathcal{T}-1} - \alpha(\frac{\partial \hat{F}_i(\bar{Q}^{1\mathcal{T}})}{\partial q_{ij}^u} + \frac{\partial \hat{C}_{ij}^u(\bar{q}_{ij}^{u\mathcal{T}})}{\partial q_{ij}^u} + \theta_i(\frac{\partial VF_i(\bar{Q}^{1\mathcal{T}})}{\partial q_{ij}^u}$$

$$+\frac{\partial VC_{ij}^u(\bar{q}_{ij}^{u\mathcal{T}})}{\partial q_{ij}^u}) + \frac{\partial \hat{C}_j^1(\bar{Q}^{1\mathcal{T}}, \bar{Q}^{2\mathcal{T}})}{\partial q_{ij}^u} + \varpi_j \frac{\partial VC_j^1(\bar{Q}^{1\mathcal{T}}, \bar{Q}^{2\mathcal{T}})}{\partial q_{ij}^u} - \bar{\gamma}_j^{\mathcal{T}})\},\ \forall i, j, u; \tag{5.23}$$

$$q_{jk}^{v\mathcal{T}} = \max\{0, q_{jk}^{v\mathcal{T}-1} - \alpha(\frac{\partial \hat{C}_j^1(\bar{Q}^{1\mathcal{T}}, \bar{Q}^{2\mathcal{T}})}{\partial q_{jk}^v} + \varpi_j \frac{\partial VC_j^1(\bar{Q}^{1\mathcal{T}}, \bar{Q}^{2\mathcal{T}})}{\partial q_{jk}^v}$$

$$+\bar{\gamma}_j^{\mathcal{T}} + c_{jk}^v(\bar{Q}^{2\mathcal{T}}) - \bar{\rho}_{3k}^{\mathcal{T}})\},\ \forall j, k, v. \tag{5.24}$$

The prices are also computed explicitly as follows

$$\gamma_j^{\mathcal{T}} = \max\{0, \gamma_j^{\mathcal{T}-1} - \alpha(\sum_{i=1}^{m}\sum_{u=1}^{g} \bar{q}_{ij}^{u\mathcal{T}} - \sum_{k=1}^{o}\sum_{v=1}^{h} \bar{q}_{jk}^{v\mathcal{T}})\},\ \forall j; \tag{5.25}$$

$$\rho_{3k}^{\mathcal{T}} = \max\{0, \rho_{3k}^{\mathcal{T}-1} - \alpha(\bar{q}_{jk}^{v\mathcal{T}} - \hat{d}_k(\bar{\rho}_3^{\mathcal{T}}))\},\ \forall k. \tag{5.26}$$

In the next section, we use this algorithm to compute solutions to numerical supply chain network equilibrium examples.

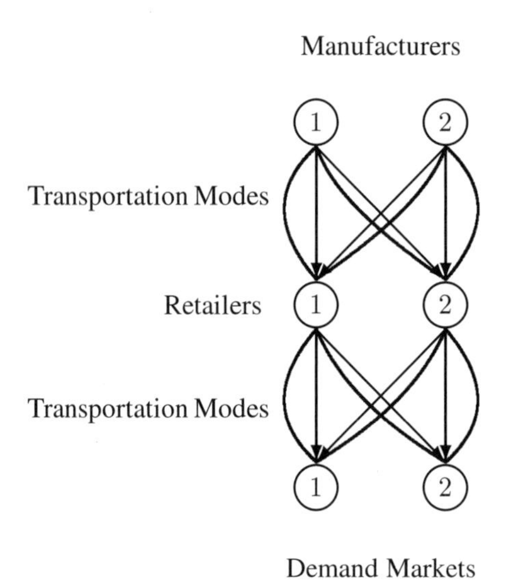

Figure 5.3: The supply chain network for the numerical examples

5.5 EXAMPLES

In the numerical examples, the supply chain consists of two manufacturers, two retailers, and two demand markets, as depicted in Figure 5.3. There are two modes of transportation available between each manufacturer and retailer pair and between each retailer and demand market pair. These examples are solved by the modified projection method, as discussed in Section 5.4. We set α equal to .5 and the convergence tolerance $\epsilon = .001$. The initial product flows and prices were all set to 1. The modified projection method, as described in Section 5.4, was implemented in MATLAB (www.mathworks.com) on an IBM laptop T61 and used to solve the following numerical examples.

5.5.1 Example 5.1

In the first example, for illustration purposes, we assumed that all the random parameters followed uniform distributions. The relevant parameters were as follows $\alpha_i \sim [0, 2]$ for $i = 1, 2$; $\beta_{ij}^u \sim [0, 1]$ for $i = 1, 2$; $j = 1, 2$; $u = 1, 2$; $\eta_j \sim [0, 3]$ for $j = 1, 2$.

We further assumed that the demand functions followed a uniform distribution given by $[-2\rho_{3k} + 200, -2\rho_{3k} + 600]$, for $k = 1, 2$. Hence the expected demand functions were

$$\hat{d}_k(\rho_3) = -2\rho_{3k} + 400, \quad \text{for } k = 1, 2.$$

The production cost functions for the manufacturers were

$$f_1(Q^1, \alpha_1) = 2.5(\sum_{j=1}^{2}\sum_{u=1}^{2} q_{1j}^u)^2 + (\sum_{j=1}^{2}\sum_{u=1}^{2} q_{1j}^u)(\sum_{j=1}^{2}\sum_{u=1}^{2} q_{2j}^u) + 2\alpha_1(\sum_{j=1}^{2}\sum_{u=1}^{2} q_{1j}^u),$$

$$f_2(Q^1, \alpha_2) = 2.5(\sum_{j=1}^{2}\sum_{u=1}^{2} q_{2j}^u)^2 + (\sum_{j=1}^{2}\sum_{u=1}^{2} q_{1j}^u)(\sum_{j=1}^{2}\sum_{u=1}^{2} q_{2j}^u) + 2\alpha_2(\sum_{j=1}^{2}\sum_{u=1}^{2} q_{2j}^u).$$

The expected production cost functions for the manufacturers were

$$\hat{F}_1(Q^1) = 2.5(\sum_{j=1}^{2}\sum_{u=1}^{2} q_{1j}^u)^2 + (\sum_{j=1}^{2}\sum_{u=1}^{2} q_{1j}^u)(\sum_{j=1}^{2}\sum_{u=1}^{2} q_{2j}^u) + 2(\sum_{j=1}^{2}\sum_{u=1}^{2} q_{1j}^u),$$

$$\hat{F}_2(Q^1) = 2.5(\sum_{j=1}^{2}\sum_{u=1}^{2} q_{2j}^u)^2 + (\sum_{j=1}^{2}\sum_{u=1}^{2} q_{1j}^u)(\sum_{j=1}^{2}\sum_{u=1}^{2} q_{2j}^u) + 2(\sum_{j=1}^{2}\sum_{u=1}^{2} q_{2j}^u).$$

The variances of the production cost functions for the manufacturers were

$$VF_1(Q^1) = \frac{4}{3}(\sum_{j=1}^{2}\sum_{u=1}^{2} q_{1j}^u)^2, \quad VF_2(Q^1) = \frac{4}{3}(\sum_{j=1}^{2}\sum_{u=1}^{2} q_{2j}^u)^2.$$

The transaction cost functions faced by the manufacturers and associated with transacting with the retailers were

$$c_{ij}^1(q_{ij}^1, \beta_{ij}^1) = .5(q_{ij}^1)^2 + 3.5\beta_{ij}^1 q_{ij}^1, \text{ for } i = 1, 2; j = 1, 2,$$

$$c_{ij}^2(q_{ij}^2, \beta_{ij}^2) = (q_{ij}^2)^2 + 5.5\beta_{ij}^2 q_{ij}^2, \text{ for } i = 1, 2; j = 1, 2.$$

The expected transaction cost functions faced by the manufacturers and associated with transacting with the retailers were

$$\hat{C}_{ij}^1(q_{ij}^1) = .5(q_{ij}^1)^2 + 1.75q_{ij}^1, \text{ for } i = 1, 2; j = 1, 2,$$

$$\hat{C}_{ij}^2(q_{ij}^2) = .5(q_{ij}^2)^2 + 2.75q_{ij}^2, \text{ for } i = 1, 2; j = 1, 2.$$

The variances of the transaction cost functions faced by the manufacturers and associated with transacting with the retailers were given by

$$VC_{ij}^1(q_{ij}^1) = 1.0208(q_{ij}^1)^2, \quad VC_{ij}^2(q_{ij}^2) = 2.5208(q_{ij}^2)^2, \text{ for } i = 1, 2; j = 1, 2.$$

The handling costs of the retailers, in turn, were

$$c_j(Q^1, Q^2, \eta_j) = .5(\sum_{i=1}^{2}\sum_{u=1}^{2} q_{ij}^u)^2 + \eta_j(\sum_{i=1}^{2}\sum_{u=1}^{2} q_{ij}^u), \text{ for } j = 1, 2.$$

The expected handling costs of the retailers were given by

$$\hat{C}_j^1(Q^1, Q^2) = .5(\sum_{i=1}^{2}\sum_{u=1}^{2} q_{ij}^u)^2 + 1.5(\sum_{i=1}^{2}\sum_{u=1}^{2} q_{ij}^u), \text{ for } j = 1, 2.$$

The variances of the handling costs of the retailers were

$$VC_j(Q^1, Q^2) = \frac{3}{4}(\sum_{i=1}^{2}\sum_{u=1}^{2} q_{ij}^u)^2, \text{ for } j = 1, 2.$$

The unit transaction costs from the retailers to the demand markets were

$$c_{jk}^1(Q^2) = .3q_{jk}^1, \quad c_{jk}^2(Q^2) = .6q_{jk}^2, \quad \text{for } j = 1, 2;\ k = 1, 2.$$

We assumed that the manufacturers and the retailers placed zero weights on the disruption risks, as discussed in Section 5.3.1, when computing $\mathcal{E}_{SCN}^0$.

With the above expected costs and demands, the computed equilibrium shipments between manufacturers and retailers were $q_{ij}^{1*} = 8.5022$, for $i = 1, 2$; $j = 1, 2$; $q_{ij}^{2*} = 3.7511$, for $i = 1, 2$; $j = 1, 2$; whereas the computed equilibrium shipments between the retailers and the demand markets were $q_{jk}^{1*} = 8.1767$, for $j = 1, 2$; $k = 1, 2$; $q_{jk}^{2*} = 4.0767$, for $j = 1, 2$; $k = 1, 2$. Finally, the computed equilibrium prices were $\rho_{31}^* = \rho_{32}^* = 187.7466$, and the expected equilibrium demands were $\hat{d}_1 = \hat{d}_2 = 24.5068$. The supply chain performance measure $\mathcal{E}_{SCN}^0 = .1305$.

Now, assume that $w = .95$, that is, all the random cost parameters are fixed at a 95% probability level. The resulting supply chain performance measure was computed as $\mathcal{E}_w$=.1270. When we set $\omega = .5$ [cf. (5.16)], which means that we placed equal emphasis on the performance and on the robustness of the supply chain, the weighted supply chain performance measure was then $\hat{\mathcal{E}}^{SCN} = .0635$.

5.5.2 Example 5.2

For the same network structure and cost and demand functions, we then assumed that the relevant parameters were changed as follows: $\alpha_i \sim [0, 4]$ for $i = 1, 2$; $\beta_{ij}^u \sim [0, 2]$ for $i = 1, 2$; $j = 1, 2$; $u = 1, 2$; $\eta_j \sim [0, 6]$ for $j = 1, 2$.

With the revised expected costs and demands, the computed equilibrium shipments between manufacturers and retailers were now $q_{ij}^{1*} = 8.6008$, for $i = 1, 2$; $j = 1, 2$; $q_{ij}^{2*} = 3.3004$, for $i = 1, 2$; $j = 1, 2$; whereas the equilibrium shipments between the retailers and the demand markets were $q_{jk}^{1*} = 7.9385$, for $j = 1, 2$; $k = 1, 2$; $q_{jk}^{2*} = 3.9652$, for $j = 1, 2$; $k = 1, 2$. Finally, the equilibrium prices were $\rho_{31}^* = \rho_{32}^* = 188.0963$ and the expected equilibrium demands were $\hat{d}_1 = \hat{d}_2 = 23.8074$. The supply chain performance measure $\mathcal{E}_{SCN}^0 = .1266$.

Similar to Example 5.1, we then assumed that $w = .95$, that is, all the random cost parameters were fixed at a 95% probability level. The resulting supply chain performance measure $\mathcal{E}_w$=.1194. When we set $\omega = .5$, the weighted supply chain performance measure $\hat{\mathcal{E}}^{SCN} = .0597$.

Observe that the first example led to a better measure of performance since the uncertain parameters do not have as great of an impact as in the second one for the cost functions under the given threshold level.

5.6 SUMMARY AND CONCLUSIONS

In this chapter, we developed a novel supply chain network model to study demand-side as well as supply-side risks, with the demand assumed to be random and the supply-side risks modeled as uncertain parameters in the underlying cost functions. This supply chain model generalizes several existing models by including such features as multiple transportation modes from the manufacturers to the retailers and from the retailers to the demand markets. We also proposed a weighted supply chain performance and robustness measure based on an extended version of the unified network performance/efficiency measure $\mathcal{E}$ [cf. (3.2)] described in Section 3.2. We illustrated the supply chain network model through numerical examples for which the equilibrium prices and product shipments were computed and robustness analyses conducted.

The application of the foundational constructs in this chapter to supply chains in different industries is now possible. For definiteness, we now describe the scale and scope of supply chains of particular firms in different industries.

Hewlett-Packard (HP) is a leading provider of printing and imaging products, desktop and laptop computers, and computer servers and storage products. The company operates in more than 170 countries and generated annual revenues of more than $90 billion from the third quarter of 2005 to the second quarter of 2006. In 2005, HP shipped more than 50 million printers, 30 million personal computers (PCs), and two million servers through its global supply chain network [Dow Theory Forecasts (2006)].

Cardinal Health Inc., one of the largest providers of supply chain services in the healthcare industry, has more than $70 billion in annual sales. Its logistics network processes a third of all pharmaceutical, medical, and laboratory products in the industry, which represents 50,000 deliveries every day to 40,000 acute care hospitals, retail pharmacy chains, independent and mail-order pharmacies, clinical laboratories, and other providers of care. The company expects to increase its markets in the healthcare industry in the United States, which has an annual market value of approximately $300 billion. [Drug Week (2005)].

Unilever, with 2006 net sales of $50 billion in the Americas, Europe, Asia, and Africa, operates a truly global supply chain. The corporation has a presence in 150 countries around the world, with some 179,000 employees, and offers a product portfolio that spans grocery stores with well-known trademarked brand names [Monahan and Nardone (2007)].

Target Corporation (2008) operates large-format general merchandise discount stores in the United States, which include Target and SuperTarget stores. As of the end of 2007, the company owned 1,352 stores, leased 73 stores, and operated 32 distribution centers, with a presence in 47 states. Target's total revenue in 2007 was $63,367 million.

As one of the largest private-sector energy companies in the world, Shell is involved in more than 40 refineries globally and has 45,000 retail gas stations across the world operating under a single brand name. Its products include low-sulphur diesel; lead replacement fuel; liquefied petroleum gas (LPG); and differentiated fuels, including Shell Pura, Optimax, V-Power, V-Power Racing, and V-Power Diesel. It has convenience stores at more than 10,000 locations. A range of products (including fuels, lubricants, and specialty products) are sold to a wide range of business customers through five subsegments of the company: Shell Aviation, Shell Marine Products, Shell Gas LPG, and Commercial Fuels and Lubricants [cf. Royal Dutch Shell plc (2008)].

According to the Vlasic and Wayne (2008), the economic and financial troubles of the automobile companies in the United States among the Big Three (General Motors, Chrysler, and Ford) are creating a domino effect throughout the supply chain and the vast network of auto supplier firms. For example, General Motors alone has approximately 2,000 suppliers, whereas Ford has about 1,600 suppliers, and Chrysler about 900 suppliers. According to the same article, suppliers make most of the 15,000 parts that make up a car, with more than 70% of a car's value being composed of parts (from the seats to the electronics and bumpers) sold to automakers by suppliers.

Large supply chains of a conglomerate size are not the signature solely of multinational firms any more. In developing countries, with globalization, supply chains have also undergone tremendous growth. For example, two Chinese home appliance manufacturers, Kelong and Little Swan, teamed up with the logistics arm of state-owned China Ocean Shipping (Group) to form An Tai Da Logistics to consolidate and integrate their domestic distribution. By centralizing Kelong's transport fleet and rationalizing its network of warehouses, the firm's requirement for freight transport fell 9.6% in the first year. Furthermore, the mainland market for third-party logistics providers is expected to be worth $32 billion by 2010, against sales of $11 billion in 2004 [Barling (2005)].

The development of theoretical frameworks to capture the complexity of global supply chain networks has been a growing area of research. Recently, Cruz, Nagurney, and Wakolbinger (2006) constructed a model of the integration of social networks with global supply chains along with risk management. The multilevel network model allowed for multicriteria decision-making and demonstrated the dynamic evolution of product flows and prices, along with the relationship levels between the various supply chain decision-makers.

5.7 SOURCES AND NOTES

This chapter is based on the paper by Qiang, Nagurney, and Dong (2009). Here we have expanded the motivation and have also fully explicated the algorithm applied to solve the numerical examples. For further background on supply chain network equilibrium models, applications, and the associated dynamics, see Nagurney (2006a).

For a discussion of the distinction between supply-side versus demand-side risk, see Snyder (2003). Tang (2006a) discussed how to deploy certain strategies to enhance the robustness and the resiliency of supply chains. Kleindorfer and Saad (2005), in turn, provided an overview of strategies for mitigating supply chain disruption risks, which were exemplified by a case study in a chemical product supply chain. For a comprehensive review of supply chain risk management models to that date, refer to Tang (2006b).

It is also important to recall here that supply chain network equilibrium models in which decision-makers compete across a tier of the supply chain but cooperate, if appropriate, between tiers, have been shown to be transformable into transportation network equilibrium problems with elastic demands [cf. Nagurney (2006b)] over an appropriately expanded network or supernetwork [see also Nagurney and Dong (2002a)]. This connection further supports the relevance of the network equilibrium paradigm, emphasized in both Chapters 2 and 3, for a variety of complex network-based decision-making problems in which there is no single controller of the network operating in a centralized manner. Such a transformation, besides enabling the transferral of theoretical results and algorithms from the transportation domain to different applications, also provides for richer interpretations of equilibrium conditions in terms of path "costs" and path flows from appropriately defined origins to destinations. Finally, it supports the commonality of structure as well as the behavior of seemingly distinct network systems that underpin our societies and economies.

Increasingly, it is supply chains that compete with other supply chains in today's globalized economy. Zhang, Dong, and Nagurney (2003) and Zhang (2006) developed network-based models for supply chain versus supply chain competition that capture the transformation processes associated with production and distribution. The models consider costs associated with both operation (production, storage, distribution) and interface (coordination, communication) links. The governing equilibrium conditions may be viewed as an extension of Wardrop's first principle from used paths to used or "active," that is, "winning" chains. Importantly, the unified network performance/efficiency measure can also be used to identify the most important nodes and links in the Zhang, Dong, and Nagurney (2003) and Zhang (2006) frameworks through a simple adaptation of (5.14). Indeed, we are seeing today that disruptions in the supply chains of auto manufacturers may affect even pharmaceutical supply chains. Furthermore, disruptions in food supply chains due to adulterated inputs have been major news both in the United States and China this past year alone. Hence our holistic, system-wide approach to supply chain network modeling and vulnerability analysis is very timely.

CHAPTER 6

CRITICAL NODES AND LINKS IN FINANCIAL NETWORKS

6.1 INTRODUCTION

The study of financial networks dates to Quesnay (1758), who, in his *Tableau Economique*, conceptualized the circular flow of financial funds in an economy as a network. Copeland (1952), subsequently, studied the relationships among financial funds as a network and asked the question, "Does money flow like water or electricity?" The advances in information technology and globalization have further shaped today's financial world into a complex network, which is characterized by distinct sectors, the proliferation of new financial instruments, and increasing international diversification of portfolios. Recently, financial networks have been studied using network models with multiple tiers of decision-makers, including intermediaries.

Because today's financial networks may be highly interconnected and interdependent, any disruptions that occur in one part of the network may produce consequences in other parts of the network, which may not only be in the same region but miles away in other countries. For example, the unforgettable 1987 stock market crash was, in effect, a chain reaction throughout the world; it originated in Hong Kong, then propagated to Europe, and, finally, the United States. More recently, notable financial crises have included the Mexican meltdown in 1994 - 1995, sometimes referred to as

Fragile Networks. By Anna Nagurney and Qiang Qiang

the "Tequila Effect," and the Asian flu crisis of 1997 - 1998, which spread financial *contagion*, that is, the transmission and impact of financial crises [see, e.g., Kali and Reyes (2005)]. The losses due to this Asian financial crisis, have been estimated at $3 trillion in GDP and $2 trillion in equity on financial markets [see, e.g., Winters (1998), Garten (1999), Wade (2000)].

The Severe Acute Respiratory Syndrome (SARS) crisis, which began in November 2002 and was pronounced by the World Health Organization to be contained on July 5, 2003, claimed 812 lives with more than 8,400 people infected. According to deLisle (2003), SARS seemed to parallel the Asian financial crisis of the 1990s, and just like the "Asian contagion" SARS spread quickly along the pathways made possible by globalization. However, rather than international capital mobility that allowed for possible attacks by speculators on vulnerable currencies, it was the global mobility of humans through air travel that made the quick spread of SARS possible and, as noted by deLisle (2003), unpredictable. Toronto, Canada was the area most affected by SARS outside of East Asia, with more than 30 deaths among approximately 200 infections. SARS cost Canada, during its duration, tens of millions of dollars per day in economic terms [see, e.g., Krauss (2003), Catto (2003)].

In 2008 and 2009, the world reeled from the effects of the financial credit crisis; leading financial services and banks closed (including the investment bank Lehman Brothers), others merged, and the financial landscape was changed for forever. The domino effect of the U.S. economic troubles rippled through overseas markets and pushed countries such as Iceland to the verge of bankruptcy [see also Chari, Christiano, and Kehoe (2008)]. The root of the financial problems was considered to have stemmed from the U.S. housing market and the huge number of bad loans that could not be repaid. Many of the subprime loans had been bundled and then sold to investment banks, some of whom had, in turn, borrowed money for these transactions, thus further complexifying the number of decision-makers or agents and the financial linkages. Ultimately, businesses could not obtain financial resources because so many banks and financial institutions were failing, which caused millions of people lost their homes. Figure 6.1 shows the increasing number of home foreclosures over the years.

It is, therefore, crucial for the decision-makers in financial systems (managers, executives, and regulators) to be able to identify a financial network's vulnerable components to protect the functionality of the network. As an illustration, the management at Merrill Lynch well understood the criticality of its operations in the World Trade Center and established contingency plans. Directly after the 9/11 terrorist attacks, management was able to switch its operations from the World Trade Center to the backup centers and the redundant trading floors near New York City. Consequently, the company managed to mitigate the losses for both its customers and itself [see Sheffi (2005)].

In Chapter 3, we proposed a unified network performance measure [cf. (3.2)] that can be used to assess the network performance in the case of either fixed or elastic demands. The measure captures flow information and user/decision-maker behavior and allows one to determine the criticality of various nodes (as well as links) through the identification of their importance and ranking. In Chapter 4, we demonstrated

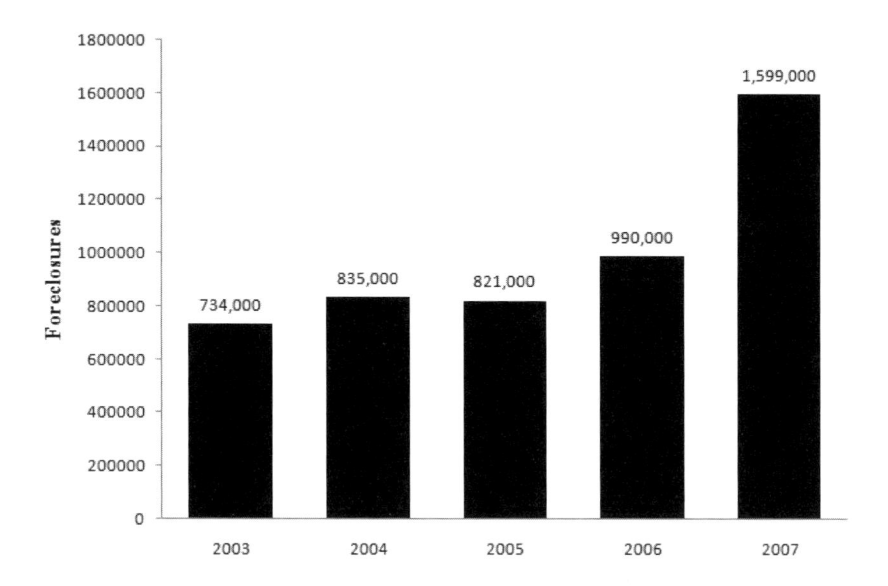

Figure 6.1: Number of homes entering foreclosure [Source: Office of the Speaker of the U.S. House of Representatives (2008)]

the applicability of the new measure to transportation networks, and in Chapter 5, we developed an extension to assess supply chains in the case of disruptions and uncertainty.

Financial networks, as extremely important infrastructure networks, have a great impact on the global economy, and their study has recently also attracted attention of researchers in the area of complex networks. For example, Onnela, Kaski, and Kertész (2004) studied a financial network in which the nodes are stocks and the edges are the correlations among the prices of stocks [see also, Kim and Jeong (2005)]. Caldarelli et al. (2004) studied different financial networks, namely, board and director networks and stock ownership networks, and discovered that all these networks displayed scale-free properties [see also Boginski, Butenko, and Pardalos (2003)]. Several recent studies in finance, in turn, have analyzed the local consequences of catastrophes and the design of risk sharing/management mechanisms since the occurrence of disasters such as 9/11 and Hurricane Katrina [see, for example, Doherty (1997), Loubergé, Këllezi, and Gilli (1999), Niehaus (2002), Gilli and Këllezi (2006), and the references therein].

Nevertheless, there is very little literature that addresses the vulnerability of financial networks. Robinson, Woodard, and Varnado (1998) discussed, from the policymaking point of view, how to protect the critical infrastructure in the United States, including financial networks. Odell and Phillips (2001) conducted an empirical study to analyze the effect of the 1906 San Francisco earthquake on bank loan rates in the financial network within that city. To the best of our knowledge, however, there is no network performance measure to-date that has been applied to financial networks that captures both economic behavior and the underlying network/graph structure. A related relevant network study to ours is that by Jackson and Wolinsky (1996), which defines a value function for the network topology and proposes the network efficiency concept based on the value function from the point of view of network formation.

More recently, there has appeared a stream of literature that begins to further explore financial network structure and vulnerability; see Allen and Gale (1998, 2000) and Boss et al. (2004). Kali and Reyes (2005), motivated by such complex network measures (see Chapter 2) as node degree centrality, node importance, and maximum flow [see Hanneman (2001)], which measures the number of nodes in the neighborhood of a source that lead to pathways to a target, ranked countries that were behind various financial crises and contagion effects using international trade network data. Marquez Diez-Canedo and Martinez-Jaramillo (2007) proposed a network model to capture financial contagion in the banking system to characterize the dynamics of contagion and the corresponding losses in the network. They utilized a systemic risk framework.

In this chapter, we introduce a novel financial network performance measure, which is motivated by the unified network measure in Chapter 3 and that evaluates the network performance in the context in which there is noncooperative competition among source fund agents and among financial intermediaries. Our measure, as we demonstrate in this chapter, can be further applied to identify the importance and the ranking of financial network components.

This chapter is organized as follows. In Section 6.2, we recall the Liu and Nagurney's (2007) financial network model with intermediation and electronic transactions, which is based on Nagurney and Ke's (2003) model but assumes that the demand price functions, rather than the demand functions, are given at the demand markets. Liu and Nagurney (2007) proved that such financial networks could be reformulated and solved as network equilibrium problems (cf. Chapter 2). Nagurney et al. (2007), on the other hand, demonstrated that complex electric power generation and distribution networks could also be reformulated as network equilibrium problems. Hence, in these two papers, the authors demonstrated, respectively, that money flows like transportation flows and so does electric power, answering questions posed decades ago by Copeland (1952) and Beckmann, McGuire, and Winsten (1956).

The financial network performance measure is developed in Section 6.3, along with the associated definition of the importance of network components. Section 6.4 demonstrates the realization of the Euler method discussed in Chapter 2 for the computation of solutions of the financial network equilibrium model. Section 6.5 presents two financial network examples for which the proposed performance measure is computed and the node and link importance rankings determined. The results in this chapter are summarized in Section 6.6.

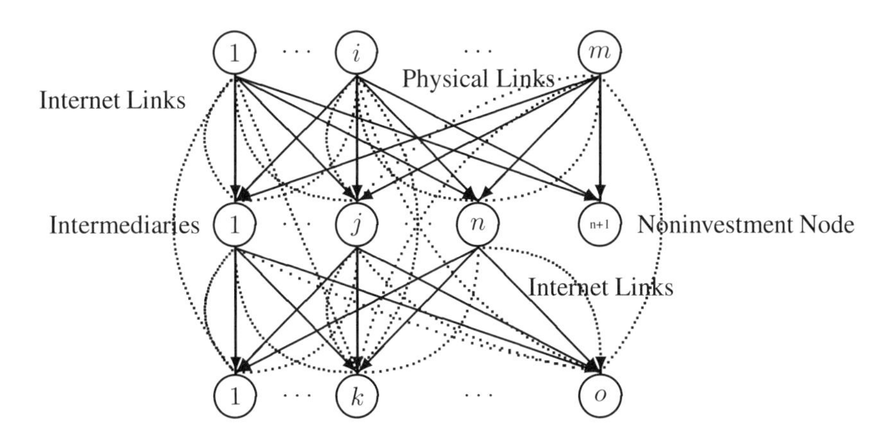

Figure 6.2: The structure of the financial network with intermediation and with electronic transactions

6.2 THE FINANCIAL NETWORK MODEL

In this section, we present the financial network model with intermediation and with electronic transactions in the case of known inverse demand functions associated with the financial products at the demand markets. The financial network consists of m sources of financial funds, n financial intermediaries, and o demand markets, as depicted in Figure 6.2. In the financial network figure, the financial transactions are denoted by the links with the transactions representing electronic transactions delineated by hatched links.

As in the other chapters in this book, all vectors are assumed to be column vectors. The equilibrium solutions in this chapter, as throughout this book, are denoted by *.

The m agents or sources of funds at the top tier of the financial network in Figure 6.2 seek to determine the optimal allocation of their financial resources transacted either physically or electronically with the intermediaries or electronically with the demand markets. Examples of source agents include households and businesses. The financial intermediaries, in turn, which can include banks, insurance companies, and investment companies, in addition to transacting with the source agents, determine how to allocate the incoming financial resources among the distinct uses or financial products associated with the demand markets, which correspond to the nodes at the bottom tier of the financial network in Figure 6.2. Examples of demand markets are the markets for real estate loans, household loans, and business loans. The transactions between the financial intermediaries and the demand markets can also take place physically or electronically via the Internet.

We denote a typical source agent by i; a typical financial intermediary by j, and a typical demand market by k. The mode of transaction is denoted by l with $l = 1$ denoting the physical mode and with $l = 2$ denoting the electronic mode.

We now describe the behavior of the decision-makers with sources of funds. We then discuss the behavior of the financial intermediaries and, finally, the consumers at the demand markets. Subsequently, we state the financial network equilibrium conditions and derive the variational inequality formulation governing the equilibrium conditions.

6.2.1 The Behavior of the Source Agents

The behavior of the decision-makers with sources of funds, also referred to as source agents, is now described.

Because there is the possibility of noninvestment allowed, the node $n+1$ in the second tier in Figure 6.2 represents the "sink" to which the uninvested portion of the financial funds flows from the particular source agent or source node.

Let S^i denote the amount of financial funds held by source agent i; $i = 1, \ldots, m$. Let q_{ijl} denote the volume of financial transactions transacted between i and j using mode of transaction l. We group all the financial transaction/flows between all source agents, intermediaries, and modes into the vector Q^1. Similarly, let q_{ik} denote the amount of financial flows from i to demand market k, and we group all such flows into the vector Q^2. We let q_i denote then the $(2n+o)$-dimensional vector associated with source agent i; $i = 1, \ldots, m$, with components $\{q_{ijl}; j = 1, \ldots, n; l = 1, 2; q_{ik}; k = 1, \ldots, o\}$.

Let q_{jkl} denote the financial flow between intermediary j and demand market k transacted via mode l, and we group all such flows into the $2no$-dimensional vector Q^3. Finally, let q_j denote the $(2m+2o)$-dimensional vector associated with financial intermediary j; $j = 1, \ldots, n$ with components $\{q_{ijl};\ i = 1, \ldots, m;\ l = 1, 2;\ q_{jkl};\ k = 1, \ldots, o; l = 1, 2\}$.

We then have the following conservation of flow equations

$$\sum_{j=1}^{n} \sum_{l=1}^{2} q_{ijl} + \sum_{k=1}^{o} q_{ik} \leq S^i, \quad i = 1, \ldots, m, \tag{6.1}$$

that is, the amount of financial funds available at source agent i and given by S^i cannot exceed the amount transacted physically and electronically with the intermediaries plus the amount transacted electronically with the demand markets. Note that the "slack" associated with constraint (6.1) for a particular source agent i is given by $q_{i(n+1)}$ and corresponds to the uninvested amount of funds.

Let ρ_{1ijl} denote the price charged by source agent i to intermediary j for a transaction via mode l and let ρ_{1ik} denote the price charged by source agent i for the electronic transaction with demand market k. The ρ_{1ijl} and ρ_{1ik} are endogenous variables and their equilibrium values ρ^*_{1ijl} and ρ^*_{1ik}; $i = 1, \ldots, m$; $j = 1, \ldots, n$; $l = 1, 2$; $k = 1, \ldots, o$ are determined once the complete financial network model is solved. As noted earlier, we assume that each source agent seeks to maximize its net revenue and to minimize its risk. We assume that the risk for source agent i is represented by the $(2n+o) \times (2n+o)$ positive definite variance-covariance matrix V^i so that the optimization problem faced by source agent i can be expressed as

$$\text{Maximize} \quad U^i(q_i) = \sum_{j=1}^{n}\sum_{l=1}^{2} \rho_{1ijl}^* q_{ijl} + \sum_{k=1}^{o} \rho_{1ik}^* q_{ik} - \sum_{j=1}^{n}\sum_{l=1}^{2} c_{ijl}(q_{ijl})$$

$$- \sum_{k=1}^{o} c_{ik}(q_{ik}) - q_i^T V^i q_i \tag{6.2}$$

subject to:

$$\sum_{j=1}^{n}\sum_{l=1}^{2} q_{ijl} + \sum_{k=1}^{o} q_{ik} \leq S^i$$

$$q_{ijl} \geq 0, \quad j = 1, \ldots, n; l = 1, 2,$$

$$q_{ik} \geq 0, \quad k = 1, \ldots, o,$$

$$q_{i(n+1)} \geq 0.$$

The first four terms in the objective function (6.2) represent the net revenue of source agent i, and the last term is the variance of the return of the portfolio, which represents the risk associated with the financial transactions.

We now discuss the transaction cost functions. The transaction incurred by source agent i in transacting with intermediary j using mode l is denoted by c_{ijl}, where we assume that $c_{ijl} = c_{ijl}(q_{ijl})$ for all i, j, l. The transaction cost incurred by source agent i in transacting with demand market k via mode l is denoted by c_{ik} where $c_{ik} = c_{ik}(q_{ik})$, for all i, k.

We assume that the transaction cost functions for each source agent are continuously differentiable and convex and that the source agents compete in a noncooperative manner in the sense of Nash (1950, 1951). The optimality conditions for all decision-makers with source of funds simultaneously coincide with the solution of the following variational inequality: Determine $(Q^{1*}, Q^{2*}) \in \mathcal{K}^0$ such that

$$\sum_{i=1}^{m}\sum_{j=1}^{n}\sum_{l=1}^{2} [2V_{z_{jl}}^i \cdot q_i^* + \frac{\partial c_{ijl}(q_{ijl}^*)}{\partial q_{ijl}} - \rho_{1ijl}^*] \times [q_{ijl} - q_{ijl}^*]$$

$$+ \sum_{i=1}^{m}\sum_{k=1}^{o} [2V_{z_{2n+k}}^i \cdot q_i^* + \frac{\partial c_{ik}(q_{ik}^*)}{\partial q_{ik}} - \rho_{1ik}^*] \times [q_{ik} - q_{ik}^*] \geq 0, \quad \forall (Q^1, Q^2) \in \mathcal{K}^0, \tag{6.3}$$

where $V_{z_{jl}}^i$ denotes the z_{jl}-th row of V^i and z_{jl} is defined as the indicator: $z_{jl} = (l-1)n + j$. Similarly, $V_{z_{2n+k}}^i$ denotes the z_{2n+k}-th row of V^i but with z_{2n+k} defined as the $2n+k$-th row and the feasible set $\mathcal{K}^0 \equiv \{(Q^1, Q^2) | (Q^1, Q^2) \in R_+^{2mn+mo}$ and (6.1) holds for all $i\}$.

6.2.2 The Behavior of the Financial Intermediaries

The behavior of the intermediaries in the financial network model is now described.

Let the endogenous variable ρ_{2jkl} denote the product price charged by intermediary j, with ρ^*_{2jkl} denoting the equilibrium price, where $j = 1, \dots, n$; $k = 1, \dots, o$, and $l = 1, 2$. We assume that each financial intermediary also seeks to maximize its net revenue while minimizing his risk. Note that a financial intermediary, by definition, may transact either with decision-makers in the top tier of the financial network and with consumers associated with the demand markets in the bottom tier.

V^j is the $(2m+2o) \times (2m+2o)$-dimensional positive definite variance-covariance matrix associated with intermediary j; $j = 1, \dots, n$. The transaction cost $c_{jkl}(q_{jkl})$ is the transaction cost incurred by intermediary j in transacting with demand market k via mode l, where we assume that $c_{jkl} = c_{jkl}(q_{jkl})$ for all j, k, l. Finally, the conversion/handling cost associated with financial intermediary j is denoted by c_j and we assume that $c_j = c_j(Q^1)$.

Noting the conversion/handling cost and the various transaction costs faced by a financial intermediary and recalling that the variance-covariance matrix associated with financial intermediary j is given by V^j, we have that the financial intermediary is faced with the following optimization problem

$$\text{Maximize} \quad U^j(q_j) = \sum_{k=1}^{o} \sum_{l=1}^{2} \rho^*_{2jkl} q_{jkl} - c_j(Q^1) - \sum_{i=1}^{m} \sum_{l=1}^{2} \hat{c}_{ijl}(q_{ijl})$$

$$- \sum_{k=1}^{o} \sum_{l=1}^{2} c_{jkl}(q_{jkl}) - \sum_{i=1}^{m} \sum_{l=1}^{2} \rho^*_{1ijl} q_{ijl} - q_j^T V^j q_j \tag{6.4}$$

subject to:

$$\sum_{k=1}^{o} \sum_{l=1}^{2} q_{jkl} \le \sum_{i=1}^{m} \sum_{l=1}^{2} q_{ijl}, \tag{6.5}$$

$$q_{ijl} \ge 0, \quad i = 1, \dots, m; l = 1, 2,$$

$$q_{jkl} \ge 0, \quad k = 1, \dots, o; l = 1, 2.$$

The first five terms in the objective function (6.4) denote the net revenue, whereas the last term is the variance of the return of the financial allocations, which represents the risk to each financial intermediary. Constraint (6.5) guarantees that an intermediary cannot reallocate more of its financial funds among the demand markets than it has available.

Let γ_j be the Lagrange multiplier associated with constraint (6.5) for intermediary j. We assume that the cost functions are continuously differentiable and convex and that the intermediaries compete in a noncooperative manner. Hence the optimality conditions for all intermediaries simultaneously can be expressed as the following variational inequality: Determine $(Q^{1*}, Q^{3*}, \gamma^*) \in R_+^{2mn+2no+n}$ satisfying

$$\sum_{i=1}^{m} \sum_{j=1}^{n} \sum_{l=1}^{2} [2V^j_{zil} \cdot q^*_j + \frac{\partial c_j(Q^{1*})}{\partial q_{ijl}} + \rho^*_{1ijl} + \frac{\partial \hat{c}_{ijl}(q^*_{ijl})}{\partial q_{ijl}} - \gamma^*_j] \times [q_{ijl} - q^*_{ijl}]$$

$$+\sum_{j=1}^{n}\sum_{k=1}^{o}\sum_{l=1}^{2}[2V^j_{z_{kl}}\cdot q^*_j+\frac{\partial c_{jkl}(q^*_{jkl})}{\partial q_{jkl}}-\rho^*_{2jkl}+\gamma^*_j]\times\left[q_{jkl}-q^*_{jkl}\right]$$

$$+\sum_{j=1}^{n}[\sum_{i=1}^{m}\sum_{l=1}^{2}q^*_{ijl}-\sum_{k=1}^{o}\sum_{l=1}^{2}q^*_{jkl}]\times\left[\gamma_j-\gamma^*_j\right]\geq 0,\quad \forall(Q^1,Q^3,\gamma)\in R^{2mn+2no+n}_{+},\tag{6.6}$$

where $V^j_{z_{il}}$ denotes the z_{il}-th row of V^j and z_{il} is defined as the indicator $z_{il} = (l-1)m+i$. Similarly, $V^j_{z_{kl}}$ denotes the z_{kl}-th row of V^j and z_{kl} is defined as the indicator $z_{kl} = 2m+(l-1)o+k$.

Additional background on risk management in finance can be found in Nagurney and Siokos (1997); see also Rustem and Howe (2002).

6.2.3 The Consumers at the Demand Markets and the Equilibrium Conditions

We now assume, as given, the inverse demand functions $\rho_{3k}(d)$; $k = 1,\ldots,o$, associated with the demand markets at the bottom tier of the financial network, where d_k denotes the demand at demand market k and d is the vector of demands at all the demand markets. Recall that the demand markets correspond to distinct financial products. Of course, if the demand functions are invertible, then one may obtain the price functions simply by inversion.

The following conservation of flow equations must hold

$$d_k=\sum_{j=1}^{n}\sum_{l=1}^{2}q_{jkl}+\sum_{i=1}^{m}q_{ik},\quad k=1,\ldots,o.\tag{6.7}$$

Equations (6.7) state that the demand for the financial product at each demand market is equal to the financial transactions from the intermediaries to that demand market plus those from the source agents.

Let $\hat{c}_{jkl}$ denote the unit transaction cost associated with transacting between intermediary j and demand market k using mode l for all j,k,l, where we assume that $\hat{c}_{jkl} = \hat{c}_{jkl}(Q^2,Q^3)$. Similarly, let $\hat{c}_{ik}$ denote the unit cost associated with transacting between source agent i and demand market k, for all i,k, where we assume that $\hat{c}_{ik} = \hat{c}_{ik}(Q^2,Q^3)$.

The equilibrium condition for the consumers at demand market k are as follows: For each intermediary j; $j = 1,\ldots,n$ and mode of transaction l; $l = 1,2$

$$\rho^*_{2jkl}+\hat{c}_{jkl}(Q^{2*},Q^{3*})\begin{cases}=\rho_{3k}(d^*), & \text{if}\quad q^*_{jkl}>0\\ \geq\rho_{3k}(d^*), & \text{if}\quad q^*_{jkl}=0.\end{cases}\tag{6.8}$$

In addition, we must have that, in equilibrium, for each source of funds i; $i = 1,\ldots,m$

$$\rho^*_{1ik}+\hat{c}_{ik}(Q^{2*},Q^{3*})\begin{cases}=\rho_{3k}(d^*), & \text{if}\quad q^*_{ik}>0\\ \geq\rho_{3k}(d^*), & \text{if}\quad q^*_{ik}=0.\end{cases}\tag{6.9}$$

Conditions (6.8) state that, in equilibrium, if consumers at demand market k purchase the financial product from intermediary j via mode l, then the price the consumers pay is exactly equal to the price charged by the intermediary plus the unit transaction cost via that mode. However, if the sum of price charged by the intermediary and the unit transaction cost is greater than the price the consumers are willing to pay at the demand market, there will be no transaction between this intermediary/demand market pair via that mode. Conditions (6.9) state the analogue but for the case of electronic transactions with the source agents.

In equilibrium, conditions (6.8) and (6.9) must hold for all demand markets. We can also express these equilibrium conditions using the following variational inequality: Determine $(Q^{2*}, Q^{3*}, d^*) \in \mathcal{K}^1$, such that

$$\sum_{j=1}^{n}\sum_{k=1}^{o}\sum_{l=1}^{2}[\rho_{2jkl}^* + \hat{c}_{jkl}(Q^{2*}, Q^{3*})] \times \left[q_{jkl} - q_{jkl}^*\right]$$

$$+\sum_{i=1}^{m}\sum_{k=1}^{o}\left[\rho_{1ik}^* + \hat{c}_{ik}(Q^{2*}, Q^{3*})\right] \times \left[q_{ik} - q_{ik}^*\right] - \sum_{k=1}^{o}\rho_{3k}(d^*) \times \left[d_k - d_k^*\right] \geq 0,$$

$$\forall (Q^2, Q^3, d) \in \mathcal{K}^1, \tag{6.10}$$

where $\mathcal{K}^1 \equiv \{(Q^2, Q^3, d) | (Q^2, Q^3, d) \in R_+^{2no+mo+o}$ and (6.7) holds.$\}$

6.2.4 The Equilibrium Conditions for the Financial Network with Electronic Transactions

In equilibrium, the optimality conditions for all decision-makers with source of funds, the optimality conditions for all the intermediaries, and the equilibrium conditions for all the demand markets must be simultaneously satisfied so that no decision-maker has any incentive to alter his or her decision. We recall the equilibrium conditions [cf. Liu and Nagurney (2007)] for the entire financial network with intermediation and electronic transactions as follows.

Definition 6.1: Financial Network Equilibrium with Intermediation and with Electronic Transactions
The equilibrium state of the financial network with intermediation is one in which the financial flows between tiers coincide and the financial flows and prices satisfy the sum of conditions (6.3), (6.6), and (6.10).

Let $\mathcal{K}^2 \equiv \{(Q^1, Q^2, Q^3, \gamma, d) | (Q^1, Q^2, Q^3, \gamma, d) \in R_+^{m+2mn+2no+n+o}$ and (6.1) and (6.7) hold$\}$ and state the following theorem.

Theorem 6.1: Variational Inequality Formulation
The equilibrium conditions governing the financial network model with intermediation are equivalent to the solution to the variational inequality problem given by: determine $(Q^{1^*}, Q^{2^*}, Q^{3^*}, \gamma^*, d^*) \in \mathcal{K}^2$ satisfying

$$\sum_{i=1}^{m}\sum_{j=1}^{n}\sum_{l=1}^{2}[2V_{z_{jl}}^i \cdot q_i^* + 2V_{z_{il}}^j \cdot q_j^* + \frac{\partial c_{ijl}(q_{ijl}^*)}{\partial q_{ijl}} + \frac{\partial c_j(Q^{1*})}{\partial q_{ijl}} + \frac{\partial \hat{c}_{ijl}(q_{ijl}^*)}{\partial q_{ijl}} - \gamma_j^*]$$

$$\times \left[q_{ijl} - q^*_{ijl}\right]$$

$$+\sum_{i=1}^{m}\sum_{k=1}^{o}[2V^i_{z_{2n+k}} \cdot q^*_i + \frac{\partial c_{ik}(q^*_{ik})}{\partial q_{ik}} + \hat{c}_{ik}(Q^{2*}, Q^{3*})] \times [q_{ik} - q^*_{ik}]$$

$$+\sum_{j=1}^{n}\sum_{k=1}^{o}\sum_{l=1}^{2}[2V^j_{z_{kl}} \cdot q^*_j + \frac{\partial c_{jkl}(q^*_{jkl})}{\partial q_{jkl}} + \hat{c}_{jkl}(Q^{2*}, Q^{3*}) + \gamma^*_j] \times \left[q_{jkl} - q^*_{jkl}\right]$$

$$+\sum_{j=1}^{n}[\sum_{i=1}^{m}\sum_{l=1}^{2} q^*_{ijl} - \sum_{k=1}^{o}\sum_{l=1}^{2} q^*_{jkl}] \times \left[\gamma_j - \gamma^*_j\right] - \sum_{k=1}^{o} \rho_{3k}(d^*) \times [d_k - d^*_k] \geq 0,$$

$$\forall (Q^1, Q^2, Q^3, \gamma, d) \in \mathcal{K}^2. \tag{6.11}$$

Proof: We first prove that an equilibrium according to Definition 6.1 satisfies variational inequality (6.11). Summation of (6.3), (6.6), and (6.10), after algebraic simplifications, yields (6.11).

We now prove the converse, that is, that a solution to variational inequality (6.11) satisfies the sum of conditions (6.3), (6.6), and (6.10) and is, therefore, a financial network equilibrium pattern. First, we add the term $-\rho^*_{1ijl} + \rho^*_{1ijl}$ to the term in the first set of brackets in (6.11). Then, we add the term $-\rho^*_{1ik} + \rho^*_{1ik}$ to the term before the second multiplication sign, and finally, we add the term $-\rho^*_{2jkl} + \rho^*_{2jkl}$ to the term preceding the third multiplication sign in (6.11). Note that these terms are identically equal to zero and do not change the variational inequality. Hence we obtain the following inequality

$$\sum_{i=1}^{m}\sum_{j=1}^{n}\sum_{l=1}^{2}[2V^i_{z_{jl}} \cdot q^*_i + 2V^j_{z_{il}} \cdot q^*_j + \frac{\partial c_{ijl}(q^*_{ijl})}{\partial q_{ijl}} + \frac{\partial c_j(Q^{1*})}{\partial q_{ijl}} + \frac{\partial \hat{c}_{ijl}(q^*_{ijl})}{\partial q_{ijl}} - \gamma^*_j$$

$$-\rho^*_{1ijl} + \rho^*_{1ijl}] \times [q_{ijl} - q^*_{ijl}]$$

$$+\sum_{i=1}^{m}\sum_{k=1}^{o}[2V^i_{z_{2n+k}} \cdot q^*_i + \frac{\partial c_{ik}(q^*_{ik})}{\partial q_{ik}} + \hat{c}_{ik}(Q^{2*}, Q^{3*}) - \rho^*_{1ik} + \rho^*_{1ik}] \times [q_{ik} - q^*_{ik}]$$

$$\sum_{j=1}^{n}\sum_{k=1}^{o}\sum_{l=1}^{2}[2V^j_{z_{kl}} \cdot q^*_j + \frac{\partial c_{jkl}(q^*_{jkl})}{\partial q_{jkl}} + \hat{c}_{jkl}(Q^{2*}, Q^{3*}) + \gamma^*_j - \rho^*_{2jkl} + \rho^*_{2jkl}]$$

$$\times[q_{jkl} - q^*_{jkl}] + \sum_{j=1}^{n}[\sum_{i=1}^{m}\sum_{l=1}^{2} q^*_{ijl} - \sum_{k=1}^{n}\sum_{l=1}^{2} q^*_{jkl}] \times [\gamma_j - \gamma^*_j]$$

$$-\sum_{k=1}^{o} \rho_{3k}(d^*) \times [d_k - d^*_k] \geq 0, \quad \forall (Q^1, Q^2, Q^3, \gamma, d) \in \mathcal{K}^2, \tag{6.12}$$

which, can be rewritten as

$$\sum_{i=1}^{m}\sum_{j=1}^{n}\sum_{k=1}^{o}[2V^i_{z_{jl}} \cdot q_i^* + \frac{\partial c_{ijl}(q_{ijl}^*)}{\partial q_{ijl}} - \rho_{1ijl}^*] \times [q_{ijl} - q_{ijl}^*]$$

$$+\sum_{i=1}^{m}\sum_{k=1}^{o}[2V^i_{z_{2n+k}} \cdot q_i^* + \frac{\partial c_{ik}(q_{ik}^*)}{\partial q_{ik}} - \rho_{1ik}^*] \times [q_{ik} - q_{ik}^*]$$

$$+\sum_{i=1}^{m}\sum_{j=1}^{n}\sum_{l=1}^{2}[2V^j_{z_{il}} \cdot q_j^* + \frac{\partial c_j(Q^{1*})}{\partial q_{ijl}} + \rho_{1ijl}^* + \frac{\partial \hat{c}_{ijl}(q_{ijl}^*)}{\partial q_{ijl}} - \gamma_j^*] \times [q_{ijl} - q_{ijl}^*]$$

$$+\sum_{j=1}^{n}\sum_{k=1}^{o}\sum_{l=1}^{2}[2V^j_{z_{kl}} \cdot q_j^* + \frac{\partial c_{jkl}(q_{jkl}^*)}{\partial q_{jkl}} - \rho_{2jkl}^* + \gamma_j^*] \times [q_{jkl} - q_{jkl}^*]$$

$$+\sum_{j=1}^{n}[\sum_{i=1}^{m}\sum_{l=1}^{2} q_{ijl}^* - \sum_{k=1}^{o}\sum_{l=1}^{2} q_{jkl}^*] \times [\gamma_j - \gamma_j^*]$$

$$+\sum_{j=1}^{n}\sum_{k=1}^{o}\sum_{l=1}^{2}[\rho_{2jkl}^* + \hat{c}_{jkl}(Q^{2*}, Q^{3*})] \times [q_{jkl} - q_{jkl}^*]$$

$$+\sum_{i=1}^{m}\sum_{k=1}^{o}[\rho_{1ik}^* + \hat{c}_{ik}(Q^{2*}, Q^{3*})] \times [q_{ik} - q_{ik}^*]$$

$$-\sum_{k=1}^{o}\rho_{3k}(d^*) \times [d_k - d_k^*] \geq 0, \quad \forall (Q^1, Q^2, Q^3, \gamma, d) \in \mathcal{K}^2. \tag{6.13}$$

Obviously, the solution to inequality (6.13) satisfies the sum of the conditions (6.3), (6.6), and (6.10). The proof is complete.

The variables in the variational inequality problem (6.11) are the financial flows from the source agents to the intermediaries Q^1, the direct financial flows via electronic transaction from the source agents to the demand markets Q^2, the financial flows from the intermediaries to the demand markets Q^3, the shadow prices associated with handling the product by the intermediaries γ, and the prices at demand markets ρ_3. The solution to the variational inequality problem (6.11), $(Q^{1^*}, Q^{2^*}, Q^{3^*}, \gamma^*, d^*)$, coincides with the equilibrium financial flow and price pattern according to Definition 6.1.

Variational inequality (6.11) is distinct from the variational inequality derived in Nagurney and Ke (2003) in which the demand functions at the markets were assumed known and given.

6.3 THE FINANCIAL NETWORK PERFORMANCE MEASURE AND THE IMPORTANCE OF COMPONENTS

In this section, we propose a novel financial network performance measure and the associated network component importance definition. For completeness, we also discuss the difference between our measure and a standard efficiency measure in economics.

6.3.1 The Financial Network Performance Measure

As stated in Section 6.1, the financial network performance measure is motivated by the unified network performance measure detailed in Chapter 3. In the case of the financial network performance measure, we state the definitions directly within the context of financial networks, without making use of the transformation of the financial network model into a network equilibrium model with defined origin/destination pairs and paths, as in Liu and Nagurney (2007).

Definition 6.2: The Financial Network Performance Measure
The financial network performance measure, $\mathcal{E}^{FN}$, for a given network topology $\mathcal{G}$ (cf. Figure 6.2), and demand price functions $\rho_{3k}(d)$; $k = 1, 2, \ldots, o$, and available funds held by source agents S, where S is the vector of financial funds held by the source agents, is defined as follows

$$\mathcal{E}^{FN} = \frac{\sum_{k=1}^{o} \frac{d_k^*}{\rho_{3k}(d^*)}}{o}, \tag{6.14}$$

where o is the number of demand markets in the financial network, and d_k^* and $\rho_{3k}(d^*)$ denote the equilibrium demand and the equilibrium price for demand market k, respectively.

The financial network performance measure $\mathcal{E}^{FN}$ defined in (6.14) is actually the average demand to price ratio. It measures the overall (economic) functionality of the financial network. When the network topology $\mathcal{G}$, the demand price functions, and the available funds held by source agents are given, a financial network is considered performing better if it can satisfy higher demands at lower prices.

By referring to the equilibrium conditions (6.8) and (6.9), we assume that if there is a positive transaction between a source agent or an intermediary with a demand market at equilibrium, the price charged by the source agent or the intermediary plus the respective unit transaction costs is always positive. Furthermore, we assume that if the equilibrium demand at a demand market is zero, the demand market price (i.e., the inverse demand function value) is positive. Hence the demand market prices will always be positive and the network performance measure is well-defined.

In the definition, we assume that all the demand markets are given the same weight when aggregating the demand to price ratio, which can be interpreted as all the demand markets are of equal strategic importance. Of course, it may be interesting and appropriate to weight demand markets differently by incorporating managerial or governmental factors into the measure. For example, one could give more preference to the markets with large demands. Furthermore, it would also be interesting to explore different functional forms associated with the definition of the performance measure to ascertain different aspects of network performance. However, in this chapter, we focus on the definition in the form of (6.12) and these issues will be considered for future research. Finally, the performance measure in (6.14) is based on the "pure" cost incurred between different tiers of the financial network.

6.3.2 Network Efficiency vs. Network Performance

It is worth pointing out further relationships between our network performance measure and other measures in economics, in particular, an *efficiency* measure. In economics, the total utility gained (or cost incurred) in a system may be used as an efficiency measure. Such a criterion is basically the underlying rationale for the concept of *Pareto efficiency*, which plays a very important role in the evaluation of economic policies in terms of social welfare. As is well-known, a Pareto efficient outcome indicates that there is no alternative way to organize the production and distribution of goods that makes some economic agent better off without making another worse off [see, e.g., Samuelson (1983), Mas-Colell, Whinston, and Green (1995)]. Under certain conditions, which include that externalities are not present in an economic system, the equilibrium state ensures that the system is Pareto efficient and that the social welfare is maximized. The concept of *Kaldor-Hicks efficiency*, in turn, relaxes the requirement of Pareto efficiency by incorporating the compensation principle: an outcome is efficient if those that are made better off could, in theory, compensate those that are made worse off and leads to a Pareto optimal outcome [see, e.g., Chipman (1987) and Buchanan and Musgrave (1999)].

These economic efficiency concepts have important implications for government and/or central planners, for example, by suggesting and enforcing policies that ensure that the system is running cost efficiently. For instance, in the transportation literature, these efficiency concepts have been used to model the "system-optimal" objective, as outlined in Chapter 2. Moreover, a toll policy can be implemented to guarantee that the minimum total travel cost for the entire network (cf. Beckmann, McGuire, and Winsten (1956), Dafermos (1973), Nagurney (2000), and the references therein) is achieved. It is worth noting that the system-optimal concept in transportation networks has stimulated a tremendous amount of interest also, recently, among computer scientists, which has led to the study of the price of anarchy [cf. (3.11)]. Recall that, as discussed in Chapter 3, the price of anarchy is defined as the ratio of the system-optimization objective function evaluated at the user-optimized solution divided by that objective function evaluated at the system-optimized solution. It has been used to study a variety of noncooperative games on such networks as telecommunication networks and the Internet. Notably, the aforementioned principles are mainly used to access the tenability of the resource allocation policies from a societal point of view. However, we believe that, in addition to evaluating an economic system in the sense of optimizing the resource allocation, there should also be a measure that can assess the network performance and functionality. Although in such networks as the Internet and certain transportation networks, the assumption of having a central planner to ensure the minimization of the total cost may, in some instances, be natural and reasonable (as in the case of system-optimized networks, discussed in Chapters 2 and 3, along with appropriate robustness analyses), the same assumption faces difficulty when extended to the larger and more complex networks as in the case of financial networks, where the control by a "central planner" is not realistic.

The purpose of this chapter is not to study the efficiency of a certain market mechanism or policy, which can be typically analyzed via the Pareto criterion and the Kaldor-Hicks test. Instead, we would like to address the following question: Given a certain market mechanism, network structure, objective functions, and demand price and cost functions, how should one evaluate the performance and the functionality of the network? In the context of a financial network where there exists noncooperative competition among the source agents as well as among the financial intermediaries, if, on the average and across all demand markets, a large amount of financial funds can reach the consumers, through the financial intermediaries, at low prices, we consider the network as performing well. Thus, instead of studying the efficiency of an economic policy or market mechanism, we evaluate the functionality and the performance of a financial network in a given environment. The proposed performance measure of the financial network is based on the financial equilibrium model outlined in Section 6.2. However, our measure can be applied to other economic networks, as well, and has been done so in the case of transportation and supply chain networks and other critical infrastructure networks, as discussed in Chapters 3 through 5. Notably, we believe that such a financial network equilibrium model is general and relevant and, moreover, it also has deep theoretic foundations [see, for example, Judge and Takayama (1973)].

Furthermore, three points merit discussion as to the need for a network performance measure besides solely looking at the total cost of the network. First, the function of an economic network is to serve the demand markets at a reasonable price. Hence it is reasonable and important to have a performance measure targeted at the functionality perspective. Second, when faced with network disruptions whereby certain parts of the network may be destroyed or becoming inoperable, the cost of providing services/products through the dysfunctional/disconnected part reaches infinity. Therefore, the total cost of the system is also equal to infinity and hence it becomes undefined as a measure. However, since the remaining network components may still be functioning, it is valid to analyze the network performance in such a situation. Finally, it has been shown in Qiang and Nagurney (2008) that the total system cost measure is not appropriate as a means of evaluating the performance of a network with elastic demands and hence a unified network measure is needed.

Based on the discussion in this section, we denote our proposed measure $\mathcal{E}^{FN}$ as the "financial network performance measure" in order to avoid confusion with efficiency measures in economics and elsewhere.

6.3.3 The Importance of a Financial Network Component

The importance of the network components is analyzed, in turn, by studying the effect on the network performance measure through their removal. The financial network performance is expected to deteriorate when a critical network component is eliminated from the network. Such a component can include a link or a node or a subset of nodes and links, depending on the financial network problem under investigation. Furthermore, the removal of a critical network component will cause more severe damage than that caused by the removal of a trivial component. Hence the importance of a network component is defined as follows.

Definition 6.3: Importance of a Financial Network Component
The importance of a financial network component $g \in \mathcal{G}$, $I^{FN}(g)$, is measured by the relative financial network performance drop after g is removed from the network

$$I^{FN}(g) = \frac{\triangle \mathcal{E}^{FN}}{\mathcal{E}^{FN}} = \frac{\mathcal{E}^{FN}(\mathcal{G}) - \mathcal{E}^{FN}(\mathcal{G} - g)}{\mathcal{E}^{FN}(\mathcal{G})} \tag{6.15}$$

where $\mathcal{G} - g$ is the resulting financial network after component g is removed from network $\mathcal{G}$.

It is worth pointing out that the importance of the network components is well-defined even in a financial network with disconnected source agent/demand market pairs. In our financial network performance measure, the elimination of a transaction link is treated by removing that link from the network while the removal of a node is managed by removing the transaction links entering or exiting that node. In the case that the removal results in no transaction path connecting a source agent/demand market pair, we simply assign the demand for that source agent/demand market pair to an abstract transaction path with an associated cost of infinity. The above procedure for handling disconnected agent/demand market pairs is illustrated with numerical examples in Section 6.4, when we compute the importance of the financial network components and their associated rankings.

6.4 THE ALGORITHM

In this section, we present the Euler method, which is applied to solve the numerical examples in the subsequent section. We first, however, simplify variational inequality (6.11). Specifically, we note that, in view of (6.7), we may define the functions

$$\hat{\rho}_{3k}(Q^2, Q^3) \equiv \rho_{3k}(d), \quad k = 1, \ldots, o. \tag{6.16}$$

Let $\mathcal{K}^3 \equiv \{(Q^1, Q^2, Q^3, \gamma) | (Q^1, Q^2, Q^3, \gamma) \in R_+^{m+2mn+2no+n}$ and (6.1) holds$\}$. Therefore, we may rewrite variational inequality (6.11) as: Determine the vector $(Q^{1*}, Q^{2*}, Q^{3*}, \gamma^*) \in \mathcal{K}^3$ satisfying

$$\sum_{i=1}^{m} \sum_{j=1}^{n} \sum_{l=1}^{2} [2V^i_{z_{jl}} \cdot q_i^* + 2V^j_{z_{il}} \cdot q_j^* + \frac{\partial c_{ijl}(q^*_{ijl})}{\partial q_{ijl}} + \frac{\partial c_j(Q^{1*})}{\partial q_{ijl}} + \frac{\partial \hat{c}_{ijl}(q^*_{ijl})}{\partial q_{ijl}} - \gamma_j^*]$$

$$\times[q_{ijl} - q^*_{ijl}]$$

$$+\sum_{i=1}^{m}\sum_{k=1}^{o}[2V^i_{z_{2n+k}} \cdot q^*_i + \frac{\partial c_{ik}(q^*_{ik})}{\partial q_{ik}} + \hat{c}_{ik}(Q^{2*}, Q^{3*}) - \hat{\rho}_{3k}(Q^{2*}, Q^{3*})] \times [q_{ik} - q^*_{ik}]$$

$$+\sum_{j=1}^{n}\sum_{k=1}^{o}\sum_{l=1}^{2}[2V^j_{z_{kl}} \cdot q^*_j + \frac{\partial c_{jkl}(q^*_{jkl})}{\partial q_{jkl}} + \hat{c}_{jkl}(Q^{2*}, Q^{3*}) + \gamma^*_j - \hat{\rho}_{3k}(Q^{2*}, Q^{3*})]$$

$$\times[q_{jkl} - q^*_{jkl}]$$

$$+\sum_{j=1}^{n}[\sum_{i=1}^{m}\sum_{l=1}^{2} q^*_{ijl} - \sum_{k=1}^{o}\sum_{l=1}^{2} q^*_{jkl}] \times [\gamma_j - \gamma^*_j] \geq 0, \quad \forall (Q^1, Q^2, Q^3, \gamma) \in \mathcal{K}^3. \tag{6.17}$$

We now present the explicit statement of the Euler method [cf. (2.54)] for the solution of the variational inequality (6.17), which is equivalent to (6.11).

The Euler Method

Step 0: Initialization Step
Set $(Q^{10}, Q^{20}, Q^{30}, \gamma^0) \in \mathcal{K}^3$. Let $\mathcal{T} = 1$, and set the sequence $\{\alpha_{\mathcal{T}}\}$ so that $\sum_{\mathcal{T}=1}^{\infty} \alpha_{\mathcal{T}} = \infty$, $\alpha_{\mathcal{T}} > 0$, $\alpha_{\mathcal{T}} \to 0$, as $\mathcal{T} \to \infty$.

Step 1: Computation Step
Compute $(Q^{1\mathcal{T}}, Q^{2\mathcal{T}}, Q^{3\mathcal{T}}, \gamma^{\mathcal{T}}) \in \mathcal{K}^3$ by solving the variational inequality subproblem

$$\sum_{i=1}^{m}\sum_{j=1}^{n}\sum_{l=1}^{2}[q^{\mathcal{T}}_{ijl} + \alpha_{\mathcal{T}}(2V^i_{z_{jl}} \cdot q^{\mathcal{T}-1}_i + 2V^j_{z_{ik}} \cdot q_j + \frac{\partial c_{ijl}(q^{\mathcal{T}-1}_{ijl})}{\partial q_{ijl}} + \frac{\partial c_j(Q^{1\mathcal{T}-1})}{\partial q_{ijl}}$$

$$+\frac{\partial \hat{c}_{ijl}(q^{\mathcal{T}-1}_{ijl})}{\partial q_{ijl}} - \gamma^{\mathcal{T}-1}_j) - q^{\mathcal{T}-1}_{ijl}] \times [q_{ijl} - q^{\mathcal{T}}_{ijl}]$$

$$+\sum_{i=1}^{m}\sum_{k=1}^{o}[q^{\mathcal{T}}_{ik} + \alpha_{\mathcal{T}}(2V^i_{z_{2n+k}} \cdot q^{\mathcal{T}-1}_i + \frac{\partial c_{ik}(q^{\mathcal{T}-1}_{ik})}{\partial q_{ik}} + \hat{c}_{ik}(Q^{2\mathcal{T}}, Q^{3\mathcal{T}})$$

$$-\hat{\rho}_{3k}(Q^{2\mathcal{T}-1}, Q^{3\mathcal{T}-1})) - q^{\mathcal{T}-1}_{ik}] \times [q_{ik} - q^{\mathcal{T}}_{ik}]$$

$$+\sum_{j=1}^{n}\sum_{k=1}^{o}\sum_{l=1}^{2}[q^{\mathcal{T}}_{jkl} + \alpha_{\mathcal{T}}(2V^j_{z_{kl}} \cdot q^{\mathcal{T}-1}_j + \frac{\partial c_{jkl}(q^{\mathcal{T}-1}_{jkl})}{\partial q_{jkl}} + \hat{c}_{jkl}(Q^{2\mathcal{T}-1}, Q^{3\mathcal{T}-1})$$

$$+\gamma^{\mathcal{T}-1}_j - \hat{\rho}_{3k}(Q^{2\mathcal{T}-1}, Q^{3\mathcal{T}-1})) - q^{\mathcal{T}-1}_{jkl}] \times [q_{jkl} - q^{\mathcal{T}}_{jkl}]$$

$$+\sum_{j=1}^{n}[\gamma^{\mathcal{T}}_j + \alpha_{\mathcal{T}}(\sum_{i=1}^{m}\sum_{l=1}^{2} q^{\mathcal{T}-1}_{ijl} - \sum_{k=1}^{o}\sum_{l=1}^{2} q^{\mathcal{T}-1}_{jkl}) - \gamma^{\mathcal{T}-1}_j] \times [\gamma_j - \gamma^{\mathcal{T}}_j] \geq 0,$$

$$\forall (Q^{1\mathcal{T}}, Q^{2\mathcal{T}}, Q^{3\mathcal{T}}, \gamma^{\mathcal{T}}) \in \mathcal{K}^3. \tag{6.18}$$

Step 2: Convergence Verification
If $|q_{ijl}^{\mathcal{T}} - q_{ijl}^{\mathcal{T}-1}| \leq \epsilon$, $|q_{ik}^{\mathcal{T}} - q_{ik}^{\mathcal{T}-1}|$, $|q_{jkl}^{\mathcal{T}} - q_{jkl}^{\mathcal{T}-1}| \leq \epsilon$, $|\gamma_j^{\mathcal{T}} - \gamma_j^{\mathcal{T}-1}| \leq \epsilon$, $|\rho_{3k}^{\mathcal{T}} - \rho_{3k}^{\mathcal{T}-1}| \leq \epsilon$, for all $i = 1, \ldots, m$; $j = 1, \ldots, n$; $k = 1, \ldots, o$; $l = 1, 2$, with $\epsilon > 0$, a prespecified tolerance, then stop; otherwise, set $\mathcal{T} = \mathcal{T} + 1$, and go to Step 1.

Note that the variational inequality subproblem (6.18) encountered at each iteration of the discrete-time algorithm can be solved explicitly and in closed form because it is actually a quadratic programming problem and the feasible set is a Cartesian product consisting of the constraints (6.1), with a simple network structure and the nonnegative orthants, R_+^{2no} and R_+^o, corresponding to the variables Q^3 and γ, respectively.

Computation of Financial Flows
In fact, the subproblem in (6.18) in the (Q^1, Q^2) variables can be solved using exact equilibration (see Chapter 2), whereas the remainder of the variables in (6.18) can be obtained by explicit formulae, which are provided below for convenience.

In particular, compute, at iteration $\mathcal{T}$, the financial flows, $q_{jkl}^{\mathcal{T}}$s, according to

$$q_{jkl}^{\mathcal{T}} = \max\{0, q_{jkl}^{\mathcal{T}-1} - \alpha_{\mathcal{T}}(2V_{z_{kl}}^j \cdot q_j^{\mathcal{T}-1} + \frac{\partial c_{jkl}(q_{jkl}^{\mathcal{T}-1})}{\partial q_{jkl}} + \hat{c}_{jkl}(Q^{2\mathcal{T}-1}, Q^{3\mathcal{T}-1})$$
$$+\gamma_j^{\mathcal{T}-1} - \hat{\rho}_{3k}(Q^{2\mathcal{T}-1}, Q^{3\mathcal{T}-1}))\}, \quad \forall j, k, l. \tag{6.19}$$

Computation of the Prices
At iteration $\mathcal{T}$, compute the prices, $\gamma_j^{\mathcal{T}}$ s, according to

$$\gamma_j^{\mathcal{T}} = \max\{0, \gamma_j^{\mathcal{T}-1} - \alpha_{\mathcal{T}}(\sum_{i=1}^{m}\sum_{l=1}^{2} q_{ijl}^{\mathcal{T}-1} - \sum_{k=1}^{o}\sum_{l=1}^{2} q_{jkl}^{\mathcal{T}-1})\}, \quad \forall j. \tag{6.20}$$

Note that in the discrete-time adjustment process [cf. (6.19) and (6.20)], the financial flows and the prices can be updated simultaneously at each iteration.

This algorithm hence tracks the dynamic trajectory of the financial flows and prices until the stationary point; equivalently, the equilibrium point is reached.

6.5 NUMERICAL EXAMPLES

To further demonstrate the applicability of the financial network performance measure proposed in Section 6.3, we present two numerical financial network examples. For each example, our network performance measure is computed and the importance and the rankings of links and the nodes are also reported.

The examples consisted of two source agents, two financial intermediaries, and two demand markets. These examples had the financial network structure depicted in Figure 6.3. For simplicity, we excluded the electronic transactions. The transaction links between the source agents and the intermediaries were denoted by a_{ij} where $i = 1, 2$; $j = 1, 2$. The transaction links between the intermediaries and the demand

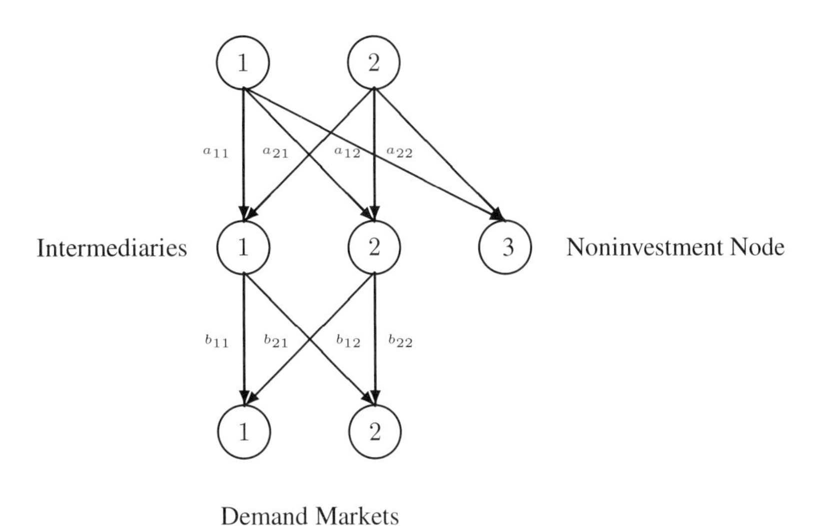

Figure 6.3: The financial network structure of the numerical examples

markets were denoted by b_{jk} where $j = 1, 2; k = 1, 2$. Since the noninvestment portions of the funds do not participate in the actual transactions, we do not discuss the importance of the links and the nodes related to the noninvestment funds. The examples below were solved using the Euler method [cf. Section 2.2.4; see also Nagurney and Zhang (1996, 1997), Nagurney and Ke (2003), and Nagurney, Wakolbinger, and Zhao (2006)]. The convergence tolerance ϵ is set to be equal to 10^{-4}.

Example 6.1
The financial holdings for the two source agents in the first example were $S^1 = 10$ and $S^2 = 10$. The variance-covariance matrices V^i and V^j were identity matrices for all the source agents $i = 1, 2$. We suppressed the subscript l associated with the transaction cost functions because we assumed a single (physical) mode of transaction being available.

The transaction cost function of Source Agent 1 associated with his transaction with Intermediary 1 was given by

$$c_{11}(q_{11}) = 4q_{11}^2 + q_{11} + 1.$$

The other transaction cost functions of the source agents associated with the transactions with the intermediaries were given by

$$c_{ij}(q_{ij}) = 2q_{ij}^2 + q_{ij} + 1, \quad i = 1, 2; j = 1, 2$$

with i and j not equal to 1 at the same time.

The transaction cost functions of the intermediaries associated with transacting with the sources agents were

$$\hat{c}_{ij}(q_{ij}) = 3q_{ij}^2 + 2q_{ij} + 1, \quad i = 1, 2; j = 1, 2.$$

The handling cost functions of the intermediaries were

$$c_1(Q^1) = .5(q_{11} + q_{21})^2,$$

$$c_2(Q^1) = .5(q_{12} + q_{22})^2.$$

We assumed that in the transactions between the intermediaries and the demand markets, the transaction costs perceived by the intermediaries were all equal to zero, that is

$$c_{jk}(q_{jk}) = 0, \quad j = 1, 2; k = 1, 2.$$

The transaction costs between the intermediaries and the consumers at the demand markets, in turn, were given by

$$\hat{c}_{jk}(Q^2, Q^3) = q_{jk} + 2, \quad j = 1, 2; k = 1, 2.$$

The demand price functions at the demand markets were

$$\rho_{3k}(d) = -2d_k + 100, \quad k = 1, 2.$$

The equilibrium financial flow pattern, the equilibrium demands, and the incurred equilibrium demand market prices are as follows.

The computed components of Q^{1*} were

$$q_{11}^* = 3.27, \; q_{12}^* = 4.16,$$

$$q_{21}^* = 4.36, \; q_{22}^* = 4.16.$$

For Q^{2*}, we obtained

$$q_{11}^* = 3.81, \; q_{12}^* = 3.81$$

$$q_{21}^* = 4.16, \; q_{22}^* = 4.16.$$

Also, the equilibrium demands and incurred prices were

$$d_1^* = 7.97, \; d_2^* = 7.97,$$

$$\rho_{31}(d^*) = 84.06, \; \rho_{32}(d^*) = 84.06.$$

Table 6.1: Importance and ranking of links in Example 6.1

Link	Importance Value	Ranking
a_{11}	.1574	3
a_{12}	.2003	2
a_{21}	.2226	1
a_{22}	.2003	2
b_{11}	.0304	5
b_{12}	.0304	5
b_{21}	.0359	4
b_{22}	.0359	4

Table 6.2: Importance and ranking of nodes in Example 6.1

Node	Importance Value	Ranking
Source Agent 1	.4146	4
Source Agent 2	.4238	3
Intermediary 1	.4759	2
Intermediary 2	.5159	1
Demand Market 1	.0566	5
Demand Market 2	.0566	5

The value of the financial network performance measure [cf. (6.12)] was

$$\mathcal{E}^{FN} = \frac{\frac{7.97}{84.06} + \frac{7.97}{84.06}}{2} = .0949.$$

The importance of the links and the nodes and their ranking are reported in Table 6.1 and 6.2, respectively.

We show the link and node importance values and rankings for Example 6.1 in Figures 6.4 and 6.5, respectively.

Discussion

First note that, in Example 6.1, both source agents chose not to invest a portion of their financial funds. Given the cost structure and the demand price functions in the network of Example 6.1, the transaction link between Source Agent 2 and Intermediary 1 is the most important link because it carries a large amount of financial flow, in equilibrium, and the removal of the link causes the highest performance drop assessed by the financial network performance measure.

Similarly, because Intermediary 2 handles the largest amount of financial input from the source agents, it is ranked as the most important node in the above network. On the other hand, because the transaction links between Intermediary 1 to Demand

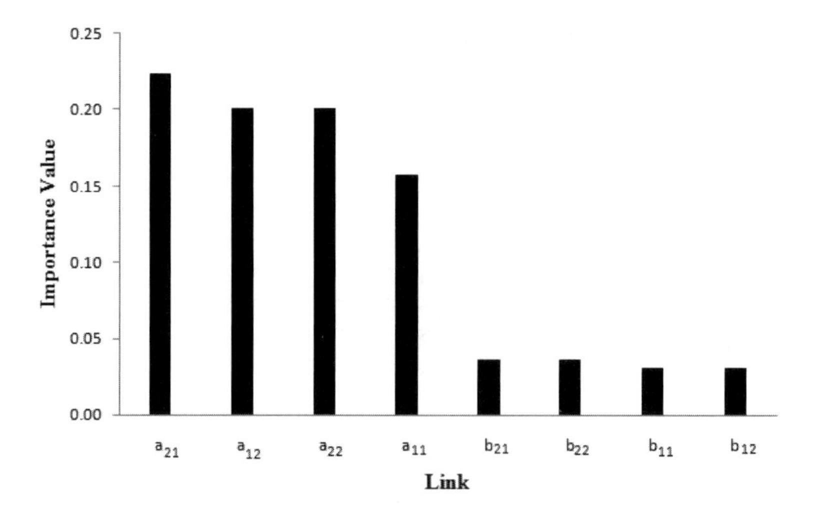

Figure 6.4: Link importance rankings for Example 6.1

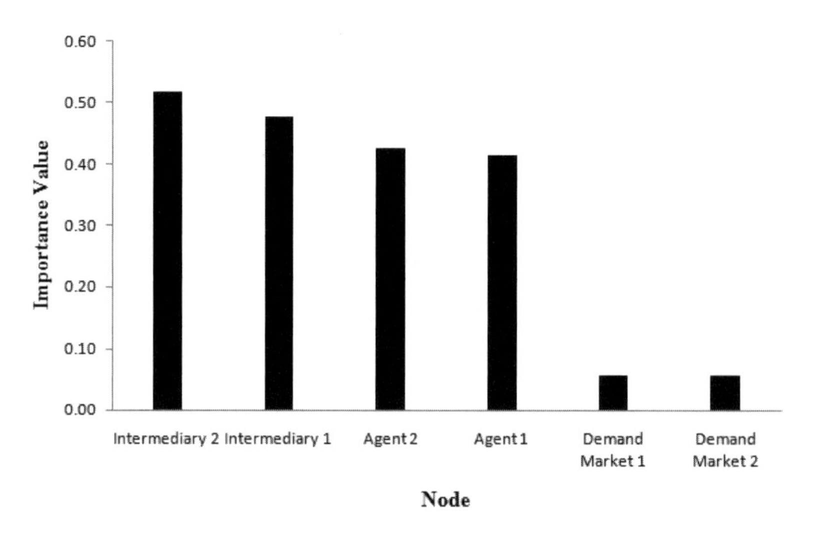

Figure 6.5: Node importance rankings for Example 6.1

Markets 1 and 2 carry the least amount of equilibrium financial flow, they are the least important links.

Example 6.2

In the second example, the parameters are identical to those in Example 6.1, except for the following changes.

Table 6.3: Importance and ranking of links in Example 6.2

Link	Importance Value	Ranking
a_{11}	.0917	2
a_{12}	.0917	2
a_{21}	.3071	1
a_{22}	.3071	1
b_{11}	.0211	3
b_{12}	.0211	3
b_{21}	.0211	3
b_{22}	.0211	3

The transaction cost function of Source Agent 1 associated with his transaction with intermediary 1 was changed to

$$c_{11}(q_{11}) = 2q_{11}^2 + q_{11} + 1$$

and the financial holdings of the source agents were changed, respectively, to $S_1 = 6$ and $S_2 = 10$.

The equilibrium financial flow pattern, the equilibrium demands, and the incurred equilibrium demand market prices are as follows.

The computed components of Q^{1*} were

$$q_{11}^* = 3.00, \; q_{12}^* = 3.00,$$

$$q_{21}^* = 4.48, \; q_{22}^* = 4.48.$$

The computed components of Q^{2*} were

$$q_{11}^* = 3.74, \; q_{12}^* = 3.74,$$

$$q_{21}^* = 3.74, \; q_{22}^* = 3.74.$$

The computed equilibrium demands and incurred prices were

$$d_1^* = 7.48, \; d_2^* = 7.48,$$

$$\rho_{31}(d^*) = 85.04, \; \rho_{32}(d^*) = 85.04.$$

The value of the financial network performance measure [cf. (6.12)] was

$$\mathcal{E}^{FN} = \frac{\frac{7.48}{85.04} + \frac{7.48}{85.04}}{2} = .0880.$$

The importance of the links and the nodes and their ranking are reported in Table 6.3 and 6.4, respectively.

Table 6.4: Importance and ranking of links in Example 6.2

Node	Importance Value	Ranking
Source Agent 1	.3687	3
Source Agent 2	.6373	1
Intermediary 1	.4348	2
Intermediary 2	.4348	2
Demand Market 1	-.0085	4
Demand Market 2	-.0085	4

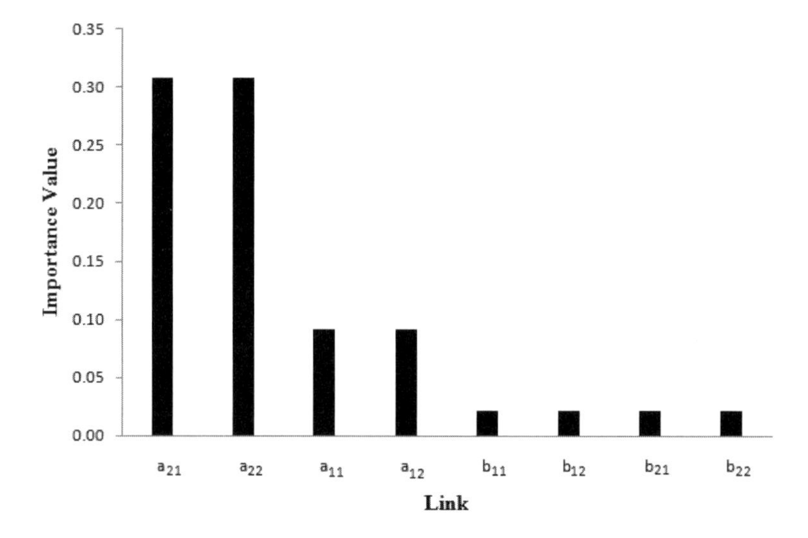

Figure 6.6: Link importance rankings for Example 6.2

We display the link and node importance values and rankings for Example 6.2 in Figures 6.6 and 6.7 respectively.

Discussion

Note that, in Example 6.2, the first source agent had no funds noninvested. Given the cost structure and the demand price functions, since the transaction links between Source Agent 2 and Intermediaries 1 and 2 carried the largest amount of equilibrium financial flow, they were ranked the most important. In addition, because Source Agent 2 allocated the largest amount of financial flow in equilibrium, it was ranked as the most important node. The negative importance value for Demand Markets 1 and 2 is due to the fact that the existence of each demand market brings extra flows on the transaction links and nodes and, therefore, increases the marginal transaction cost. The removal of one demand market had two effects: first, the contribution to

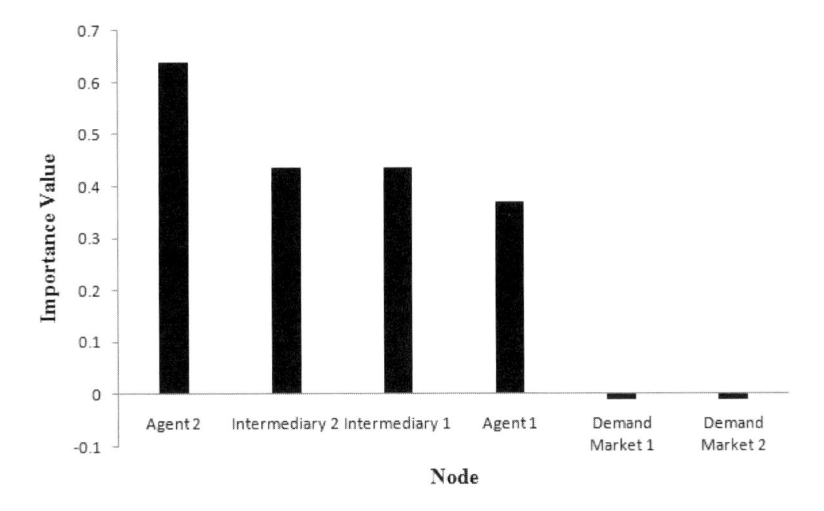

Figure 6.7: Node importance rankings for Example 6.2

the network performance of the removed demand market became zero; second, the marginal transaction cost on links/nodes decreased, which decreased the equilibrium prices and increased the demand at the other demand markets. If the performance drop caused by the removal of the demand markets is overcompensated by the improvement of the demand-price ratio of the other demand markets, the removed demand market will have a negative importance value. It simply implies that the "negative externality" caused by the demand market had a larger impact than the performance drop due to its removal.

6.6 SUMMARY AND CONCLUSIONS

In this chapter, we proposed a novel financial network performance measure, which is motivated by the unified network performance measure described in Chapter 3. The financial network measure examines the network performance by incorporating the economic behavior of the decision-makers, with the resultant equilibrium prices and transaction flows, coupled with the network topology. The financial network performance measure, and the network component importance definition provide valuable methodological tools for evaluating the financial network vulnerability and reliability. Furthermore, our measure is shown to be able to evaluate the importance of nodes and links in financial networks even when the source agent/demand market pairs become disconnected.

We believe that our network performance measure is a good starting point from which to begin to analyze the functionality of an economic network, in general, and a financial network, in particular. Especially in a network in which agents compete in a noncooperative manner in the same tier and coordinate between different tiers without the intervention from the government or a central planner, our proposed measure examines the network on a functional level other than in the traditional Pareto sense. We believe that the proposed measure has natural applicability in such networks as those studied in this chapter. Specifically, with our measure, we are also able to study the robustness and vulnerability of different networks with partially disrupted network components, as discussed in Chapter 3. In the future, additional criteria and perspectives can be incorporated to analyze the network performance more comprehensively. Moreover, with a sophisticated and informative network performance measure, financial administrators and regulators can implement effective policies to enhance the network security and begin to enhance the system robustness.

6.7 SOURCES AND NOTES

This chapter is based on the chapter by Nagurney and Qiang (2008b). Here we have provided additional motivation and background material, including the explicit statement of the Euler method for the financial network equilibrium model with intermediation. In addition, we have also included a proof of Theorem 6.1, which is due to Liu and Nagurney (2007). It is worth noting that the financial network model with intermediation and electronic transactions with known demand (rather than demand price functions) was developed by Nagurney and Ke (2003). When the demand functions are invertible, the two models are equivalent. Hence it is important to note that the financial network measure can also be applied to financial networks with given demand functions. Indeed, for the sake of completeness, we now state the following.

Definition 6.4: The Financial Network Performance Measure
The financial network performance measure, $\mathcal{E}^{FN}$, for a given network topology $\mathcal{G}$ (cf. Figure 6.2) and either given demand price functions $\rho_{3k}(d)$; $k = 1, 2, \ldots, o$, or given demand functions $d_k(\rho)$; $k = 1, 2, \ldots, o$ and available funds held by source agents S, where S is the vector of financial funds held by the source agents, is defined, respectively, as follows

$$\mathcal{E}^{FN} = \frac{\sum_{k=1}^{o} \frac{d_k^*}{\rho_{3k}(d^*)}}{o} = \frac{\sum_{k=1}^{o} \frac{d_k(\rho^*)}{\rho_{3k}^*}}{o}, \tag{6.21}$$

where o is the number of demand markets in the financial network, d_k^* and $\rho_{3k}(d^*)$ denote the equilibrium demand and the incurred equilibrium price for demand market k, respectively, in the case of the model with given demand price functions; similarly, $d_k(\rho^*)$ and ρ_{3k}^* denote the equilibrium incurred demand and equilibrium demand price for the model with given demand functions for demand market k.

For a detailed literature review of financial networks, refer to Nagurney (2003) (see also Fei (1960), Charnes and Cooper (1967), Thore (1969, 1980), Thore and Kydland (1972), Christofides, Hewins, and Salkin (1979), Crum and Nye (1981), Mulvey (1987), Nagurney and Hughes (1992), Nagurney, Dong and Hughes (1992),

Nagurney and Siokos (1997), Nagurney and Ke (2001, 2003), Boginski, Butenko, and Pardalos (2003), Geunes and Pardalos (2003), Nagurney and Cruz (2003a, 2003b), Nagurney, Wakolbinger, and Zhao (2006), and references therein). Furthermore, for a detailed discussion of optimization, risk modeling, and network equilibrium problems in finance and economics, refer to the chapters in the book edited by Kontoghiorghes, Rustem, and Siokos (2002).

By capturing the network structure of economic linkages and transactions from supply chains to financial networks, we can further explore such issues as "traceability." For example, Nagurney (2006b) established that supply chain network equilibrium problems can be reformulated and solved as transportation network equilibrium problems and hence path flow information and path cost information are available through such a transformation. It is interesting that Liu and Nagurney (2007) established an analogous result but in the context of financial networks with intermediation. With path flow information, one can then identify the most "likely" paths or routes that the flows took, be they ingredients or products or financial instruments. Clearly, such information can assist in identifying sources of toxins, whether in food supply chains or financial networks, as in the case of toxic assets.

CHAPTER 7

DYNAMIC NETWORKS, THE INTERNET, AND ELECTRIC POWER

7.1 INTRODUCTION

The Internet, as a global telecommunications network, with linkages to other network systems, including transportation, financial and energy systems, is fundamental to the functioning of our modern societies and economies. Hence an appropriate efficiency measure and a means of identifying its critical nodes and links is essential to not only the management of such a dynamic network but also to its very security. Clearly, those nodes and links that are deemed most important are those that also merit greatest protection.

According to Coffman and Odlyzko (2002) Internet traffic is approximately doubling each year, and such a rate of growth represents extremely fast growth. Figure 7.1 illustrates this rapid growth in number of Internet users in the world from 1995 to 2008.

At the same time, the number of security attacks against the Internet [cf. Bagchi and Tang (2005)] is also growing rapidly; the number of global security incidents rose to 73,359 in 2002, far surpassing the number of 52,658 in the previous year. In August 2003, a variant of the computer worm named Blaster affected over 330,000 computers globally and resulted in more than $320 million in damage [cf. Messmer

Fragile Networks. By Anna Nagurney and Qiang Qiang

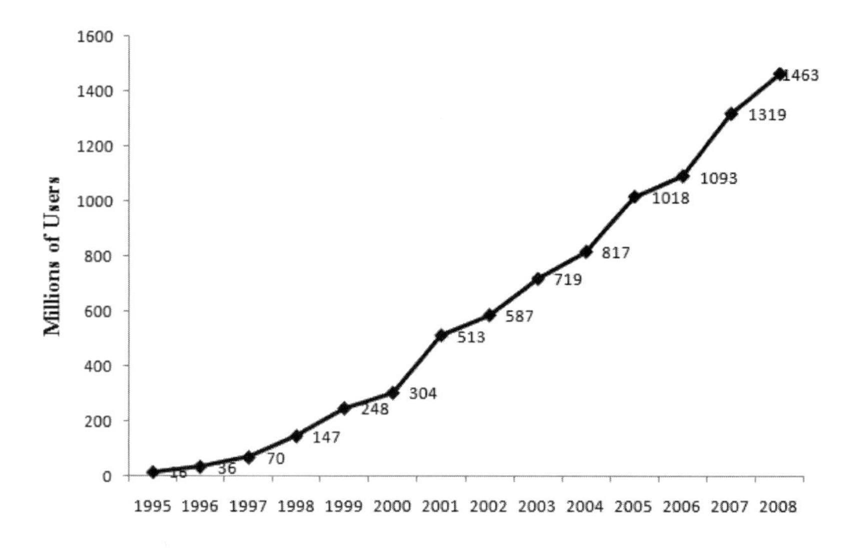

Figure 7.1: Internet users in the world [Data Source: Internet World Stats (2009)]

(2003)]. A 2002 CSI/FBI report [see Power (2002)] documents the variety of Internet attacks and attack types.

Other recent incidents have further highlighted the impacts of Internet disruptions, which need not be due to terrorist attacks but may occur due to accidents or natural disasters. For example, the severance of two cables in the Mediterranean, on January 30, 2008, apparently by ship anchors, caused major Internet service problems for days in the Middle East and parts of Asia [see Reed (2008)]. On December 27, 2006, a 7.1 magnitude earthquake off the southern coast of Taiwan severely disrupted Taiwan's Internet access [see Lemon (2006)]. Although the earthquake caused little structural damage, it disrupted telecommunication networks in Asia and, according to Wolfe (2007), exposed a vulnerable "choke" point in the network. As noted by Wolfe (2007), the physical layer of the Internet consists of several redundant lines, but many of the redundancies run through the same locations with the consequence that a single incident at one of these locations can affect a large area. In terms of the financial impact of a major Internet disruption, a report by the Business Roundtable [see Insurance Journal (2007)] estimated a cost to the global economy of $250 billion.

In this chapter, we propose a network efficiency measure for dynamic networks, which have been modeled as evolutionary variational inequalities. Such dynamic networks include the Internet [cf. Nagurney, Parkes, and Daniele (2007)], time-varying supply chains [see Nagurney and Liu (2007)], and electric power generation and distribution networks [cf. Nagurney et al. (2007)]. As mentioned in Chapter 3, the network efficiency measure proposed by Latora and Marchiori (2001), which is the sum of the inverses of the shortest paths based on geodesic distances between nodes, multiplied by one over a nodal factor, has been applied to several network systems, including the MBTA subway system in Boston and the Internet [cf. Latora and Marchiori (2003, 2004)]. However, their measure does not include behavior that would be associated with rerouting in the case of nodal or link failures, nor does it explicitly include traffic flows. Furthermore, it does not capture the time dimension. Bienenstock and Bonacich (2003) focused on social networks and their efficiency versus vulnerability and proposed a network efficiency measure, which actually coincides with that of Latora and Marchiori (2001). More recently, Jenelius, Petersen, and Mattsson (2006) proposed a network efficiency measure for congested transportation networks, but the measure requires that the network remain connected after disruptions. Furthermore, it does not measure the usage of the network as demand varies over time.

The recent theories of scale-free and small-world networks in complex network research [cf. Watts and Strogatz (1998) and Barabási and Albert (1999)] have also been applied to assess the vulnerability of real-world infrastructure networks, including the Internet. Indeed, due to the fact that the Internet exhibits the scale-free property [see, e.g., Faloutsos, Faloutsos, and Faloutsos (1999)] and a dynamic nature, more and more researchers in different disciplines have begun to address its resilience [cf. Doyle et al. (2005) and Cohen et al. (2000a, 2000b)] and reliability [cf. O'Kelly, Kim, and Kim (2006)].

We use as the basis for our dynamic network measurement framework the time-dependent, evolutionary variational inequality (EVI) model of the Internet proposed by Nagurney, Parkes, and Daniele (2007). Indeed, because the demand for Internet resources itself is dynamic, an underpinning modeling framework must be able to handle time-dependent constraints. Roughgarden (2005) notes that "A network like the Internet is volatile. Its traffic patterns can change quickly and dramatically. The assumption of a static model is therefore particularly suspect in such networks."

It has been shown [cf. Roughgarden (2005) and the references therein] that distributed routing, which is common in computer networks and, in particular, the Internet, and "selfish" routing (or "source" routing in computer networks), as occurs in the case of user-optimized transportation networks (see Chapter 3) in which travelers select the minimum cost route between an origin and destination, are one and the same if the cost functions associated with the links that make up the paths/routes coincide with the lengths used to define the shortest paths. In this chapter, we assume that the costs on the links are congestion-dependent, that is, they depend on the volume of the flow on the link. Note that network efficiency measurements based on shortest path geodesic distances ignore this property of congested networks. We permit the cost on a link to represent travel delay but we utilize "cost" functions because these are more general conceptually than delay (or latency) functions, and they can include, for example, tolls associated with pricing.

In addition, we emphasize that in the case of transportation networks it is travelers that make the decisions as to the route selection between O/D pairs of nodes, whereas in the case of the Internet, it is algorithms, implemented in software, that determine the shortest paths. Furthermore, we emphasize again that analogues exist between transportation networks and telecommunication networks and, in particular, the Internet, in terms of decentralized decision-making, flows and costs, and even the Braess paradox, which allow us to take advantage of such a connection [cf. Beckmann, McGuire, and Winsten (1956), Beckmann (1967), Braess (1968), Dafermos and Sparrow (1969), Cantor and Gerla (1974), Gallager (1977), Bertsekas and Tsitsiklis (1989), Bertsekas and Gallager (1987), Ran and Boyce (1996), Korilis, Lazar, and Orda (1999), Boyce, Mahmassani, and Nagurney (2005), Resende and Pardalos (2006), Nagurney, Parkes, and Daniele (2007)].

The evolutionary variational inequality methodology that Nagurney, Parkes, and Daniele (2007) used for the formulation and analysis of the Internet is quite natural for several reasons. First, historically, finite-dimensional variational inequality theory (cf. Smith (1979), Dafermos (1980), Nagurney (1999), Patriksson (1994), and the references therein) has been used to generalize static transportation network equilibrium models dating to the classic work of Beckmann, McGuire, and Winsten (1956), which also forms the foundation for selfish routing and decentralized decision-making on the Internet [see, e.g., Roughgarden (2005)]. Second, there has been much research activity devoted to the development of models for dynamic transportation problems, and it makes sense to exploit the connections between transportation networks and the Internet [see also Nagurney and Dong (2002a) and Boyce, Mahmassani, and Nagurney (2005)]. In addition, EVIs, which are infinite-dimensional, have been used to model a variety of time-dependent applications, including spatial price problems, financial network problems, dynamic supply chains, and electric power networks [cf. Daniele, Maugeri, and Oettli (1999), Daniele (2003a,b, 2004), Daniele (2006), Nagurney et al. (2007), and Nagurney (2006a)]. Hence we believe that they are a rather natural formalism for dynamic networks and the measurement of their efficiency/performance.

The structure of the chapter is as follows. In Section 7.2 we briefly recall the evolutionary variational inequality formulation of the Internet according to Nagurney, Parkes, and Daniele (2007). In Section 7.3, we propose the efficiency measure for dynamic networks in both continuous time and discrete time forms. In Section 7.4, we apply the measure to the Braess paradox network with time-varying demands to determine the network efficiency as well as the importance rankings of the nodes and links. In Section 7.5, we describe analogues between the dynamic model of the Internet and that of electric power generation and distribution networks.

7.2 EVOLUTIONARY VARIATIONAL INEQUALITIES AND THE INTERNET

For definiteness, in this section, we first present the evolutionary variational inequality model of the Internet, proposed by Nagurney, Parkes, and Daniele (2007), but, for simplicity, we consider the single class, rather than the multiclass version.

In particular, the Internet, a dynamic network par excellence, is modeled as a network $\mathcal{G} = [\mathcal{N}, \mathcal{L}]$, consisting of the set of nodes $\mathcal{N}$ and the set of directed links $\mathcal{L}$. The set of links $\mathcal{L}$ consists of n elements. The set of O/D pairs of nodes is denoted by W and consists of n_W elements. We denote the set of routes (with a route consisting of links) joining the O/D pair w by P_w. We assume that the routes are acyclic. Let P with n_P elements denote the set of all routes connecting all the O/D pairs in the

Internet. Links are denoted by a, b, etc; routes by r, q, etc., and O/D pairs by w_1, w_2, etc. We assume that the Internet is traversed by a single class of "jobs" or "tasks."

Let $d_w(t)$ denote the demand, that is, the traffic generated between O/D pair w at time t. The flow on route r at time t, which is assumed to be nonnegative, is denoted by $x_r(t)$, and the flow on link a at time t by $f_a(t)$.

Because the demands over time are assumed to be known, the following conservation of flow equations must be satisfied at each t

$$d_w(t) = \sum_{r \in P_w} x_r(t), \quad \forall w \in W, \tag{7.1}$$

that is, the demand associated with an O/D pair must be equal to the sum of the flows on the routes that connect that O/D pair. Also, we must have

$$0 \leq x_r(t) \leq \mu_r(t), \quad \forall r \in P, \tag{7.2}$$

where $\mu_r(t)$ denotes the capacity on route r at time t.

We group the demands at time t for all the O/D pairs into the n_W-dimensional vector $d(t)$ and the route flows at time t into the n_P-dimensional vector $x(t)$. The upper bounds/capacities on the routes at time t are grouped into the n_P-dimensional vector $\mu(t)$.

The link flows are related to the route flows, in turn, through the following conservation of flow equations

$$f_a(t) = \sum_{r \in P} x_r(t)\delta_{ar}, \quad \forall a \in \mathcal{L}, \tag{7.3}$$

where $\delta_{ar} = 1$ if link a is contained in route r, and $\delta_{ar} = 0$, otherwise. Hence the flow on a link is equal to the sum of the flows on routes that contain that link. All the link flows at time t are grouped into the vector $f(t)$, which is of dimension n.

The cost on route r at time t is denoted by $C_r(t)$, and the cost on a link a at time t by $c_a(t)$. We allow the cost on a link, in general, to depend on the entire vector of link flows at time t, so that

$$c_a(t) = c_a(f(t)), \quad \forall a \in \mathcal{L}. \tag{7.4}$$

In view of (7.3), we may write the link costs as a function of route flows, that is

$$c_a(x(t)) \equiv c_a(f(t)), \quad \forall a \in \mathcal{L}. \tag{7.5}$$

The costs on routes are related to costs on links through the following equations

$$C_r(x(t)) = \sum_{a \in \mathcal{L}} c_a(x(t))\delta_{ar}, \quad \forall r \in P, \tag{7.6}$$

which means that the cost on a route at a time t is equal to the sum of costs on links that make up the route at time t. We group the route costs at time t into the vector $C(t)$, which is of dimension n_P.

We consider the Hilbert space $\mathcal{H} = L^2([0,T], R^{n_P})$ (see the glossary) where T denotes the time interval under consideration and define the feasible set $\hat{\mathcal{K}}$ as follows

$$\hat{\mathcal{K}} \equiv \{x \in L^2([0,T], R^{n_P}) : 0 \leq x(t) \leq \mu(t) \text{ a.e. in } [0,T];$$
$$\sum_{p \in P_w} x_p(t) = d_w(t), \forall w, \text{ a.e. in } [0,T]\}. \tag{7.7}$$

We assume that the capacities $\mu_r(t)$, for all r, are in $\mathcal{H}$ and that the demands $d_w \geq 0$, for all w, are also in $\mathcal{H}$. Further, we assume that

$$0 \leq d(t) \leq \Phi\mu(t), \text{ a.e. on } [0,T], \tag{7.8}$$

where Φ is the $n_W \times n_P$-dimensional O/D pair-route incidence matrix, with element (w,r) equal to 1 if route r is contained in P_w, and 0, otherwise. Due to (7.8), the feasible set $\mathcal{K}$ is nonempty. As noted in Nagurney, Parkes, and Daniele (2007), $\hat{\mathcal{K}}$ is also convex, closed, and bounded. We are not restricted as to the form that the time-varying demands for the O/D pairs take because convexity is guaranteed even if the demands have a step-wise structure, or are piecewise continuous.

The dual space of $\mathcal{H}$ is denoted by $\mathcal{H}^*$. On $\mathcal{H} \times \mathcal{H}^*$ we define the canonical bilinear form by

$$\langle\langle \hat{G}^T, x \rangle\rangle \equiv \int_0^T \langle \hat{G}(t)^T, x(t) \rangle dt, \quad \hat{G} \in \mathcal{H}^*, \quad x \in \mathcal{H}. \tag{7.9}$$

Furthermore, the cost mapping $C : \hat{\mathcal{K}} \mapsto \mathcal{H}^*$, assigns to each flow trajectory $x(\cdot) \in \hat{\mathcal{K}}$ the cost trajectory $C(x(\cdot)) \in \mathcal{H}^*$.

We now state the dynamic network equilibrium conditions governing the Internet, assuming shortest path routing. These conditions are a generalization of the Wardropian (1952) first principle of user behavior [cf. (2.19)] to include the time dimension and capacities on the route flows. Of course, if the capacities are very large and exceed the demand at each t, then the upper bounds are never attained by the route flows and the conditions below will collapse, in the case of fixed time t, to the well-known static network equilibrium conditions. We next present a special case of the multiclass network equilibrium conditions stated in Nagurney, Parkes, and Daniele (2007); see also Daniele (2006).

Definition 7.1: Dynamic Network Equilibrium
A route flow pattern $x^* \in \hat{\mathcal{K}}$ is said to be a dynamic network equilibrium (according to the generalization of Wardrop's first principle), if, at each time t, only the minimum cost routes not at their capacities are used (i.e., have positive flow) for each O/D pair unless the flow on a route is at its upper bound (in which case those routes' costs can be lower than those on the routes not at their capacities). The state can be expressed by the following equilibrium conditions, which must hold for every O/D pair $w \in W$, every route $r \in P_w$, and a.e. on $[0,T]$

$$C_r(x^*(t)) - \lambda_w^*(t) \begin{cases} \leq 0, & \textit{if} \quad x_r^*(t) = \mu_r(t), \\ = 0, & \textit{if} \quad 0 < x_r^*(t) < \mu_r(t), \\ \geq 0, & \textit{if} \quad x_r^*(t) = 0. \end{cases} \tag{7.10}$$

Hence conditions (7.10) state that all used routes not at their capacities connecting an O/D pair have equal and minimal costs at each time t in $[0, T]$. If a route flow is at its capacity then its cost can be lower than the minimal cost for that O/D pair. Of course, if we have that $\mu_r = \infty$, for all routes $r \in P$, then the dynamic equilibrium conditions state that all used routes connecting an O/D pair of nodes have equal and minimal route costs at each time t. For fixed t, the latter conditions coincide with a single-class version of Wardrop's first principle [cf. (2.19)] governing static network equilibrium problems. Note that this concept has also been applied to static models of the Internet (cf. Roughgarden (2005) and the references therein).

In this discussion, we assume that the general "cost" on links and paths is "instantaneous," which solely depends on the current network flows and demands. Such an assumption is relatively general and can be easily extended to the discrete-time case, which will be discussed in the following section. Indeed, in the discrete-time case, demand is assumed to be constant during each time interval and, therefore, the link and path costs can be interpreted as the average costs based on the flows in that particular time interval. Furthermore, similar data and computer communication network models and analysis, but in the discrete-time case, have also been studied by Bertsekas and Gallager (1987).

The standard form of the EVI that we work with is

$$\text{determine } X^* \in \hat{\mathcal{K}} \text{ such that } \langle\langle F(X^*)^T, X - X^* \rangle\rangle \geq 0, \ \forall X \in \hat{\mathcal{K}}. \tag{7.11}$$

We now state the following theorem, which is an adaptation of Theorem 1 in Nagurney, Parkes, and Daniele (2007) to the single-class case; see that reference for a proof.

Theorem 7.1
$x^* \in \hat{\mathcal{K}}$ is an equilibrium flow according to Definition 7.1 if and only if it satisfies the evolutionary variational inequality

$$\int_0^T \langle C(x^*(t))^T, x(t) - x^*(t) \rangle dt \geq 0, \quad \forall x \in \hat{\mathcal{K}}. \tag{7.12}$$

Furthermore, for completeness, we also provide some existence results.

Theorem 7.2 [cf. Daniele, Maugeri, and Oettli (1999) and Daniele (2006)]
If C in (7.12) satisfies any of the following conditions

1. *C is hemicontinuous with respect to the strong topology on $\hat{\mathcal{K}}$, and there exist $A \subseteq \hat{\mathcal{K}}$ nonempty, compact, and $B \subseteq \hat{\mathcal{K}}$ compact such that, for every $y \in \hat{\mathcal{K}} \setminus A$, there exists $x \in B$ with $\langle\langle C(x)^T, y - x \rangle\rangle < 0$;*

2. *C is hemicontinuous with respect to the weak topology on $\hat{\mathcal{K}}$;*

3. *C is pseudomonotone and hemicontinuous along line segments,*

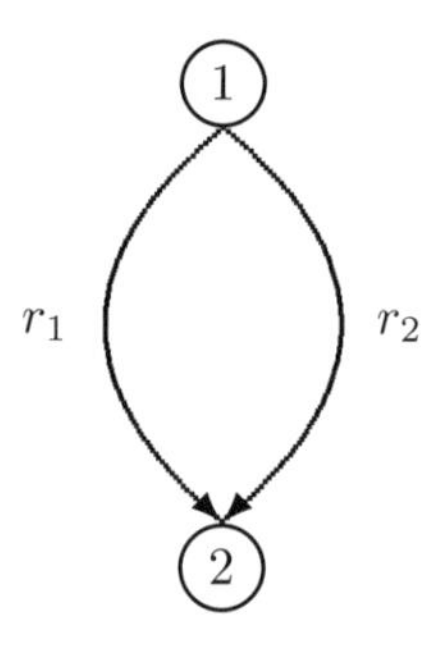

Figure 7.2: Network structure of Example 7.1

then the EVI problem (7.12) admits a solution over the constraint set $\hat{\mathcal{K}}$.

Recall that $C : \hat{\mathcal{K}} \mapsto \mathcal{H}^*$, where $\hat{\mathcal{K}}$ is convex, is said to be

pseudomonotone if and only if, for all $x, y \in \hat{\mathcal{K}}$

$$\langle\langle C(x)^T, y - x\rangle\rangle \leq 0 \Rightarrow \langle\langle C(y)^T, x - y\rangle\rangle \leq 0;$$

hemicontinuous if and only if, for all $y \in \hat{\mathcal{K}}$, the function $\xi \mapsto \langle\langle C(\xi)^T, y - \xi\rangle\rangle$ is upper semicontinuous on $\hat{\mathcal{K}}$; and

hemicontinuous along line segments if and only if, for all $x, y \in \hat{\mathcal{K}}$, the function $\xi \mapsto \langle\langle C(\xi)^T, y - x\rangle\rangle$ is upper semicontinuous on the line segment $[x, y]$.

Moreover, if C is strictly monotone, then the solution of (7.12) is unique [see, e.g., Kinderlehrer and Stampacchia (1980)].

Daniele, Maugeri, and Oettli (1999) presented dynamic network equilibrium conditions for transportation networks. Here we state the dynamic equilibrium conditions in a manner that is more transparent [cf. (7.10)], noting that the lower bounds on the route flows on the Internet will be zero.

The solution of finite-dimensional variational inequalities is at quite an advanced state (cf. Chapter 3). The solution of evolutionary variational inequalities, which are infinite-dimensional, is a topic discussed in the books by Daniele (2006) and Nagurney (2006a) and in the papers by Cojocaru, Daniele, and Nagurney (2005, 2006, 2007) and Barbagallo (2007). In particular, under certain conditions, an evolutionary variational inequality of the form (7.12) can be solved at discrete points in time and the solutions then interpolated. Hence, the advances in the computation of finite-dimensional variational inequalities, with many network equilibrium problems solved to-date, may be exploited for the effective solution of the EVI for dynamic networks.

We now present a small dynamic network numerical example.

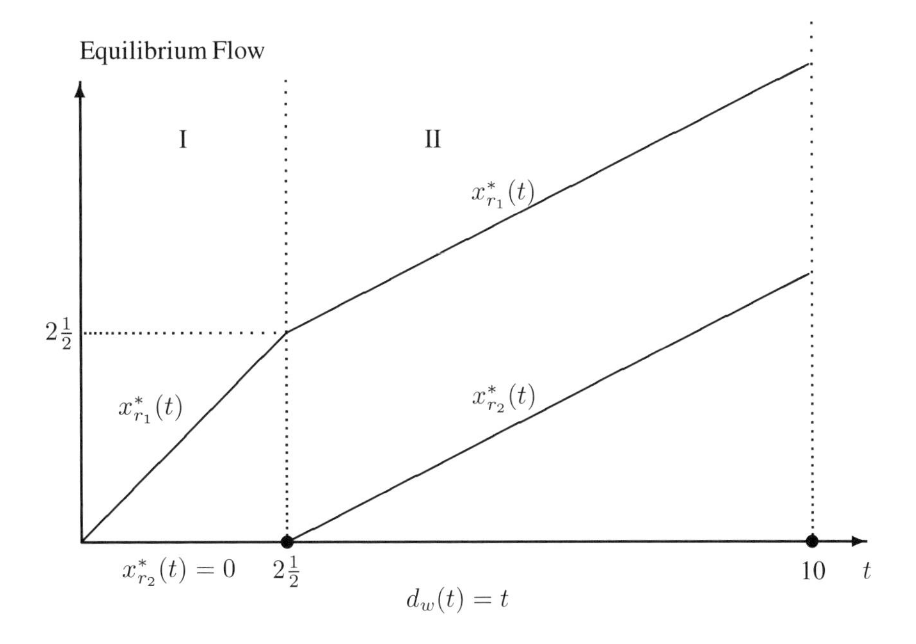

Figure 7.3: Equilibrium trajectories of the simple numerical example with time-dependent demands

Example 7.1: A Simple Numerical Example
Consider a network (small subnetwork of the Internet) consisting of two nodes and two links as in Figure 7.2. Because the routes consist of single links we work with the routes directly as in Figure 7.2. The route costs are

$$C_{r_1}(x(t)) = 2x_{r_1}(t) + 5, \quad C_{r_2}(x(t)) = 2x_{r_2}(t) + 10.$$

Also, the route capacities are: $\mu_{r_1}(t) = \mu_{r_2}(t) = \infty$.

The time horizon is $[0, 10]$. There is a single O/D pair $w = (1, 2)$ and the time-varying demand is assumed to be $d_w(t) = t$.

Given the special structure of the network in Figure 7.2 and the linearity and separability of the route cost functions, it is easy to determine the equilibrium route trajectories satisfying (7.10) or, equivalently, (7.12), and they are for $t \in [0, 2\frac{1}{2}]$, $x^*_{r_1}(t) = t$ and $x^*_{r_2}(t) = 0$, and for $t \in [2\frac{1}{2}, 10]$, $x^*_{r_1}(t) = \frac{1}{2}t + 1\frac{1}{4}$ and $x^*_{r_2}(t) = \frac{1}{2}t - 1\frac{1}{4}$.

The equilibrium flow trajectories of the above network are depicted in Figure 7.3.

7.3 THE EFFICIENCY MEASURE FOR DYNAMIC NETWORKS MODELED AS EVOLUTIONARY VARIATIONAL INEQUALITIES

In this section, we propose an efficiency measure for dynamic networks, modeled as evolutionary variational inequalities, which we denote by $\mathcal{E}^{DN}(\mathcal{G}, d, T)$.

Definition 7.2a: Dynamic Network Efficiency: Continuous Time Version
The network efficiency for the network $\mathcal{G}$ with time-varying demand d for $t \in [0,T]$, denoted by $\mathcal{E}^{DN}(\mathcal{G}, d, T)$, is defined as follows

$$\mathcal{E}^{DN}(\mathcal{G}, d, T) = \frac{\int_0^T [\sum_{w \in W} \frac{d_w(t)}{\lambda_w(t)}]/n_W \, dt}{T}. \tag{7.13}$$

The dynamic network efficiency measure $\mathcal{E}^{DN}$ defined in (7.13) is actually the average demand to price ratio over time. It measures the overall (economic) functionality of the network with time-varying demands. When the network topology $\mathcal{G}$, the demand pattern over time and the time span are given, a network is considered to be more efficient if it can satisfy higher demands at lower costs over time. We assume that the integral in (7.13) is well-defined.

The network efficiency measure (7.13) can be easily adapted to dynamic networks in which the demands change at discrete points in time, as we now demonstrate. Let $d_w^1, d_w^2, ..., d_w^H$ denote demands for O/D pair w in H discrete time intervals, given, respectively, by: $[t_0, t_1], (t_1, t_2], ..., (t_{H-1}, t_H]$, where $t_H \equiv T$. We assume that the demand is constant in each such time interval for each O/D pair. Moreover, we denote the corresponding minimal costs for each O/D pair w [see (7.10)] at the H different time intervals by: $\lambda_w^1, \lambda_w^2, ..., \lambda_w^H$. The demand vector d, in this special discrete case, is a vector in $R^{n_W \times H}$. The dynamic network efficiency measure in this case is as follows.

Definition 7.2b: Dynamic Network Efficiency: Discrete Time Version
The network efficiency for the network $(\mathcal{G}, d)$ over H discrete time intervals: $[t_0, t_1], (t_1, t_2], ..., (t_{H-1}, t_H]$, where $t_H \equiv T$, and with the respective constant demands: $d_w^1, d_w^2, ..., d_w^H$ for all $w \in W$ is defined as follows

$$\mathcal{E}^{DN}(\mathcal{G}, d, t_H = T) = \frac{\sum_{i=1}^H [(\sum_{w \in W} \frac{d_w^i}{\lambda_w^i})(t_i - t_{i-1})/n_W]}{t_H}. \tag{7.14}$$

We now provide the relationship between the dynamic network efficiency measure (7.13) and the unified network efficiency measure (3.2) for static networks, and we focus on the case of fixed demands.

Theorem 7.3
Assume that $d_w(t) = d_w$, for all O/D pairs $w \in W$ and for $t \in [0,T]$. Then, the dynamic network efficiency measure (7.13) collapses to the network measure (3.2)

$$\mathcal{E} = \frac{1}{n_W} \sum_{w \in W} \frac{d_w}{\lambda_w}. \tag{7.15}$$

Proof: Because the d_ws are fixed over the time horizon $[0,T]$, the minimal equilibrium route costs for each O/D pair w, denoted by λ_w, $\forall w \in W$, are also fixed over the time horizon. Substituting these terms into (7.13), we obtain

$$\mathcal{E}^{DN}(\mathcal{G}, d, T) = \frac{\int_0^T [\sum_{w \in W} \frac{d_w}{\lambda_w}]/n_W \, dt}{T} = \frac{T}{T}[\sum_{w \in W} \frac{d_w}{\lambda_w}]/n_W = [\sum_{w \in W} \frac{d_w}{\lambda_w}]/n_W. \tag{7.16}$$

But the right-most expression in (7.16) is precisely the network efficiency/performance measure for network equilibrium problems given by (3.2).

Several points merit some discussion here. First, one of the advantages of this measure is that it allows us to study the network efficiency when there are disconnected O/D pairs. Such a feature is extremely useful in assessing the network functionality under partial disruptions, whereas traditional measures, such as the total network cost over time, are no longer applicable [cf. Roughgarden (2005)]. An illustrative example is given later in this section. Second, in this definition, we consider the network model without explicit edge, that is, link, capacity constraints. It is well-known in the transportation literature that edge capacities can be implicitly included in the link travel cost functions (cf. Section 3.3), in which case the travel time tends to infinity as the link flows approach their respective capacities [cf. Daganzo (1977a, b)]. A network model with explicit edge capacity constraints was studied by Patriksson (1994), who generalized link costs to reformulate the network as an equivalent uncapacitated network. Hence in such a model, $\lambda_w(t)$ in (7.13) can be interpreted as a generalized cost. Nevertheless, a discussion of the distinction and the relationships between the two approaches is not the focus of this chapter. In Chapters 8 through 10 we develop system-optimized network models with explicit capacities on the links and consider network synergies associated the network integration through mergers and acquisitions.

The importance of a network component in the dynamic network case is the same as that defined in (3.2), but with the static efficiency measure now replaced by the dynamic network efficiency measure given by (7.13) in the continuous case and by (7.14) in the discrete case. Hence we have the following.

Definition 7.3: Importance of a Dynamic Network Component
The importance of dynamic network component g of network G with demand d over time horizon T is defined as follows

$$I^{DN}(g, d, T) = \frac{\mathcal{E}^{DN}(\mathcal{G}, d, T) - \mathcal{E}^{DN}(\mathcal{G} - g, d, T)}{\mathcal{E}^{DN}(\mathcal{G}, d, T)} \tag{7.17}$$

where $\mathcal{E}^{DN}(\mathcal{G} - g, d, T)$ is the dynamic network efficiency after component g is removed.

In studying the importance of a network component, the elimination of a link is treated in this measure by removing that link, while the removal of a node is managed by removing the links entering and exiting that node. When the removal results in no path/route connecting an O/D pair, we simply assign the demand for that O/D pair to an abstract path with a cost of infinity. Hence our dynamic network measure is

Table 7.1: Importance and ranking of links in Example 7.1

Link	Importance Value	Importance Ranking
r_1	.3887	1
r_2	.1894	2

Table 7.2: Importance and ranking of nodes in Example 7.1

Node	Importance Value	Importance Ranking
1	1.0000	1
2	1.0000	1

well-defined even in the case of disconnected networks; see also Chapter 3, where additional theoretical properties of the static measure are discussed.

For illustration purpose, we apply the above dynamic network efficiency measure to study the importance of links and nodes in Example 7.1 in Section 7.2. For the time horizon given by $t \in [0, 10]$, with $T = 10$, the network efficiency $\mathcal{E}^{DN}(\mathcal{G}, d, 10) = .3686$. The importance and the rankings of the links and the nodes are given in Tables 7.1 and 7.2, respectively.

It is quite reasonable that link/route r_1 is more important than link/route r_2, as reported in Table 7.1, because over the time horizon it carries more traffic. Note that a measure such as the Latora and Marchiori (2001) measure does not capture either flows or the time dimension.

Furthermore, as discussed earlier in this section, the proposed dynamic network efficiency measure enables us to study the network with disconnected O/D pairs, while the other measure (e.g., total network cost over time) is not applicable. The following simple example is given to illustrate this idea.

Example 7.2: An Illustrative Network Example

Consider the following network, as shown in Figure 7.4, with three nodes and two links. Nodes are denoted by 1, 2, and 3, and links are denoted by a and b. Similar to Example 7.1, since the routes consist of single links, we work with the routes directly, so that $r_1 = a$ and $r_2 = b$. The route costs were

$$C_{r_1}(x(t)) = x_{r_1}(t) + 1, \quad C_{r_2}(x(t)) = x_{r_2}(t) + 2.$$

Also, we assume that the route capacities were $\mu_{r_1}(t) = \mu_{r_1}(t) = \infty$.

The time horizon is $[0, 10]$. There are two O/D pairs, namely, $w_1 = (1, 2)$ and $w_2 = (1, 3)$. The time-varying demands are $d_{w_1}(t) = 3t + 10$ and $d_{w_2}(t) = 3t$.

Given the special structure of the network, it is easy to determine the equilibrium route trajectories satisfying (7.10) or, equivalently, (7.12), and they are $x^*_{r_1}(t) = 3t + 10$ and $x^*_{r_2}(t) = 3t$.

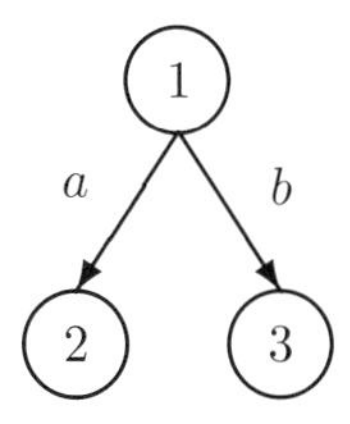

Figure 7.4: Network structure of Example 7.2

Table 7.3: Importance and ranking of links in Example 7.2

Link	Importance Value	Importance Ranking
a	.5398	1
b	.4602	2

Table 7.4: Importance and ranking of nodes in Example 7.2

Node	Importance Value	Importance Ranking
1	1.0000	1
2	.5398	2
3	.4602	3

The network efficiency measure is given by $\mathcal{E}^{DN}(\mathcal{G}, d, 10) = .8857$.

The importance and the rankings of the links and the nodes are given in Tables 7.3 and 7.4, respectively. In this example, link a/route r_1 carries more flow over time and, therefore, it is ranked higher than link b/route r_2.

The proposed measure provides important insights into network vulnerability and security analysis, especially because the total cost of the network over time, after a link or a node is destroyed, becomes infinite in Example 7.2. Hence this undesirable feature of a total cost measure, unlike the proposed measure, prevents one from assessing the network efficiency under disruptions and, as a consequence, the importance and the rankings of the network component. In particular, when a disruption causes partial damage to a large network and a certain demand cannot be satisfied, the remaining network components still may function properly. Hence it still makes sense to analyze network performance in such a situation and, obviously, those network components whose removal causes a large performance drop should receive more attention.

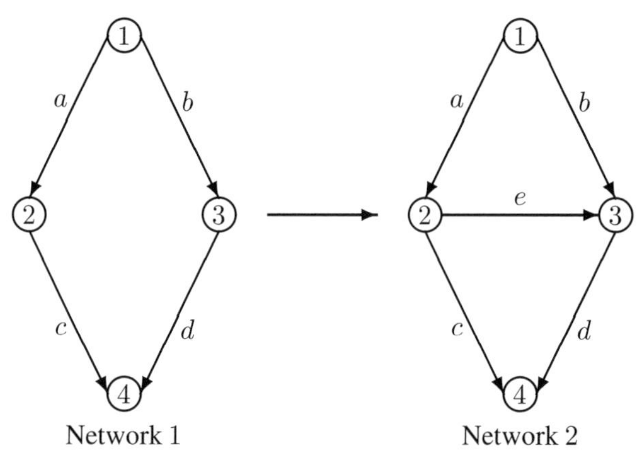

Figure 7.5: The time-dependent Braess network example with relevance to the Internet

7.4 AN APPLICATION TO THE TIME-DEPENDENT BRAESS NETWORK

In this section, we use the proposed dynamic network measure to the time-dependent Braess example [cf. Nagurney, Parkes, and Daniele (2007), Nagurney (2006a), and, also, Pas and Principio (1997)], but, first, for completeness and easy reference, we revisit the Braess (1968) paradox [see also, Boyce, Mahmassani, and Nagurney (2005) and Braess, Nagurney, and Wakolbinger (2005)]. Recall that in the Braess paradox, which is an example of a fixed demand network equilibrium problem, the addition of a new link, which yields a new route, makes all the "users" in the network worse off. As emphasized in Korilis, Lazar, and Orda (1999), this is also relevant to the Internet. We present an EVI formulation with instantaneous costs that deepens the understanding of the Braess paradox and illustrates dramatically the importance of time-varying demands and the associated equilibrium flows and costs. Here, in a sense, we reinterpret the results in Section 3.2.4 in a dynamic, time dimension setting.

The Time-Dependent Braess Paradox

Assume a network as the first network depicted in Figure 7.5 in which there are four nodes 1, 2, 3, and 4; four links a, b, c, and d; and a single O/D pair $w = (1, 4)$. There are hence two routes available between this O/D pair: $r_1 = (a, c)$ and $r_2 = (b, d)$.

The networks given in Figure 7.5 are due to Braess (1968). We now construct time-dependent link costs, route costs, and demand for $t \in [0, T]$. It is important to emphasize that when time t is discrete, that is, $t = 0, 1, 2, \ldots, T$, is trivially included in the equilibrium conditions (7.10) and captured in the EVI formulation (7.12).

We considered, to start, the first network in Figure 7.5, consisting of links: a, b, c, d. We assumed that the capacities $\mu_{r_1}(t) = \mu_{r_2}(t) = \infty$ for all $t \in [0, T]$. The link cost functions were assumed to be given and as follows for time $t \in [0, T]$

$$c_a(f_a(t)) = 10 f_a(t), \quad c_b(f_b(t)) = f_b(t) + 50,$$

$$c_c(f_c(t)) = f_c(t) + 50, \quad c_d(f_d(t)) = 10 f_d(t).$$

We assumed a time-varying demand $d_w(t) = t$ for $t \in [0, T]$.

Observe that at time $t = 6$, $d_w(6) = 6$, and it is easy to verify that the equilibrium route flows at time $t = 6$ were $x^*_{r_1}(6) = 3$, $x^*_{r_2}(6) = 3$, the equilibrium link flows were $f^*_a(6) = 3$, $f^*_b(6) = 3$, $f^*_c(6) = 3$, $f^*_d(6) = 3$, with associated equilibrium route costs

$$C_{r_1}(6) = c_a(6) + c_c(6) = 83, \quad C_{r_2} = c_b(6) + c_d(6) = 83,$$

and hence equilibrium conditions (7.10) are satisfied for time $t = 6$. This is the solution to the classical (static) Braess (1968) network without the route addition.

We now construct and solve EVI (7.12) for the dynamic network equilibrium problem over $t \in [0, T]$. We first expressed the route costs in terms of route flows for Network 1 in Figure 7.5, where we have that, because of the conservation of flow equations (7.3), $f_a(t) = f_c(t) = x_{r_1}(t)$ and $f_b(t) = f_d(t) = x_{r_2}(t)$. That is, we must have that $C_{r_1}(t) = 11x_{r_1}(t) + 50$, $\; C_{r_2}(t) = 11x_{r_2}(t) + 50$, with the route conservation of flow equations (7.1), yielding $d_w(t) = t = x_{r_1}(t) + x_{r_2}(t)$, and hence we may write $x_{r_2}(t) = t - x_{r_1}(t)$.

Similarly, we must have, because of the feasible set $\hat{\mathcal{K}}$ [cf. (7.7)], the simplicity of the network topology, and the cost structure

$$x^*_{r_1}(t) = x^*_{r_2}(t). \tag{7.18}$$

Hence we may write EVI (7.12) for this problem as: Determine $x^* \in \hat{\mathcal{K}}$ satisfying

$$\int_0^T (11x^*_{r_1}(t)+50) \times (x_{r_1}(t) - x^*_{r_1}(t)) + (11x^*_{r_1}(t)+50) \times (x_{r_2}(t) - x^*_{r_2}(t))dt \geq 0,$$

$$\forall x \in \hat{\mathcal{K}}, \tag{7.19}$$

which, in view of the feasibility condition $x_{r_1}(t) + x_{r_2}(t) = t$ implies that (7.19) can be expressed as

$$\int_0^T (11x^*_{r_1}(t)+50) \times (x_{r_1}(t) - x^*_{r_1}(t)) + (11x^*_{r_1}(t)+50) \times (t - x_{r_1}(t) - x^*_{r_1}(t))dt \geq 0,$$

$$\forall x \in \hat{\mathcal{K}}, \tag{7.20}$$

which, after algebraic simplification, is

$$\int_0^T (11x^*_{r_1}(t) + 50) \times (t - 2x^*_{r_1}(t))dt \geq 0, \quad \forall x \in \hat{\mathcal{K}}. \tag{7.21}$$

But (7.21) implies that $2x^*_{r_1}(t) = t$; for $t \in [0, T]$ or $x^*_{r_1}(t) = \frac{t}{2}$. Thus we also have $x^*_{r_2}(t) = \frac{t}{2}$.

Moreover, the equilibrium route costs for $t \in [0, T]$ are given by $C_{r_1}(x^*_{r_1}(t)) = 5\frac{1}{2}t + 50 = C_{r_2}(x^*_{r_2}(t)) = 5\frac{1}{2}t + 50$, and, clearly, equilibrium conditions (7.10) hold for $\in [0, T]$ a.e.

Assume now that, as depicted in Figure 7.5, a new link e, joining node 2 to node 3, is added to the original network, with cost $c_e(f_e(t)) = f_e(t) + 10$ for $t \in [0, T]$. The addition of this link creates a new route $r_3 = (a, e, d)$ that is available for the Internet traffic. Assume that the time-varying demand is still given by $d_w(t) = t$. Note that for $t = 6$, for example, the original equilibrium flow distribution pattern $x_{r_1}(6) = 3$ and $x_{r_2}(6) = 3$ is no longer an equilibrium pattern because at this level of flow the cost on route r_3, $C_{r_3}(6) = 70$. Hence the traffic from routes r_1 and r_2 would be switched to route r_3.

The equilibrium flow pattern at time $t = 6$ on the new network (which would correspond to the classic Braess paradox in a static network equilibrium setting) is now $x^*_{r_1}(6) = 2$, $x^*_{r_2}(6) = 2$, $x^*_{r_3}(6) = 2$, with equilibrium link flows $f^*_a(6) = 4$, $f^*_b(6) = 2$, $f^*_c(6) = 2$, $f^*_e(6) = 2$, $f^*_d(6) = 4$, and with associated equilibrium route costs

$$C_{r_1}(6) = 92, \quad C_{r_2}(6) = 92, \quad C_{r_3}(6) = 92.$$

Indeed, one can verify that any reallocation of the route flows would yield a higher cost on a route.

Note that, with the route addition, the cost at time $t = 6$ increased for every "user" of the network from 83 to 92 without a change in the demand or traffic rate! This is the classical Braess paradox.

We now recall the solution to the EVI problem (7.12) for the second network in Figure 7.5 over the time interval $[0, T]$, which illustrates the time-dependent Braess paradox (see also Section 3.2.4).

In particular, the solution of the EVI for the second Braess network in Figure 7.5 with $d_w(t) = t$, for $t \in [0, T]$, yields three regimes, denoted by I, II, and III, respectively, and depicted in Figure 7.5, where for $d_w(t) = t \in [0, t_1 = 3\frac{7}{11}]$ (Regime I)

$$x^*_{r_1}(t) = x^*_{r_2}(t) = 0, \quad x^*_{r_3}(t) = d_w(t) = t;$$

for $d_w(t) = t \in (t_1 = 3\frac{7}{11}, 8\frac{8}{9}]$ (Regime II), we have

$$x^*_{r_1}(t) = x^*_{r_2}(t) = \frac{11}{13}t - \frac{40}{13}, \quad x^*_{r_3}(t) = -\frac{9}{13}t + \frac{80}{13}.$$

Finally, for $d_w(t) = t \in (t_2 = 8\frac{8}{9}, T < \infty]$ (Regime III), we have

$$x^*_{r_1}(t) = x^*_{r_2}(t) = \frac{d_{r_1}(t)}{2} = \frac{t}{2}, \quad x^*_{r_3}(t) = 0.$$

The curves of equilibria are depicted in Figure 7.6.

Clearly, one can see from Figure 7.6, that in the range $(0, t_1 = 3\frac{7}{11}]$, that is, in Regime I (once the demand is positive), only the new route r_3 would be used. Hence at a relatively low level of demand, up to a value of $3\frac{7}{11}$, only the new route is used. In the range of demands: $(3\frac{7}{11}, 8\frac{8}{9}]$, that is, Regime II, all three routes are used, and in this range the Braess paradox occurs. Finally, once the demand (recall that $d_w(t) = t$ here) exceeds $8\frac{8}{9}$ and we are in Regime III, then the new route is never used! Thus the use of an EVI formulation reveals that over time the Braess paradox

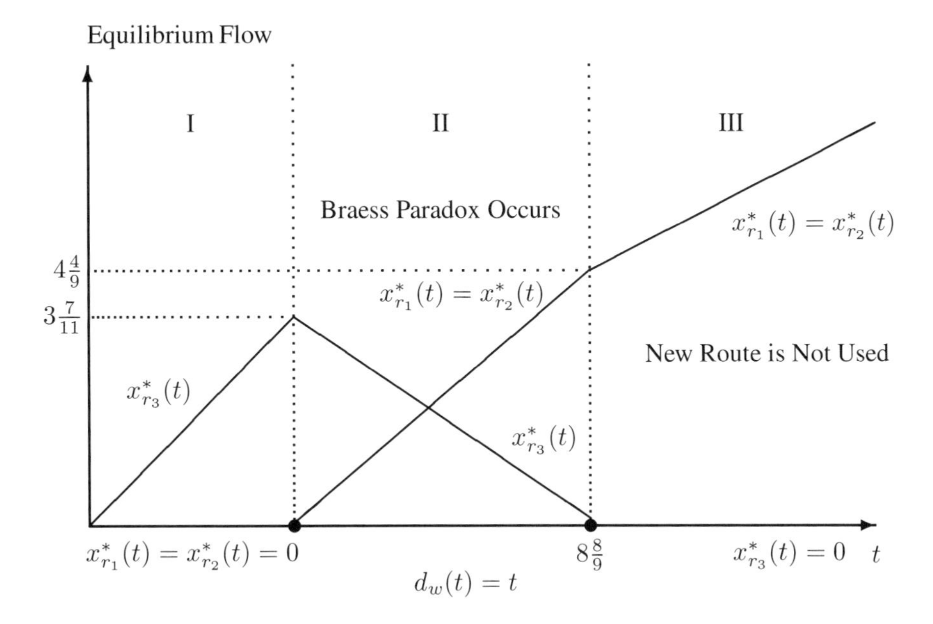

Figure 7.6: Equilibrium trajectories of the Braess network with time-dependent demands

is even more profound and the addition of a new route may result in the route never being used. Finally, if the demand lies within a particular range, then the addition of a new route may result in everyone being worse off because it results in higher costs than before the route/link was added to the network.

In particular, the classical Braess paradox, in which the addition of the route makes the travel cost higher for everyone, always occurs in Regime II. To find the minimal demand at which the Braess paradox occurs, we note that in the first network in Figure 7.5, the demand will always equally distribute itself. Hence on the original network, the equilibrium flow pattern on each route would be given by $\frac{d_w(t)}{2} = \frac{t}{2}$ for $t \in [0, T]$, with a minimal route cost over the horizon being thus equal to $11(\frac{t}{2}) + 50$.

Consider now the second network in Figure 7.5. We know that in Regime I, only the new route would be used, assuming shortest path routing; the minimal route cost hence being given by the expression in this range of demands as $21t + 10$. Setting now, $11(\frac{t}{2}) + 50 = 21t + 10$, and solving for t, which is also in this problem equal to the demand $d_w(t)$, yields $t = 2\frac{18}{31} = 2.58$. For demand in the range $2.58 < d_w(t) = t < 8\frac{8}{9} = 8.89$, the addition of the new route will result in everyone being worse off (see Figure 7.7).

For both networks in Figure 7.5, with the associated link and route cost functions, it is easy to verify that the corresponding vector of route costs $C(x)$ is strictly monotone (cf. Chapter 2) in route flows x, that is

$$\langle\langle (C(x^1) - C(x^2))^T, x^1 - x^2 \rangle\rangle > 0, \quad \forall x^1, x^2 \in \hat{\mathcal{K}}, x^1 \neq x^2,$$

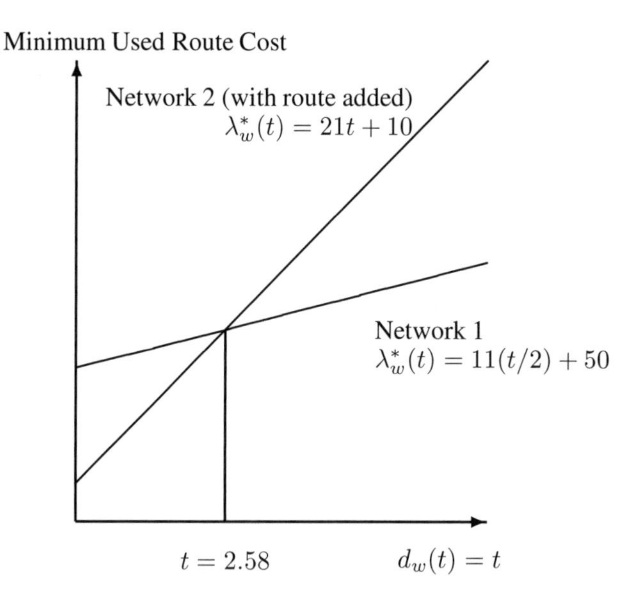

Figure 7.7: Minimum used route costs for Braess Networks 1 and 2

because the Jacobian of the route costs is strictly diagonally dominant at each t and thus positive definite. Hence the corresponding equilibrium route flow solutions $x^*(t)$ will be unique.

Network Efficiency of the Dynamic Braess Network and Importance Rankings of Nodes and Links over the Time Horizon

Let us now consider the second dynamic Braess network in Figure 7.5 for $t \in [0, 10]$. As demonstrated in Section 3.2.4, and just noted, different routes (and links) are used in different demand ranges. Therefore, it is interesting and relevant to study the network efficiency and the importance of the network components over the time horizon. Because in the above example the demand varies continuously over time, the formula (7.13) is used to compute the network efficiency. The computations for both (7.13) and for (7.17) were done using MATLAB (see www.mathworks.com).

The network efficiency $\mathcal{E}^{DN}(\mathcal{G}, d, 10)$ for this dynamic network is .5793. The importance and the rankings of the links and the nodes are summarized in Table 7.5 and Table 7.6, respectively.

From this analysis, it is clear which nodes and links are more important in the dynamic Braess network and hence should, in effect, be better protected and secured because their elimination results in a more significant drop in network efficiency or performance. Indeed, link e after $t = 8\frac{8}{9}$ is never used and in the range $t \in [3\frac{7}{11}, 8\frac{8}{9}]$ increases the cost, so the fact that link e has a *negative* importance value makes sense; over time, its removal would, on the average, improve the network efficiency! This analysis also has implications for network design because, over the time horizon, adding or building link e does not make sense.

Table 7.5: Importance and ranking of links in the dynamic Braess network

Link	Importance Value	Importance Ranking
a	.2604	1
b	.1784	2
c	.1784	2
d	.2604	1
e	-.1341	3

Table 7.6: Importance and ranking of nodes in the dynamic Braess network

Node	Importance Value	Importance Ranking
1	1.0000	1
2	.2604	2
3	.2604	2
4	1.0000	1

7.5 ELECTRIC POWER NETWORKS

Electric power generation and distribution networks provide another example of critical infrastructure. Although the research and analytical efforts have been quite impressive in this area, from August to September 2003 major outages occurred in North America and Europe. First, on August 14, 2003, large portions of the Midwest, the Northeastern United States, and Ontario, Canada, experienced an electric power blackout. The outage affected an area with an estimated 50 million people and 61,800 megawatts (MW) of electric load in the states of Ohio, Michigan, Pennsylvania, New York, Vermont, Massachusetts, Connecticut, New Jersey, and the Canadian province of Ontario. Parts of Ontario suffered rolling blackouts for more than a week before full power was restored. Estimates of the total associated costs in the United States ranged between $4 and $10 billion (U.S. dollars). In Canada, the gross domestic product was down .7% that August; there was a net loss of 18.9 million work hours, and manufacturing shipments in Ontario were down by $2.3 billion (Canadian dollars) [United States - Canada Power System Outage Task Force (2004)]. In addition, two other significant outages occurred during September 2003, one in England and one, initiated in Switzerland, which cascaded over much of Italy. These events clearly indicate that recent changes in the electric power markets require deep and thorough analyses.

For illustration purpose, we now use the dynamic network efficiency measure to study an electric power supply chain network. The example is adapted from Example 1 in Nagurney et al. (2007), where the notation and model are fully described.

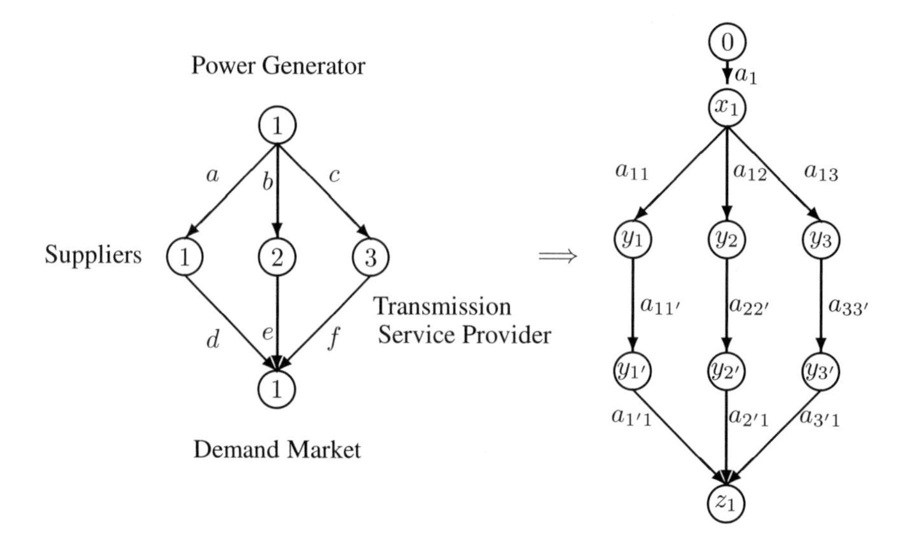

Figure 7.8: Electric power supply chain network and the corresponding supernetwork

In the example, the electric power supply chain network consists of one power generator, three power suppliers, one transmission provider, and one demand market, as depicted in Figure 7.8. The links in the first network in Figure 7.8 are labeled a, b, etc., for subsequent reference. The supernetwork representation that allows the transformation of the electric power supply chain network [cf. Wu et al. (2006) and Nagurney et al. (2007)] into a transportation network equilibrium problem is also shown in Figure 7.8.

The power generating cost function for the power generator is given by

$$f_1(q_1(t)) = 2.5(q_1(t))^2 + 2q_1(t).$$

The transaction cost functions faced by the power generator and associated with transacting with the power suppliers: $1, 2, 3$ are given, respectively, by

$$c_{11}(q_{11}(t)) = .5(q_{11}(t))^2 + 3.5q_{11}(t), \quad c_{12}(q_{12}(t)) = .5(q_{12}(t))^2 + 2.5q_{12}(t),$$

$$c_{13}(q_{13}(t)) = .5(q_{13}(t))^2 + 3.5q_{13}(t).$$

The operating costs of the three power suppliers, in turn, are given, respectively, by

$$c_1(q_{11}(t)) = .5(q_{11}(t))^2, \quad c_2(q_{12}(t)) = .5(q_{12}(t))^2, \quad c_3(q_{13}(t)) = .5(q_{13}(t))^2.$$

The unit transaction costs associated with transacting between the power suppliers and the demand market are

$$\hat{c}_{11}(q_{11}(t)) = q_{11}(t) + 1, \quad \hat{c}_{21}(q_{21}(t)) = q_{21}(t) + 5, \quad \hat{c}_{31}(q_{31}(t)) = q_{31}(t) + 10.$$

In the corresponding supernetwork representation, there are 9 nodes and 10 links with 1 O/D pair denoted by $w_1 = (0, z_1)$. There are three paths connecting w_1

$$p_1 = (a_1, a_{11}, a_{11'}, a_{1'1}), \quad p_2 = (a_1, a_{12}, a_{22'}, a_{2'1}), \quad p_3 = (a_1, a_{13}, a_{33'}, a_{3'1}).$$

The time horizon $T = 1$. The time-varying demand function is given by $d_{w_1} = d_1(t) = 41 + 10t$.

Here we present the importance and the ranking of the individual links and nodes for this example using the dynamic network efficiency measure. We conduct the analysis in the context of the supernetwork, which represents the transportation network equilibrium transformation [see Nagurney et al. (2007)]. We then translate the results back to the first network.

Recall that (cf. Figure 7.8), as discussed in Nagurney et al. (2007), $c_{a_1} = \frac{\partial f_1}{\partial q_1}$, $c_{a_{gs}} = \frac{\partial c_{gs}}{\partial q_{gs}}$; for $gs = 11, 12, 13$; $c_{a_{ss'}} = \frac{\partial c_s}{\partial q_{1s}}$, for $s = 1, 2, 3$, and $c_{a_{s'1}} = \hat{c}_{s1}$, for $s = 1, 2, 3$. The cost on a path $p_s \equiv p_{1ss'1}$, for $s = 1, 2, 3$ joining node 0 to node z_1 in the supernetwork representation is

$$C_{p_s} \equiv C_{p_{1ss'1}} = c_{a_1} + c_{a_{1s}} + c_{a_{ss'}} + c_{a_{s'1}}, \quad s = 1, 2, 3.$$

Due to the separability of the cost functions and the special structure of the supernetwork of this example, we can obtain the explicit formulae for the equilibrium path flows and the disutility over time $[0, T]$, which are shown below [cf. Nagurney et al. (2007)]

$$x^*_{p_1}(t) = 3.33t + 14.78, \quad x^*_{p_2}(t) = 3.33t + 13.78, \quad x^*_{p_3}(t) = 3.33t + 12.45.$$

$$\lambda^*_{w_1}(t) = 60t + 255.83.$$

By a simple calculation, the network efficiency for the above electric power network example is $\mathcal{E}^{DN} = .1609$. The importance and the rankings of the links and the nodes are given, respectively, in Tables 7.7 and 7.8.

7.5.1 Discussion

In this example, we note that links a and d are the most important links. The power generator and the demand market are, in turn, the most important nodes, which is reasonable. Power Supplier 1 is ranked as the second most important node. By observing the equilibrium path flows, we know that path p_1, which includes links a and d and Power Supplier 1, carries the largest amount of flow. Therefore, given the fact that three paths have the same equilibrium disutility, the removal of link a or d will have the most severe impact on the electric power supply chain network efficiency. The removal of Power Supplier 1 will have the largest impact on the network efficiency only second to that of the removal of Power Generator 1 or the demand market. Furthermore, at the imposed demand, link b is more important than link c and link e is more important (i.e., is ranked higher) than link f.

Table 7.7: Importance and ranking of links in the electric power supply chain network

Link	Importance Value	Importance Ranking
a	.0725	1
b	.0689	2
c	.0628	3
d	.0725	1
e	.0689	2
f	.0628	3

Table 7.8: Importance and ranking of nodes in the electric power supply chain network

Node	Importance Value	Importance Ranking
Power Generator 1	1.0000	1
Power Supplier 1	.0735	2
Power Supplier 2	.0689	3
Power Supplier 3	.0628	4
Demand Market 1	1.0000	1

7.6 SUMMARY AND CONCLUSIONS

In this chapter, we proposed a network efficiency measure that can be used for dynamic networks, formulated as evolutionary variational inequalities, including the Internet. The network efficiency measure captures costs/latencies on networks with time-varying demands. We provided both a continuous time version of the efficiency measure and a discrete version and then showed that the dynamic network efficiency measure in the case of fixed demands for all the origin/destination pairs over time collapses to the proposed measure given by (3.2), when such problems are formulated as network equilibrium problems in a static setting. The novelty of the measure lies in that it enables us to assess the network performance without worrying about the connectivity assumption and further, it can be used to study the importance of network components under changing demands over the time horizon of interest. To the best of our knowledge, this is the first rigorous and well-defined, dynamic network efficiency measure. Such a tool, we expect, will be quite useful for security purposes because the dynamic network itself will be most vulnerable, as measured by the drop in efficiency, if the most important nodes (or links) are removed, due to, for example, structural failures, natural disasters, and terrorist attacks.

We illustrated the formalism on several dynamic network examples, including the time-dependent Braess paradox and a stylized electric power supply chain network example. From the results of the importance rankings of individual nodes and links, we can see that when evaluating the vulnerability of a network, all the relevant information regarding the demands, the flows, and the costs over time has to be taken into consideration.

Future research will include the application of the results in this chapter to large-scale dynamic networks in telecommunication applications as well as other applications, such as large-scale electric power supply chain networks [cf. Liu and Nagurney (2009)]. In particular, Liu and Nagurney (2009) constructed an empirical electric power supply chain network model with fuel supply markets that captures both the economic network transactions in energy supply chains and the physical network transmission constraints in the electric power network. The theoretical derivation and analyses were done using the theory of finite-dimensional variational inequalities. The theoretical model was then applied to the New England electric power supply chain consisting of 6 states, 5 fuel types, 82 power generators, with a total of 573 generating units and 10 demand points. The empirical case study demonstrated that the regional electric power prices simulated by the model very well matched the actual electricity prices in New England. To solve this large-scale model, the authors used the connections between electric power supply chain networks and transportation networks [cf. Chapter 2 and Nagurney et al. (2007)].

Finally, the proposed dynamic network measure can be used to study network robustness and reliability under uncertainty as was done in Chapter 5 for supply chain networks under disruptions.

7.7 SOURCES AND NOTES

Section 7.2 through 7.4 are based on the papers by Nagurney and Qiang (2008c) and Nagurney, Parkes, and Daniele (2007). Here we have expanded the motivation and have added more references. The electric power supply chain example in Section 7.5 is based on the paper by Nagurney et al. (2007). In Nagurney and Qiang (2008a), we studied a static electric power supply chain network, which we extended in Section 7.5, by considering a dynamic version.

PART III

MERGERS AND ACQUISITIONS, NETWORK INTEGRATION, AND SYNERGIES

CHAPTER 8

A SYSTEM-OPTIMIZATION PERSPECTIVE FOR SUPPLY CHAIN NETWORK INTEGRATION

8.1 INTRODUCTION

With this chapter, we begin Part III of this book, which focuses on potential synergies from network integration, in the form of mergers and acquisitions (M&As), associated with two (or more) firms (or organizations). The novelty of our perspective is in the depiction of firms as networks of their economic activities. The presentation and the analysis are done from a supply chain perspective but the concepts are even broader, for example, they may be applied to the integration of firms in a variety of network industries from telecommunications, transportation (railways and airlines), and financial services to energy, including electric power. Indeed, the originality of our approach lies in the conceptualization of the firms or organizations of interest as networks and their integration through M&As being formed through the addition of appropriate new links (with associated costs). Flows on the networks before or after the integration correspond, depending on the particular application, to products, people, messages, energy in a variety of forms, financial transactions, and so on. The flows may be multiproduct in nature, and the costs may be generalized total costs that also capture different criteria, such as risk and/or environmental impact. The latter

Fragile Networks. By Anna Nagurney and Qiang Qiang

we explore more fully in Chapter 9. We study multiproduct supply chains in Chapter 10.

In particular, in this chapter, we quantify the strategic advantages of supply chain network integration using a system-optimization (see Section 2.3) perspective. We use this perspective because it enables one to graphically represent, to formulate, and compute the strategic advantages associated with a variety of mergers. In this chapter, we assume that the firms that merge are in the same industry. Recent examples of such mergers include the merger of Kmart and Sears in the retail industry [see Knowledge@Wharton (2005)] and the merger of Molson and Coors in the beverage industry [cf. Beverage World (2007)], with Molson Coors setting a merger synergy goal of $175 million in annual savings.

Langabeer (2003) noted that the use of M&As had been growing exponentially, with over 6,000 M&A transactions conducted world-wide in 2001, with a value of over $1 trillion. Nevertheless, many scholars argue whether mergers achieve their objectives. For example, Marks and Mirvis (2001) found that fewer than 25% of all mergers achieve their stated objectives. Langabeer and Seifert (2003), in turn, determined a direct correlation between how effectively supply chains of merged firms are integrated and how successful the merger is. Furthermore, they stated, based on the empirical findings in Langabeer (2003), who analyzed hundreds of mergers over the preceding decade, that improving supply chain integration between merging companies is the key to improving the likelihood of post-merger success.

According to Kusstatscher and Cooper (2005), there have been five major waves of Merger and Acquisition activity

The First Wave: 1898 – 1902: consisted of an increase in horizontal mergers that resulted in many U.S. industrial groups,

The Second Wave: 1926 – 1939: involved many public utilities,

The Third Wave: 1969 – 1973: had as its driving force *diversification*,

The Fourth Wave: 1983 – 1986: had as its goal *efficiency*,

The Fifth Wave: 1997 until the present: had as its motto *globalization*.

In the present uncertain economic and financial climate, it is practically imperative to quantitatively assess a priori the potential cost savings associated with a proposed merger or acquisition. Toward that end, and, for definiteness, we envision a firm as a network of economic activities consisting of manufacturing, which is conducted at the firm's plants or manufacturing facilities; distribution, which occurs between the manufacturing plants and the distribution centers, which store the product; and the ultimate distribution of the product to the retailers. Associated with each such economic activity is a link in the network with a total associated cost that depends on the flow of the product on the link. The links, be they manufacturing, shipment, or storage links have capacities on the flows. We assume, as given, the demand for the product at each retailer. We relax this assumption when we formulate mergers among oligopolistic firms in Chapter 11.

Of course, in other network economic settings, the "manufacturers" may correspond to producers of services, knowledge, etc., as appropriate, and the nodes and links that abstract the firms and their activities would be re(defined) accordingly as would the link costs and network flows.

In this chapter, the use of a *system-optimization* perspective enables the modeling of the economic activities associated with a firm as a network and hence the evaluation of the strategic advantages, often referred to as *synergy*, due to mergers (or acquisitions), in a network format. However, unlike the classical system-optimization formulation, which was originally proposed in the context of transportation networks, here we explicitly consider capacities on the links of the networks. As noted by Soylu et al. (2006) more and more companies now realize the strategic importance of controlling the supply chain as a whole (see Brown et al. (2001) for a specific corporate example). Min and Zhou (2002) emphasized the need to analyze the synergy obtained through both interfunctional and interorganizational integration. Hakkinen et al. (2004) further described the integration of logistics after M&As with a review of the literature and concluded that operational issues, in general, and logistics issues, in particular, have received insufficient attention [see also Herd, Saksena, and Steger (2005)].

This chapter is organized as follows. In Section 8.2, we develop the system-optimization models, which are nonlinear programming problems with network structure, faced by the two firms whose merger we wish to evaluate. We focus here on horizontal mergers. We present both the pre-merger optimization problem, which we refer to as the baseline case, or Case 0, and the system-optimization problems associated with three distinct horizontal mergers:

Case 1: the firms merge, and retailers associated with either original firm can now, in principle, be supplied the product produced at any manufacturing plant but still use their original distribution centers,

Case 2: the firms merge, and the retailers can now obtain the product from any distribution center but the original firms' manufacturing plants deal with their original distribution centers,

Case 3: the firms merge, and the retailers can, in principle, now obtain the product produced at any of the manufacturing facilities and distributed by any of the distribution centers.

We represent the underlying associated networks before and after the mergers and demonstrate that the solution of all the associated system-optimization problems can be obtained by solving a variational inequality problem, with a structure that can be easily exploited for computational purposes. Notably, our framework incorporates manufacturing/production activities as well as distribution and storage activities both before and after the merger (or acquisition). Of course, whether a manufacturing plant or a distribution center is used before or after a merger depends on the optimal solution.

In Section 8.3, we present a measure of strategic advantage, which allows one to evaluate the gains, if any, associated with these horizontal mergers. In Section 8.4, we provide a spectrum of examples, in which we compute the strategic advantage associated with mergers in the case of many different scenarios. In Section 8.5, we summarize the results in this chapter and present our conclusions.

8.2 THE PRE- AND POST-HORIZONTAL MERGERS SUPPLY CHAIN NETWORK MODELS

In this section, we develop the supply chain network models before and after the horizontal mergers. We consider two firms, denoted by Firm A and Firm B, which are integrated after the merger. We assume that each firm produces the same homogeneous product because the focus here is on horizontal mergers in the same industry. The explicit horizontal mergers that we model and evaluate are

Case 1: Firms A and B merge and share manufacturing plants only,
Case 2: Firms A and B merge and share distributions centers only,
Case 3: Firms A and B merge and share manufacturing plants and distribution centers.

As mentioned in Section 8.1, the formalism that we use is that of *system-optimization*, by which each of the firms is assumed to own its manufacturing facilities/plants and distribution centers, and each firm seeks to determine the optimal production of the product at each of its manufacturing plants and the optimal quantities shipped to the distribution centers, where the product is stored and from which it is shipped to the retail outlets. We assume that each firm seeks to minimize the total costs associated with the production, storage and distribution activities, subject to the demand being satisfied at the retail outlets. Of course, an appropriate time horizon for the merger is assumed.

In Section 8.2.1, we formulate the pre-merger system-optimization problem associated with each of the firms individually and together before to the merger and we consider this as the baseline case, Case 0. In Section 8.2.2, we formulate the three post-merger models, corresponding to Case 1, Case 2, and Case 3, respectively.

8.2.1 The Pre-Merger Supply Chain Network Model(s)

The optimization problem faced by Firm A and Firm B is as follows. Each firm is represented as a network of its economic activities, as depicted in Figure 8.1. Each firm i; $i = A, B$, has n^i_M manufacturing facilities/plants; n^i_D distribution centers, and serves n^i_R retail outlets. Let $\mathcal{G}_i = [\mathcal{N}_i, \mathcal{L}_i]$ for $i = A, B$ denote the graph consisting of nodes and directed links representing the economic activities associated with each firm i. Let $\mathcal{G}^0 = [\mathcal{N}^0, \mathcal{L}^0] \equiv \cup_{i=A,B}[\mathcal{N}_i, \mathcal{L}_i]$. The links from the top-tiered nodes i; $i = A, B$ in each network in Figure 8.1 are connected to the manufacturing nodes of the respective firm i, which are denoted, respectively, by: $M^i_1, \ldots, M^i_{n^i_M}$, and these links represent the manufacturing links.

The links from the manufacturing nodes, in turn, are connected to the distribution center nodes of each firm i; $i = A, B$, which are denoted by $D^i_{1,1}, \ldots, D^i_{n_D{}^i,1}$. These links correspond to the shipment links between the manufacturing plants and the distribution centers where the product is stored. The links joining nodes $D^i_{1,1}, \ldots, D^i_{n^i_D,1}$ with nodes $D^i_{1,2}, \ldots, D^i_{n^i_D,2}$ for $i = A, B$ correspond to the storage links. Finally, there are shipment links joining the nodes $D^i_{1,2}, \ldots, D^i_{n^i_D,2}$ for $i = A, B$ with the retail outlet nodes: $R^i_1, \ldots, R^i_{n^i_R}$ for each firm $i = A, B$. Note that

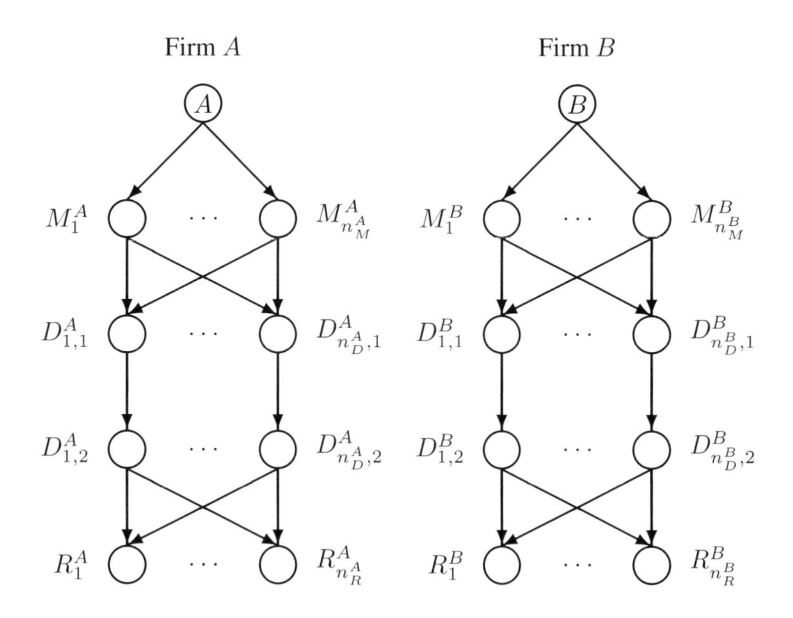

Figure 8.1: Case 0: Firms A and B before the horizontal merger

each firm i has its individual retail outlets where it sells the product, as depicted in Figure 8.1.

Assume that associated with each link (cf. Figure 8.1) of the network corresponding to each firm i; $i = A, B$ is a total cost. We denote, without any loss in generality, the links by a, b, etc., and the total cost on a link a by $\hat{c}_a$. The demands for the product are assumed as given and are associated with each firm and retailer pair. Let $d_{R_k^i}$ denote the demand for the product at retailer R_k^i associated with firm i; $i = A, B$; $k = 1, \ldots, n_R^i$. Let x_p denote the nonnegative flow of the product on path p joining (origin) node i with a (destination) retailer node of firm i; $i = A, B$. Then the following conservation of flow equations must hold for each firm i

$$d_{R_k^i} = \sum_{p \in P_{R_k^i}^0} x_p, \quad i = A, B; k = 1, \ldots, n_R^i, \tag{8.1}$$

where $P_{R_k^i}^0$ denotes the set of paths connecting (origin) node i with (destination) retail node R_k^i. In other words, the demand for the product at each retail outlet associated with each firm must be satisfied by the sum of the product flows from the firm to that retail outlet. Note that a path consists of a sequence of links representing the economic activities of manufacturing/production, distribution, storage, and final shipment to the retailer.

In addition, let f_a denote the flow of the product on link a. Hence the following conservation of flow equations must also hold

$$f_a = \sum_{p \in P^0} x_p \delta_{ap}, \quad \forall p \in P^0, \tag{8.2}$$

where $\delta_{ap} = 1$ if link a is contained in path p and $\delta_{ap} = 0$, otherwise, that is, the (product) flow on a link is equal to the sum of the (product) path flows on paths that contain that link. P^0 denotes the set of *all* paths in Figure 8.1, that is, $P^0 = \cup_{i=A,B;k=1,\ldots,n_R^i} P^0_{R_k^i}$. Because here we consider the two firms before any merger the paths associated with a given firm have no links in common with paths of the other firm. This changes when the horizontal mergers occur, in which case the sets of paths and the number of paths also change, as do the sets of links and the number of links, as we demonstrate in Section 8.2.2.

Of course, the path flows must be nonnegative, that is

$$x_p \geq 0, \quad \forall p \in P^0. \tag{8.3}$$

The total cost on a link, be it a manufacturing/production link, a shipment link, or a storage link is assumed to be a function of the flow of the product on the link. Hence we have

$$\hat{c}_a = \hat{c}_a(f_a), \quad \forall a \in \mathcal{L}^0. \tag{8.4}$$

The total cost on each link is assumed to be convex, continuously differentiable, and with a bounded third order partial derivative. The same is assumed to hold for all links that are added post the merger. Such conditions will guarantee convergence of the proposed algorithm.

There are positive capacities on the links with the capacity on link a denoted by u_a, for all links $a \in \mathcal{L}^0$. This is very reasonable because the manufacturing plants, the shipment links, and the distribution centers, which serve also as the storage facilities, can be expected to have capacities, in practice.

The total cost associated with the economic activities of both firms before the merger is minimized when the following system-optimization problem is solved

$$\text{Minimize} \quad \sum_{a \in \mathcal{L}^0} \hat{c}_a(f_a) \tag{8.5}$$

subject to: constraints (8.1) – (8.3) and

$$f_a \leq u_a, \quad \forall a \in \mathcal{L}^0. \tag{8.6}$$

The solution of the above optimization problem will minimize the total costs associated with each firm individually and both firms together because they are, before the merger, independent and share no manufacturing facilities or distribution facilities or retail outlets. Observe that this problem is, as is well-known in the literature, a system-optimization problem (see also Section 2.3) but in *capacitated* form. Under the above imposed assumptions, the optimization problem is a convex optimization problem. If we further assume that the feasible set underlying the problem represented by the constraints (8.1) through (8.3) and (8.6) is non-empty, then it follows from the standard theory of nonlinear programming (cf. Bazaraa, Sherali, and Shetty (1993) and the appendix) that the optimal solution, denoted by $f^* \equiv \{f_a^*\}, a \in \mathcal{L}^0$, exists. If the total cost functions (8.4) are strictly convex, then this link flow solution is unique.

Let $\mathcal{K}^0$ denote the feasible set: $\mathcal{K}^0 \equiv \{f | \exists x \geq 0, \text{and}\,(8.1) - (8.3), (8.6)\,\text{hold}\}$, and let f denote the vector of link flows and x the vector of path flows.

Associated with constraint (8.6) for each link a is the Lagrange multiplier β_a, with the optimal Lagrange multiplier denoted by β_a^*. This term may be interpreted as the price or value of an additional unit of capacity on link a for each $a \in \mathcal{L}^0$.

The variational inequality formulation of the problem is given in Theorem 8.1.

Theorem 8.1: Variational Inequality Formulation of the Supply Chain Network Pre-Integration Problem

The vector of link flows $f^* \in \mathcal{K}^0$ is an optimal solution to problem (8.5), subject to (8.1) through (8.3) and (8.6), if and only if it satisfies the following variational inequality problem with the vector of optimal nonnegative Lagrange multipliers β^*

$$\sum_{a \in \mathcal{L}^0} \left[\frac{\partial \hat{c}_a(f_a^*)}{\partial f_a} + \beta_a^* \right] \times [f_a - f_a^*] + \sum_{a \in \mathcal{L}^0} [u_a - f_a^*] \times [\beta_a - \beta_a^*] \geq 0, \quad \forall f \in \mathcal{K}^0,$$

$$\forall \beta_a \geq 0, \forall a \in \mathcal{L}^0. \tag{8.7}$$

Proof: See Bertsekas and Tsitsiklis (1989) and Nagurney (1999).

This variational inequality problem can be easily put into standard form (2.19). Indeed, we define $X \equiv (f^T, \beta^T)^T$, where f and β are the vectors of, respectively, link flows and Lagrange multipliers for links $a \in \mathcal{L}^0$. Also, the function $F(X)$ that enters the variational inequality problem (2.19) consists of the vector of terms $\frac{\partial \hat{c}_a(f_a)}{\partial f_a} + \beta_a$ for all links $a \in \mathcal{L}^0$ and the vector of terms $u_a - f_a$ for all links $a \in \mathcal{L}^0$. The feasible set $\mathcal{K}$ in the standard variational inequality formulation is then $\mathcal{K} \equiv \{(f, \beta) | f \in \mathcal{K}^0 \,\text{and}\, \beta_a \geq 0, \forall a \in \mathcal{L}^0\}$.

This variational inequality can be easily solved using the modified projection method (also sometimes referred to as the extragradient method); refer to Section 2.4.3. The elegance of this computational procedure in the context of variational inequality (8.7) lies in that it allows one to utilize algorithms for the solution of the *uncapacitated* system-optimization problem (for which numerous algorithms exist in the transportation science literature) with a straightforward update procedure at each iteration to obtain the Lagrange multipliers. To solve the former problem we use in Section 8.4 the well-known U-O equilibration algorithm (see Section 2.4.1.3) and embed it in the modified projection method. Indeed, we will see that the variational inequalities governing the supply chain networks post-mergers will also be of the form (8.7) and hence amenable to solution via the modified projection method. The modified projection method is guaranteed to converge to a solution of a variational inequality problem, provided that the function that enters the variational inequality problem is monotone and Lipschitz continuous and that a solution exists (cf. Theorem 2.12).

Once we have solved problem (8.7) we have the solution f^* which minimizes the total cost [cf. (8.5)] in the supply chain networks associated with the two firms. We denote this total cost given by $\sum_{a \in \mathcal{L}^0} \hat{c}_a(f_a^*)$ as TC^0, and we use this total cost value as a baseline from which to compute the strategic advantage or synergy, discussed in Section 8.3, associated with horizontal mergers that we describe next.

8.2.2 The Horizontal Merger Supply Chain Network Models

In this section, we consider three post-merger cases. In Case 1, the firms merge and retailers associated with either original firm can now get the product from any manufacturing plant but still use their original distribution centers. In Case 2, the firms merge and the retailers can obtain the product from any distribution center but the manufacturers deal with their original distribution centers, and in Case 3, the firms merge and the retailers can obtain the product produced at any of the manufacturing facilities and distributed by any of the distribution centers. Case 1 is depicted graphically in Figure 8.2; Case 2, in Figure 8.3; and Case 3, in Figure 8.4.

8.2.2.1 Case 1 In Case 1, we add to the network $\mathcal{G}^0$ depicted in Figure 8.1 a supersource node 0 and links joining node 0 to nodes $i = A, B$ to reflect the merger of the two firms. We also add new links joining each manufacturing node of each firm with the distribution center nodes of the other firm, as depicted in Figure 8.2. We denote the new network topology in Figure 8.2 by $\mathcal{G}^1 = [\mathcal{N}^1, \mathcal{L}^1]$ where $\mathcal{N}^1 = \mathcal{N}^0 \cup$ node 0 and $\mathcal{L}^1 = \mathcal{L}^0 \cup$ the additional links.

The total cost functions associated with the new links corresponding to the merger are of the form (8.4). If one wishes to focus primarily on the operational aspects of the merger, associated with the production, storage, and transportation/logistics of the product, one may wish to assign zero costs to the links emanating from the supersource node 0. On the other hand, if one wishes to include the costs associated with combining other resources, including personnel, paying legal fees, and integrating information technology, the total cost functions on the topmost links in Figure 8.2 should reflect those costs.

Let x_p, without loss of generality, now denote the flow of the product on path p joining (origin) node 0 with a (destination) retailer node in Figure 8.2. Then the following conservation of flow equations must hold

$$d_{R_k^i} = \sum_{p \in P^1_{R_k^i}} x_p, \quad i = A, B; k = 1, \ldots, n_R^i, \tag{8.8}$$

where $P^1_{R_k^i}$ denotes the set of paths connecting node 0 with retail node R_k^i. Due to the merger, the retail outlets can obtain the product from any manufacturer. The set $P^1 \equiv \cup_{i=A,B;k=1,\ldots,n_R^i} P^1_{R_k^i}$.

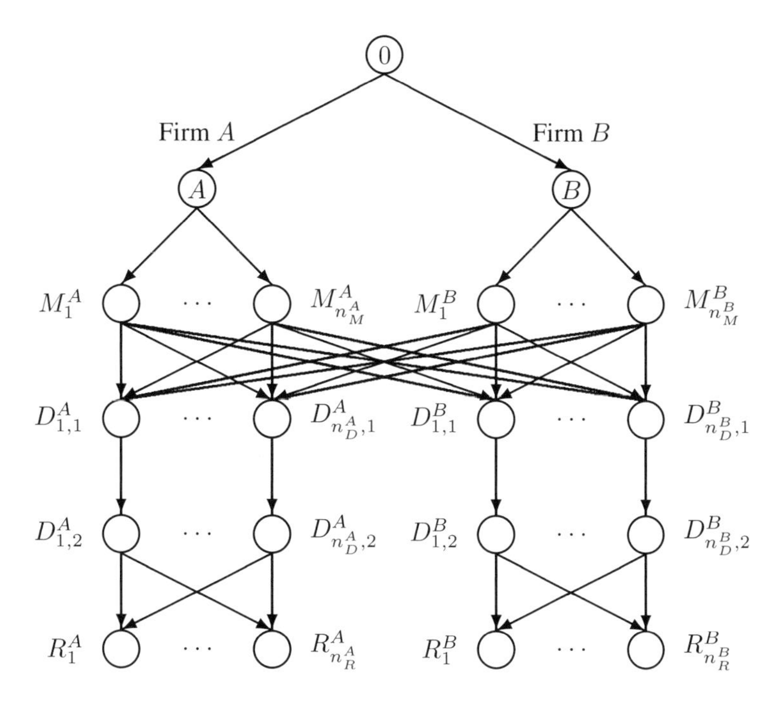

Figure 8.2: Case 1 merger

In addition, as before, let f_a denote the flow of the product on link a. Hence the following conservation of flow equations must be satisfied

$$f_a = \sum_{p \in P^1} x_p \delta_{ap}, \quad \forall p \in P^1. \tag{8.9}$$

Of course, the path flows must be nonnegative, that is

$$x_p \geq 0, \quad \forall p \in P^1. \tag{8.10}$$

The optimization problem associated with this horizontal merger, which minimizes the total cost of the network in Figure 8.2, is given by

$$\text{Minimize} \quad \sum_{a \in \mathcal{L}^1} \hat{c}_a(f_a) \tag{8.11}$$

subject to: constraints (8.8) – (8.10) and

$$f_a \leq u_a, \quad \forall a \in \mathcal{L}^1. \tag{8.12}$$

Clearly, the solution to this problem can also be obtained as a solution to a variational inequality problem akin to (8.7), where $a \in \mathcal{L}^1$ and the vectors f, f^*, x, β, and β^* have identical definitions, as before, but are redimensioned accordingly.

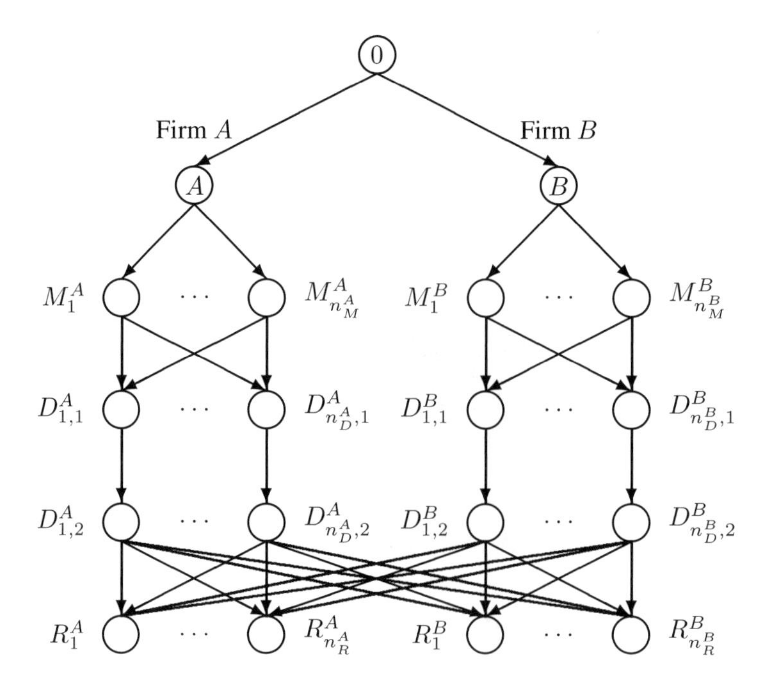

Figure 8.3: Case 2 merger

The feasible set $\mathcal{K}^0$ is replaced by the new feasible set $\mathcal{K}^1 \equiv \{f | \exists x \geq 0, \text{ and } (8.8) - (8.10) \text{ and } (8.12) \text{ hold}\}$. One can apply the modified projection problem to compute the solution to the variational inequality problem governing Case 1, as well. The optimal solution for Case 1 has an associated total cost given by $\sum_{a \in \mathcal{L}^1} \hat{c}_a(f_a^*)$, which we denote by TC^1, with f^* corresponding to the optimal link flow solution to this horizontal merger.

***8.2.2.2 Case* 2** We now formulate the merger associated with Case 2 in which Firms A and B merge and the retailers can obtain the product from any distribution center but the manufacturers deal with their original distribution centers. Figure 8.3 depicts the network topology associated with this type of horizontal merger. Specifically, to network $\mathcal{G}^0$ depicted in Figure 8.1, we add a supersource node 0 and links joining node 0 with nodes $i = A, B$ to reflect the merger (as we did for Case 1). Also, we add links connecting the distribution centers of each firm to the (other) retailers. The network underlying this merger is $\mathcal{G}^2 = [\mathcal{N}^2, \mathcal{L}^2]$, where $\mathcal{N}^2 = \mathcal{N}^1$. Associated with the new links are total cost functions as in (8.4).

Let x_p now denote the flow of the product on path p joining (origin) node 0 with a (destination) retailer node as in Figure 8.3. Then the following conservation of flow equations must hold

$$d_{R_k^i} = \sum_{p \in P^2_{R_k^i}} x_p, \quad i = A, B; k = 1, \ldots, n_R^i, \tag{8.13}$$

where $P^2_{R^i_k}$ denotes the set of paths connecting node 0 with retail node R^i_k in Figure 8.3. Due to the merger, the retail outlets can, in effect, now obtain the product from any manufacturer and any distributor. The set $P^2 \equiv \cup_{i=A,B;k=1,\ldots,n^i_R} P^2_{R^i_k}$.

In addition, as before, let f_a denote the flow of the product on link a. Hence we must now have the following conservation of flow equations satisfied

$$f_a = \sum_{p \in P^2} x_p \delta_{ap}, \quad \forall p \in P^2. \tag{8.14}$$

Of course, the path flows must be nonnegative, that is

$$x_p \geq 0, \quad \forall p \in P^2. \tag{8.15}$$

The optimization problem associated with this horizontal merger is given by

$$\text{Minimize} \quad \sum_{a \in \mathcal{L}^2} \hat{c}_a(f_a) \tag{8.16}$$

subject to: constraints (8.13) – (8.15) and

$$f_a \leq u_a, \quad \forall a \in \mathcal{L}^2. \tag{8.17}$$

The solution to this problem can also be obtained as a solution to a variational inequality problem akin to (8.7), where $a \in \mathcal{L}^2$, and the vectors: f, f^*, x, β, and β^* have the same definitions, as before, but are redimensioned accordingly. Instead of the feasible set $\mathcal{K}^1$ we now have the new feasible set $\mathcal{K}^2 \equiv \{f | \exists x \geq 0, \text{ and } (8.13) - (8.15) \text{ and } (8.17) \text{ hold}\}$. One can also apply the modified projection problem to compute the solution to the variational inequality problem governing Case 2. The total cost TC^2, which is the value of the objective function (8.16) at its optimal solution f^*, is equal to $\sum_{a \in \mathcal{L}^2} \hat{c}_a(f^*_a)$.

8.2.2.3 *Case* 3 We now formulate the merger associated with Case 3 in which Firms A and B merge and the retailers can obtain the product from any manufacturer and shipped from any distribution center. Figure 8.4 depicts the network topology associated with this type of horizontal merger. Specifically, we retain the nodes and links associated with network $\mathcal{G}^1$ depicted in Figure 8.2 but now the additional links connecting the distribution centers of each firm to the retailers of the other are added. We refer to the network underlying this merger as $\mathcal{G}^3 = [\mathcal{N}^3, \mathcal{L}^3]$ where $\mathcal{N}^3 = \mathcal{N}^1$.

Let x_p, without loss of generality, denote the flow of the product on path p joining (origin) node 0 with a (destination) retailer node as in Figure 8.4. Then the following conservation of flow equations must hold

$$d_{R^i_k} = \sum_{p \in P^3_{R^i_k}} x_p, \quad i = A, B; k = 1, \ldots, n^i_R, \tag{8.18}$$

where $P^3_{R^i_k}$ denotes the set of paths connecting node 0 with retail node R^i_k in Figure 8.4. The set of paths $P^3 \equiv \cup_{i=A,B;k=1,\ldots,n^i_R} P^3_{R^i_k}$.

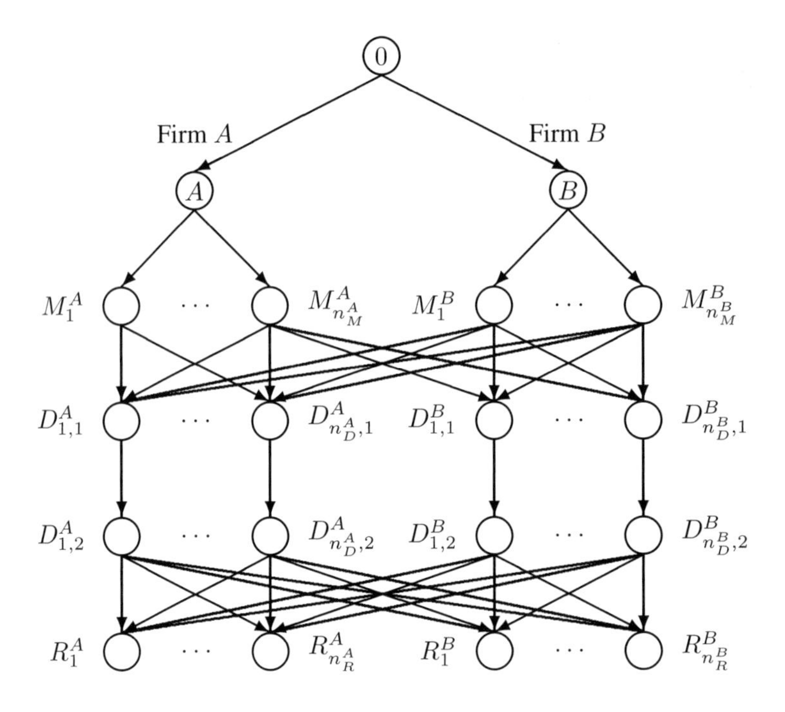

Figure 8.4: Case 3 merger

In addition, as before, let f_a denote the flow of the product on link a. Hence the following conservation of flow equations must now be satisfied

$$f_a = \sum_{p \in P^3} x_p \delta_{ap}, \quad \forall p \in P^3. \tag{8.19}$$

These path flows must also be nonnegative since they represent products, that is

$$x_p \geq 0, \quad \forall p \in P^3. \tag{8.20}$$

The optimization problem associated with this horizontal merger, which minimizes the total cost of the network in Figure 8.4, is

$$\text{Minimize} \quad \sum_{a \in \mathcal{L}^3} \hat{c}_a(f_a) \tag{8.21}$$

subject to: constraints (8.18) – (8.20) and

$$f_a \leq u_a, \quad \forall a \in \mathcal{L}^3. \tag{8.22}$$

The solution to this problem can also be obtained as a solution to a variational inequality problem akin to (8.7), where $a \in \mathcal{L}^3$, and the variable and solution vectors are as defined before but with expanded dimensions. Finally, instead of $\mathcal{K}^1$ we now have $\mathcal{K}^3 \equiv \{f | \exists x \geq 0, \text{ and } (8.18) - (8.20) \text{ and } (8.22) \text{ hold}\}$. One can also apply the modified projection problem to compute the solution to the variational inequality problem governing Case 3. The total cost TC^3, which is the value of the objective function (8.21) evaluated at its optimal solution f^*, is equal to $\sum_{a \in \mathcal{L}^3} \hat{c}_a(f_a^*)$. In the next section, we discuss how we use the total costs: TC^0, TC^1, TC^2, and TC^3, to determine the strategic advantage (or synergy) associated with the respective horizontal mergers associated with Cases 1, 2, and 3.

8.3 MEASURING THE STRATEGIC ADVANTAGE ASSOCIATED WITH HORIZONTAL MERGERS

In this section, we provide a measure for quantifying the strategic advantage or *synergy* associated with horizontal mergers. The synergy measure that we utilize to capture the gains, if any, associated with a horizontal merger Case i; $i = 1, 2, 3$ and denoted by $\mathcal{S}^i$; $i = 1, 2, 3$ is as follows

$$\mathcal{S}^i = \left[\frac{TC^0 - TC^i}{TC^0} \right] \times 100\%, \tag{8.23}$$

where TC^i is the total cost associated with the value of the objective function $\sum_{a \in \mathcal{L}^i} \hat{c}_a(f_a)$ for $i = 0, 1, 2, 3$ evaluated at the optimal solution for Case i. Note that $\mathcal{S}^i$; $i = 1, 2, 3$ may also be interpreted as *synergy*.

In the case of simple, stylized examples one may be able to derive explicit formulae for $\mathcal{S}^i$. For example, if both Firms A and B pre-merger have a single manufacturing plant, a single distribution center, and a single retailer, and identical demands at the retailers, given by d, and assuming that the total costs on each link $a \in \mathcal{L}^0$ are given by $\hat{c}_a(f_a) = gf_a^2 + hf_a$, with $g > 0$ and $h > 0$, and the capacities on the links are not less than the demand d then it is straightforward to determine (cf. Figure 8.1) that: $TC^0 = 8[gd^2 + hd]$. Assume now that new links are added to construct Case 1, Case 2, and Case 3 accordingly, where we assume that the total costs on the new links are all identically equal to zero and their capacities are greater than or equal to the demand d. Then, since the addition of the new zero cost links creates new paths and new S-O flow solutions, we obtain $TC^1 = TC^2 = 6[gd^2 + hd]$ and $TC^3 = 4[gd^2 + hd]$. It follows that

$$\mathcal{S}^1 = \mathcal{S}^2 = 25\%, \quad \mathcal{S}^3 = 50\%.$$

A slightly more general case would be as above but assume now that the manufacturing link, denoted by $a \in \mathcal{L}^0$, of either firm has an identical cost of the form $g_a f_a^2 + h_a f_a$; the first shipment link b of either firm has a total cost of the form $g_b f_b^2 + h_b f_b$; the storage link, denoted, for simplicity, by c, of either firm has a total cost of the form $g_c f_c^2 + h_c f_c$, and, finally, the total cost associated with each bottom shipment link has a total cost given by $g_d f_d^2 + h_d f_d$, where we assume that g_a, g_b, g_c, and $g_d > 0$ and h_a, h_b, h_c, and $h_d > 0$. Then, one can easily derive the following total cost formulae from which the strategic advantages can be determined according to (8.23), assuming that, as discussed earlier, the total costs associated with the new links associated with the respective mergers are all identically equal to zero

$$TC^0 = 2[g_a d^2 + h_a d] + 2[g_b d^2 + h_b d] + 2[g_c d^2 + h_c d] + 2[g_d d^2 + h_d d],$$

$$TC^1 = 2[g_a d^2 + h_a d] + 2[g_c d^2 + h_c d] + 2[g_d d^2 + h_d d],$$

$$TC^2 = 2[g_a d^2 + h_a d] + 2[g_b d^2 + h_b d] + 2[g_c d^2 + h_c d],$$

$$TC^3 = 2[g_a d^2 + h_a d] + 2[g_c d^2 + h_c d].$$

8.4 NUMERICAL EXAMPLES

In this section, we present five numerical examples for which we compute the strategic advantage measure as in (8.23) for the different cases.

We consider Firm A and Firm B, each of which has two manufacturing plants: M_1^i and M_2^i; $i = A, B$. In addition, each firm has a single distribution center to which the product is shipped from the manufacturing plants and stored. Finally, once stored, the product is shipped to the two retailers associated with each firm and denoted by R_1^i and R_2^i for $i = A, B$. A graphical depiction of the supply chain networks associated with the two firms pre-merger and representing Case 0 is given in Figure 8.5. Figure 8.6 depicts the Case 1 horizontal merger; Figure 8.7 depicts the Case 2 horizontal merger, and Figure 8.8 depicts the Case 3 horizontal merger of these two firms.

We used the modified projection method, embedded with the equilibration algorithm, as discussed in Section 8.2, to compute the solutions to the problems. We implemented the algorithm in Fortran and used a Unix system at the University of Massachusetts for the computations. The algorithm was considered to have converged to the solution when the path flows at the particular and preceding iteration differed by no more than $\epsilon = 10^{-5}$ and the same held for the Lagrange multipliers.

In Table 8.1, we define the links on the various networks, and the total link cost functions associated with the various supply chain activities of manufacturing, shipping/distribution, and storage. The merger links (emanating from node 0) are assumed to have associated total costs of zero.

The capacities on all the links in all the examples [see (8.6), (8.12), (8.17), and (8.22)] were set to 15 for all the links. The demands at the retailers, except when noted, were $d_{R_1^A} = 5$, $d_{R_2^A} = 5$, and $d_{R_1^B} = 5$, $d_{R_2^B} = 5$.

Example 8.1

Example 8.1 served as the baseline example. The total cost functions for the links are reported in Table 8.1 with the total costs associated with the optimal solutions of each of the three merger cases given in Table 8.2. Note that the strategic advantage or synergy approximately doubled when retailers could obtain the product from any manufacturer and from any distribution center (Case 3) relative to Cases 1 and 2, rising from 7.5% to 15.1%.

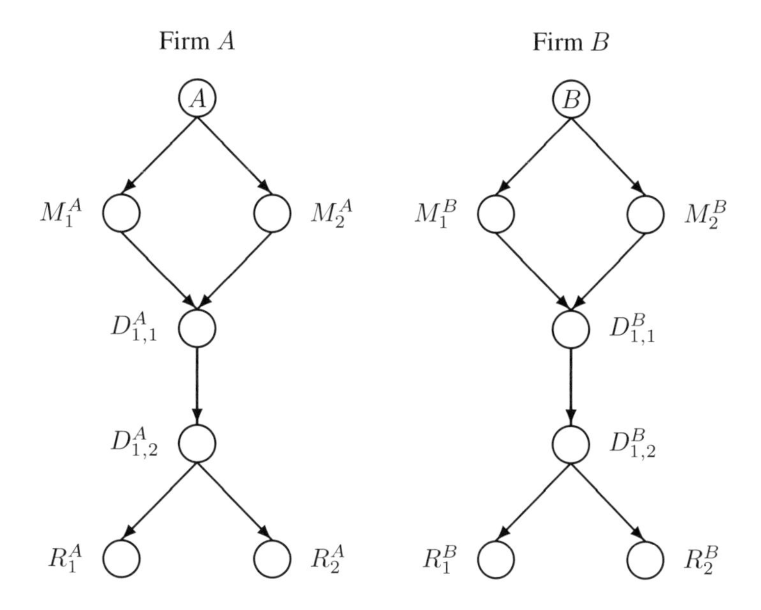

Figure 8.5: Case 0 Network topology for the numerical examples

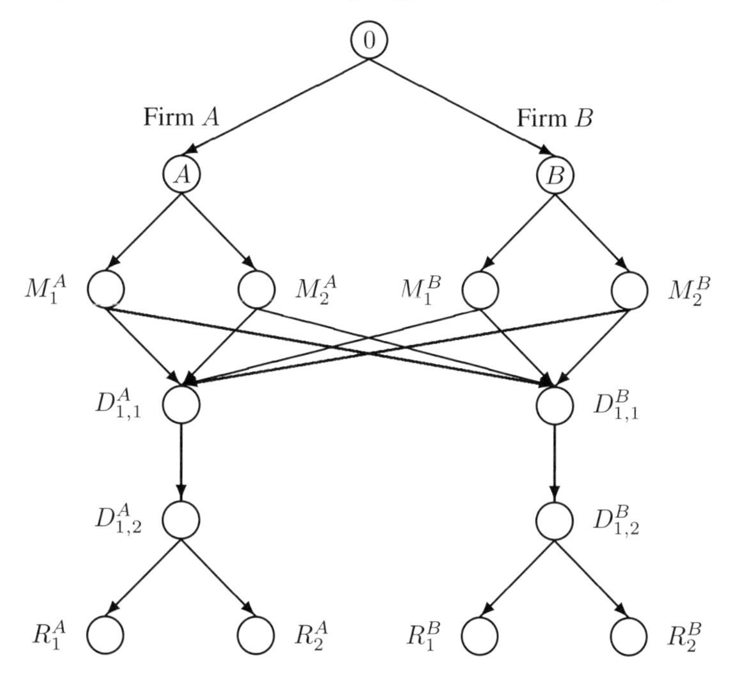

Figure 8.6: Case 1 Network topology for the numerical examples

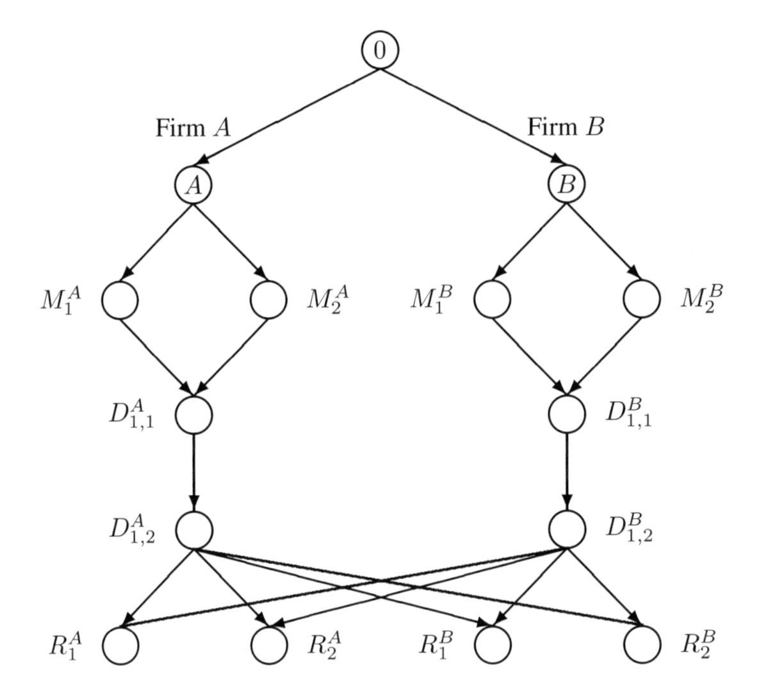

Figure 8.7: Case 2 Network topology for the numerical examples

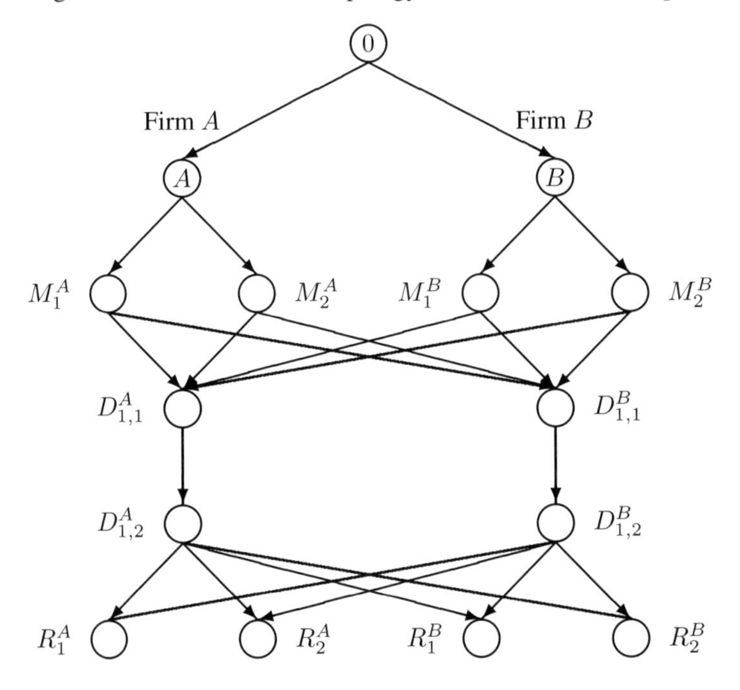

Figure 8.8: Case 3 Network topology for the numerical examples

Table 8.1: Definition of links and associated total cost functions for the numerical examples

Link a	From Node	To Node	Example 8.1	Example 8.2	Examples 8.3, 8.4	Example 8.5
1	A	M_1^A	$f_1^2+2f_1$	$f_1^2+2f_1$	$.5f_1^2+f_1$	$.5f_1^2+f_1$
2	A	M_2^A	$f_2^2+2f_2$	$f_2^2+2f_2$	$.5f_2^2+f_2$	$.5f_2^2+f_2$
3	M_1^A	$D_{1,1}^A$	$f_3^2+2f_3$	$f_3^2+2f_3$	$f_3^2+2f_3$	$f_3^2+2f_3$
4	M_2^A	$D_{1,1}^A$	$f_4^2+2f_4$	$f_4^2+2f_4$	$f_4^2+2f_4$	$f_4^2+2f_4$
5	$D_{1,1}^A$	$D_{1,2}^A$	$f_5^2+2f_5$	$.5f_5^2+f_5$	$.5f_5^2+f_5$	$.5f_5^2+f_5$
6	$D_{1,2}^A$	R_1^A	$f_6^2+2f_6$	$f_6^2+2f_6$	$f_6^2+2f_6$	$f_6^2+2f_6$
7	$D_{1,2}^A$	R_2^A	$f_7^2+2f_7$	$f_7^2+2f_7$	$f_7^2+2f_7$	$f_7^2+2f_7$
8	B	M_1^B	$f_8^2+2f_8$	$f_8^2+2f_8$	$f_8^2+2f_8$	$f_8^2+2f_8$
9	B	M_2^B	$f_9^2+2f_9$	$f_9^2+2f_9$	$f_9^2+2f_9$	$f_9^2+2f_9$
10	M_1^B	$D_{1,1}^B$	$f_{10}^2+2f_{10}$	$f_{10}^2+2f_{10}$	$f_{10}^2+2f_{10}$	$f_{10}^2+2f_{10}$
11	M_2^B	$D_{1,1}^B$	$f_{11}^2+2f_{11}$	$f_{11}^2+2f_{11}$	$f_{11}^2+2f_{11}$	$f_{11}^2+2f_{11}$
12	$D_{1,1}^B$	$D_{1,2}^B$	$f_{12}^2+2f_{12}$	$.5f_{12}^2+f_{12}$	$.5f_{11}^2+f_{11}$	$.5f_{11}^2+f_{11}$
13	$D_{1,2}^B$	R_1^B	$f_{13}^2+2f_{13}$	$f_{13}^2+2f_{13}$	$f_{13}^2+2f_{13}$	$f_{13}^2+2f_{13}$
14	$D_{1,2}^B$	R_2^B	$f_{14}^2+2f_{14}$	$f_{14}^2+2f_{14}$	$f_{14}^2+2f_{14}$	$f_{14}^2+2f_{14}$
15	M_1^A	$D_{1,1}^B$	$f_{15}^2+2f_{15}$	$f_{15}^2+2f_{15}$	$f_{15}^2+2f_{15}$	.00
16	M_2^A	$D_{1,1}^B$	$f_{16}^2+2f_{16}$	$f_{16}^2+2f_{16}$	$f_{16}^2+2f_{16}$	.00
17	M_1^B	$D_{1,1}^A$	$f_{17}^2+2f_{17}$	$f_{17}^2+2f_{17}$	$f_{17}^2+2f_{17}$	.00
18	M_2^B	$D_{1,1}^A$	$f_{18}^2+2f_{18}$	$f_{18}^2+2f_{18}$	$f_{18}^2+2f_{18}$	.00
19	$D_{1,2}^A$	R_1^B	$f_{19}^2+2f_{19}$	$f_{19}^2+2f_{19}$	$f_{19}^2+2f_{19}$	.00
20	$D_{1,2}^A$	R_2^B	$f_{20}^2+2f_{20}$	$f_{20}^2+2f_{20}$	$f_{20}^2+2f_{20}$	.00
21	$D_{1,2}^B$	R_1^A	$f_{21}^2+2f_{21}$	$f_{21}^2+2f_{21}$	$f_{21}^2+2f_{21}$	.00
22	$D_{1,2}^B$	R_2^A	$f_{22}^2+2f_{22}$	$f_{22}^2+2f_{22}$	$f_2^2+2f_{22}$	.00
23	0	A	.00	.00	.00	.00
24	0	B	.00	.00	.00	.00

Example 8.2

Example 8.2 was constructed from Example 8.1 as follows. The demands were the same as in Example 8.1, as were the capacities and total cost functions for all links, except for the total cost functions associated with the storage links, link 5 and link 12, representing the total costs associated with storing the product at the distribution centers associated with Firm A and Firm B, respectively. Rather than having these total costs be given by $\hat{c}_5(f_5) = f_5^2 + 2f_5$ and $\hat{c}_{12}(f_{12}) = f_{12} + 2f_{12}$ as they were in Example 8.1, they were now reduced to $\hat{c}_5(f_5) = .5f_5^2 + f_5$ and $\hat{c}_{12}(f_{12}) = .5f_{12} + f_{12}$, as reported in Table 8.1. The strategic advantage or synergy associated with all three horizontal mergers now increased, as reported in Table 8.2.

Example 8.3

Example 8.3 was constructed from Example 8.2 and had the same data except that now we reduced the total cost associated with the manufacturing plants belonging to Firm A as given in Table 8.1. Specifically, we changed $\hat{c}_1(f_1) = f_1^2 + 2f_1$ and

Table 8.2: Total costs and synergy of the different merger cases for the numerical examples

Example	8.1	8.2	8.3	8.4	8.5
TC^0	660.00	540.00	505.00	766.25	766.25
TC^1	610.00	490.00	447.80	687.92	573.80
$\mathcal{S}^1$	7.5%	9.2%	11.3%	10.2%	26.5%
TC^2	610.00	490.00	447.80	675.60	549.73
$\mathcal{S}^2$	7.5%	9.2%	11.3%	13.3%	28.2%
TC^3	560.00	432.00	389.80	581.30	320.20
$\mathcal{S}^3$	15.1%	20%	20.8%	24.1%	57.5%

$\hat{c}_2(f_2) = f_2 + 2f_2$ to: $\hat{c}_1(f_1) = .5f_1^2 + f_1$ and $\hat{c}_2(f_2) = .5f_2 + f_2$. The computed synergy or strategic advantage for each of the three horizontal mergers is given in Table 8.2. These values were greater than the respective ones for Example 8.2, although not substantially so.

Example 8.4
Example 8.4 was identical to Example 8.3 except that now the demand $d_{R_1^A} = 10$, that is, the demand for the product doubled at the first retailer associated with Firm A. The total costs and the strategic advantages for the different horizontal merger cases are given in Table 8.2. Note that now the synergies associated with Case 1 and Case 2 mergers were lower than those obtained for Example 8.3, suggesting that if the manufacturing costs of one plant are much higher than the other than these types of mergers are not as beneficial. However, the Case 3 merger yielded a strategic advantage of 24.1%, which was higher than that obtained for this case in Example 8.3.

Example 8.5
Example 8.5 was constructed from Example 8.4, and here we considered an idealized version in that the total cost functions (cf. Table 8.1) associated with shipment links that are added after the respective horizontal mergers are all equal to zero. The strategic advantages or synergies were now quite significant, as Table 8.2 reveals. Indeed, the strategic advantage for Case 3 was 57.5%. This example demonstrates that significant cost reductions can occur in mergers in which the costs associated with distribution between the associated manufacturing plants and distribution centers and the distribution centers and retailers are very low.

8.5 SUMMARY AND CONCLUSIONS

In this chapter, we presented a novel system-optimization approach for the representation of economic activities associated with supply chain networks, in particular, manufacturing, distribution, and storage, which we then used to quantify the strategic advantages or gains, if any, associated with the integration of supply chain networks through the horizontal merger of firms. However, unlike the classical system-optimization model, which we recalled in Chapter 2, here we presented a model with capacities on the links to represent the capacities associated with manufacturing plants, shipment/distribution routes, and storage facilities. A given firm's economic activities may be cast into this network economic form with total costs associated with the links and the demands at the retailers assumed known and given. A particular firm was assumed to be interested in determining its optimal production quantity of the product, the amounts to be shipped to its distribution centers, where the product is stored and then shipped to the retailers, to satisfy the demand and to minimize the total associated costs.

We established that the system-optimization model with capacities could be formulated and solved as a variational inequality problem. We then used this framework to explore the strategic advantages that could be obtained from the integration of supply chain networks through distinct horizontal mergers of two firms and we identified three distinct cases of horizontal mergers. In particular, we presented a measure for strategic advantage or synergy and then computed the strategic advantage for different cases of horizontal mergers for five numerical examples. In addition, in the case of certain supply chain networks with special structure, we were able to obtain explicit formulae for the total costs associated with different horizontal mergers and the associated strategic advantages.

The novelty of the framework lies in the graphical depiction of distinct types of mergers, with a focus on the involved firms' supply chain networks, and the efficient and effective computation of the total costs before and after the mergers, coupled with the determination of the associated strategic advantage or synergy of the particular merger. One can, within this framework, conduct sensitivity analyses to evaluate the synergy in the case of ranges of demand, ranges of total cost functions on the new links, etc. To capture the availability of different modes for transportation/shipment within this framework one only needs to add the additional corresponding links to the supply chain network model(s).

8.6 SOURCES AND NOTES

Some of the history of the evolution of network models for firms and numerous network-based economic models can be found in the books by Nagurney (1999, 2003, 2006a), and Nagurney and Dong (2002a). Notably, a variety of supply chain network equilibrium models, initiated by Nagurney, Dong, and Zhang (2002), have been developed that focus on competition among decision-makers (such as manufacturers, distributors, and retailers) at a tier of the supply chain but with cooperation between tiers. The relationships of such supply chain network equilibrium problems to transportation network equilibrium problems, which are characterized by user-optimizing behavior, have also been established [cf. Nagurney (2006b)]. Zhang, Dong, and Nagurney (2003) and Zhang (2006), in turn, modeled competition among supply chains in a supply chain economy context. See Nagurney (2006a) for a spectrum of supply chain network equilibrium models and applications.

Synergy in supply chains has been considered based on mixed integer linear programming models by Soylu et al. (2006), who focused on energy systems, and by Xu (2007), who developed multiperiod supply chain planning models with an emphasis on the distribution aspects, and investigated market regions, distribution configurations, as well as product characteristics and planning horizons. Juga (1996) earlier considered network synergy in logistics with a specific case study but did not present any mathematical models. Nijkamp and Reggiani (1998), in turn, investigated network synergies with links as productive factors. Gupta and Gerchak (2002) presented a model to capture operational efficiency and focused on production efficiency associated with mergers and acquisitions whereas Alptekinoglu and Tang (2005), subsequently, focused on distribution efficiency associated with mergers and acquisitions. The latter authors also established that their model was a convex nonlinear programming problem.

This chapter is based on the paper by Nagurney (2009), which introduced a system-optimization perspective for supply chain network integration in the context of horizontal mergers. However, in this chapter, we no longer assume that the total cost functions associated with the top-most merger links emanating from the supersource node 0 are zero. In this chapter, for definiteness, we focused on the mergers of two firms. The network formalism in this chapter can easily handle the merger of two or more firms from a subset of a finite number of firms. One would only need to identify the network structure of each firm as in Figure 8.1, for all the firms, and then construct the networks for the specific mergers with the inclusion of additional supersource node(s) and links. The feasible sets would be redefined accordingly. In Chapter 11, we consider such mergers, but in the case of competitive markets in the form of oligopolies. In those models, the firms supply, before and after the mergers, the same demand markets and the prices of the product at the demand markets are elastic since the demands are no longer fixed, but, rather, the consumers respond to the price of the product at the particular demand market.

Of course, the system-optimization model for supply chain networks presented in Section 8.2.1 is, in itself, of interest. For example, by formulating and studying the supply chain network of a firm at the level of detail of all the associated economic activities, one can determine the optimal product path flows, the corresponding link flows, and the incurred minimal total cost. Such a model can provide the foundation for the determination of the effects on total costs of, for example, the addition (or deletion) of a manufacturing plant, the addition (or deletion) of a distribution center, and so on. Furthermore, one can determine the effects of a decrease (or an increase) in link capacity on the associated product flows and the total costs. One can also evaluate, within this context, the robustness of a supply chain network design from a system-optimization perspective. In the transportation network applications in Chapter 4 we studied transportation networks from alternative robustness points of view, and in the case of link cost functions in which the capacities were explicitly stated. Here, in contrast, the capacities are not incorporated into the link cost functions, but, rather, are imposed on the link flows through the upper bounding constraints. Nevertheless, the robustness of a particular supply chain network design can be determined by considering uniform reductions in the link capacities, as was discussed in Chapter 3, under the assumption, that there is still a feasible solution.

The reader may have noticed the similarity between the two measures defined by (3.10b) and (8.23), respectively. Basically, both measures quantify the relative total cost difference. However, (3.10b) quantifies the relative total cost *increase* under a uniform capacity reduction ratio γ, whereas with (8.23), we examine the relative total cost *savings* for a specific merger case.

Finally, we emphasize that the network models presented in this chapter can be easily adapted and transferred to mergers and acquisitions in industries in which there are no overt supply chain networks. For example, by constructing the underlying network structure of firms in transportation (airlines, railways, etc.) or in telecommunications, before and after a proposed synergy, along with the associated total cost functions and expected demands for the services provided, one can compute the associated synergies of the M&As using the relevant formula (8.23). In addition, one can apply the ideas to the case of services, such as financial services.

In the next chapter, we consider possible environmental impacts associated with supply chain network integration and we extend the results in this chapter to capture generalized total costs.

CHAPTER 9

ENVIRONMENTAL AND COST SYNERGY IN NETWORK INTEGRATION

9.1 INTRODUCTION

In Chapter 4, Section 4.4, we focused on environmental emissions of vehicular traffic and proposed environmental impact indices for transportation networks under either U-O or S-O behavior. In this chapter, we build on Chapter 8, and we demonstrate how environmental criteria can be incorporated into a S-O modeling framework for assessing both cost and environmental synergies of network integration. Indeed, it is now well-recognized that pollution has major adverse consequences, including global warming, acid rain, rising oceanic temperatures, smog, and the resulting harmful effects on human health and wildlife. Hence the development of rigorous, quantitative measures to assess the environmental impacts of potential mergers and acquisitions of firms through network integration may yield deeper and broader knowledge regarding both cost and environmental savings. Figure 9.1 summarizes the percentage of total greenhouse gas emitted by economic sectors in the United States. We can see that the electric power industry and transportation are the top two sectors contributing to greenhouse gas emissions.

Firms are increasingly realizing the importance of their environmental impacts and the return on the bottom line for those actions expended to reduce pollution

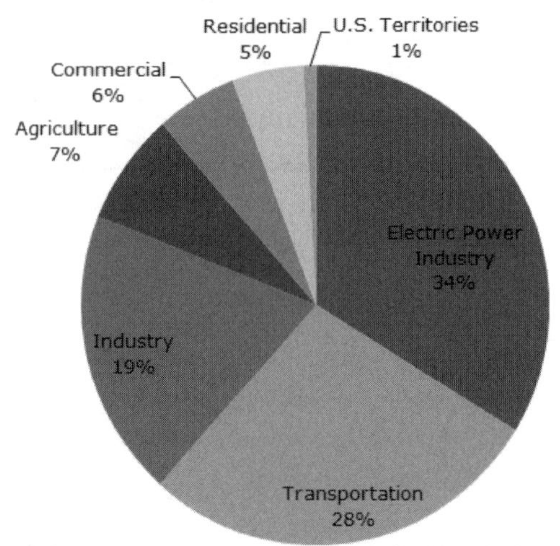

Figure 9.1: U.S. greenhouse gas emissions allocated to economic sectors (Tg CO2 Eq. and percent of total in 2006) [Data Source: U.S. Environmental Protection Agency (2008)]

[Hart and Ahuja (1996)]. For example, it has been noted that firms in the public eye have not only met but, in some cases, exceeded the required environmental mandate [Lyon (2003)]. In the United States, over 1,200 firms voluntarily participated in the Environmental Protection Agency's 33/50 program, agreeing to reduce certain chemical emissions 50% by 1995 [Arora and Cason (1996)]. It has also been argued that customers and suppliers may "punish" polluters in the marketplace that violate environmental rules. As a consequence, polluters may face lower profits, also called a "reputational penalty," which may be manifested in a lower stock price for the company [Klein and Leffler (1981), Klassen and McLaughlin (1996)]. Roper Starch Worldwide (1997) noted that more than 75% of the public would switch to a brand associated with the environment when price and quality were equal; and nearly 60% of those questioned favored organizations that support the environment. It has also been argued that sound environmental practices will reduce risk to the firm [Feldman, Soyka, and Ameer (1997)].

Due to the visibility and the number of M&As that have been occurring, it is important to understand and study the synergy results for managerial benefits from an environmental standpoint. In the first 9 months of 2007 alone, according to Thomson Financial, worldwide merger activity hit $3.6 trillion, surpassing the total from all of 2006 [Wong (2007)]. Companies merge for various reasons, some of which include such benefits as acquired technologies and greater economies of scale that improve productivity or cut costs [Chatterjee (1986)]. Successful mergers can add tremendous value; however, with a failure rate estimated to be between 74% and

83% [Devero (2004)], it is worthwhile to develop tools to better predict the associated strategic gains, which include cost savings [Eccles, Lanes, and Wilson (1999)].

With the growing investment and industrialization in developing nations, it is also important to evaluate the overall impact of merger activities at not only the operational level but also as related to environmental impacts. There is enormous potential for both developed and developing countries to adopt cleaner production, given current technologies and the levels of private capital investments. For example, between 1988 and 1995, multinational corporations invested nearly $422 billion worth of new factories, supplies, and equipment in developing countries [World Resources Institute (1998)]. Through globalization, firms of industrialized nations can acquire those firms in developing nations that offer lower production costs; however, quite possibly, combined with inferior environmental concerns. As a result of the industrialization of developing countries, the actions taken today will greatly influence the future scale of environmental and health problems.

Lambertini and Mantovani (2007) concluded that horizontal mergers result in the reduction of negative externalities related to the environment. Moreover, according to Stanwick and Stanwick (2002), if environmental issues are ignored the value of the proposed merger can be greatly compromised.

There is virtually no literature to-date that discusses the relationship between post-merger operational synergy and the effects on the environment and thus, ultimately, society. We attempt to address this issue from a quantitative perspective in this chapter. This chapter develops a multicriteria decision-making optimization framework in which each firm not only minimizes costs but also minimizes emissions. Multicriteria decision-making has been recently much-explored as related to the transportation network equilibrium problem. For example, Nagurney, Dong, and Mokhtarian (2002a) include the weighting of travel time, travel cost, and the emissions generated; see also Nagurney, Dong, and Mokhtarian (2002b, c). For general references on transportation networks and multicriteria decision-making, see Nagurney and Dong (2002a). Multicriteria decision-making within a supply chain can assist in the production and delivery of products by focusing on factors such as cost, quality, and lead times [Talluri and Baker (2002)]. Toward that end, Dong, Zhang, and Nagurney (2002) proposed a supply chain network that included multicriteria decision-makers at each tier of the supply chain, specifically, the manufacturing tier, the retailer tier, and the demand markets. Nagurney and Toyasaki (2003, 2005) focused on environmental decision-making in supply chains, with the latter paper developing a general multitiered framework for electronic recycling networks.

The proponents for a system view structure of the supply chain, which we use in this chapter and which we also used in Chapter 8, include the fostering of relationships and, possibly, enhanced coordination and management in order to achieve greater consumer satisfaction, which is necessary to be competitive in the current economic climate [Zsidisin and Siferd (2001)]. Sarkis (2003) demonstrated that environmental supply chain management, also referred to as green supply chain management, is necessary to address environmental concerns. As an illustrative example, the Ford Motor Company demanded that all of its 5000 worldwide suppliers with manufacturing plants obtain a third-party certification of an environmental management system (EMS) by 2003 [Rao (2002)].

This chapter is built on Chapter 8, but we extend the results therein to include multicriteria decision-making and environmental concerns. In particular, here we analyze the synergy effects associated with a merger in terms of the operational synergy, that is, the reduction, if any, in the cost of production, storage, and distribution as well as the environmental benefits in terms of the reduction of associated emissions (if any). The framework developed in this chapter can be expected to be especially useful and relevant in various types of manufacturing industries, including steel and cement; in the food processing industry; in electric power and other energy-related supply chains; and even in a variety of transportation sectors. In the case of the airline industry, for example, one may conceive of the underlying networks before the merger corresponding to different routes of a particular air carrier with associated demands; with the expanded network corresponding to a particular merger and the integration of routes. Similar analogies may be made to M&As associated with package delivery companies as well as trucking companies.

This chapter is organized as follows: The pre-merger supply chain network model is developed in Section 9.2. In Section 9.2 the horizontally merged (or acquired) supply chain model is also constructed. The method of quantification of the synergistic gains, if any, is provided in Section 9.3. In Section 9.4 we present numerical examples. Section 9.5 summarizes the results.

By allowing for multicriteria decision-making in the context of mergers and acquisitions and in the form of the integration of supply chain networks, we have laid the foundation for the conceptualization and analysis of the integration of other network systems. For example, another very important criterion in the new era of increasing risk and uncertainty is that of risk minimization. The theoretical formalism in this chapter can also easily capture risk, with an appropriate weight associated with the criterion of risk minimization. The network concepts associated with M&As and developed here can also be applied to financial services, including banks.

9.2 THE PRE- AND POST-MERGER SUPPLY CHAIN NETWORK MODELS

This section develops the pre- and post-merger supply chain network models with environmental concerns using a system-optimization approach. Each firm is assumed to act as a multicriteria decision-maker so as to not only to minimize costs but also to minimize the emissions generated. We formulate the pre-merger multicriteria supply chain network model in Section 9.2.1 and the post-merger one in Section 9.2.2.

9.2.1 The Pre-Merger Supply Chain Network Model with Environmental Concerns

We first formulate the pre-merger multicriteria decision-making optimization problem faced by Firm A and Firm B, and refer to this model as Case 0. As in Chapter 8, we assume that each firm is represented as a network of its economic activities, as depicted in Figure 8.1. Each firm produces a homogeneous product and the notation follows closely that of Chapter 8.

As in Chapter 8, there is a total cost associated with each link (cf. Figure 8.1) of the network corresponding to each firm i; $i = A, B$. We denote the links by a, b, etc., and the total cost on a link a by $\hat{c}_a$. The demands for the product are assumed as given and are associated with each firm and retailer pair. Let $d_{R_k^i}$ denote the demand for the product at retailer R_k^i associated with firm i; $i = A, B$; $k = 1, \ldots, n_R^i$. A path is defined as a sequence of links joining an origin node $i = A, B$ with a destination node R_k^i. Let x_p denote the nonnegative flow of the product on path p. A path consists of a sequence of economic activities comprising manufacturing, storage, and distribution. The following conservation of flow equations must hold for each firm i

$$d_{R_k^i} = \sum_{p \in P^0_{R_k^i}} x_p, \quad i = A, B; k = 1, \ldots, n_R^i, \tag{9.1}$$

where $P^0_{R_k^i}$ denotes the set of paths connecting (origin) node i with (destination) retail node R_k^i.

Let f_a denote the flow of the product on link a. We must also have the following conservation of flow equations satisfied

$$f_a = \sum_{p \in P^0} x_p \delta_{ap}, \quad \forall p \in P^0, \tag{9.2}$$

where $\delta_{ap} = 1$ if link a is contained in path p and $\delta_{ap} = 0$, otherwise. Here P^0 denotes the set of *all* paths in Figure 8.1, that is, $P^0 = \cup_{i=A,B;k=1,\ldots,n_R^i} P^0_{R_k^i}$.

The path flows must be nonnegative, that is

$$x_p \geq 0, \quad \forall p \in P^0. \tag{9.3}$$

We group the path flows into the vector x and the link flows into the vector f.

The total cost on a link, be it a manufacturing link, a shipment/distribution link, or a storage link is assumed to be a function of the flow of the product on the link, as was assumed in Chapter 8. Hence we have

$$\hat{c}_a = \hat{c}_a(f_a), \quad \forall a \in \mathcal{L}^0. \tag{9.4}$$

We assume that the total cost on each link is convex, continuously differentiable, and has a bounded third order partial derivative.

As in Chapter 8, it is assumed that there are positive capacities on the links with the capacity on link a denoted by u_a, $\forall a \in \mathcal{L}^0$.

In addition, we assume, as given, emission functions for each economic link $a \in \mathcal{L}^0$, denoted by e_a, where

$$e_a = e_a(f_a), \quad \forall a \in \mathcal{L}^0, \tag{9.5}$$

and e_a is the total amount of emissions generated by link a in processing an amount f_a of the product. We assume that the emission functions have the same properties as the total cost functions (9.4). We now discuss the units for measurement of the emissions. We propose the use of the carbon equivalent for emissions, which is commonly used in environmental modeling and research [Nagurney (2000), Wu et al. (2006)] and in practice as employed by the Kyoto Protocol [Reilly et al. (1999)] to aid in the direct comparison of environmental impacts of differing pollutants. Emissions are typically expressed in a common metric, specifically, in million metric tons of carbon equivalent (MMTCE) [U.S. Environmental Protection Agency (2005)].

It is reasonable to assume that the amount of emissions generated is a function of the flow on the associated economic link (see, e.g., Dhanda, Nagurney, and Ramanujam (1999) and the references therein).

Because the firms, pre-merger, have no links in common (cf. Figure 8.1), their individual cost minimization problems can be formulated jointly as follows

$$\text{Minimize} \quad \sum_{a \in \mathcal{L}^0} \hat{c}_a(f_a) \tag{9.6}$$

subject to: constraints (9.1) - (9.3) and

$$f_a \leq u_a, \quad \forall a \in \mathcal{L}^0. \tag{9.7}$$

In addition, because we are considering multicriteria decision-making with environmental concerns, the minimization of emissions generated can, in turn, be expressed as follows

$$\text{Minimize} \quad \sum_{a \in \mathcal{L}^0} e_a(f_a) \tag{9.8}$$

subject to: constraints (9.1) - (9.3) and (9.7).

We can now construct a weighted total cost function or generalized cost function, which we refer to as the generalized total cost [cf. Fishburn (1970), Chankong and Haimes (1983), Yu (1985), Keeney and Raiffa (1992), Nagurney and Dong (2002a)], associated with the two criteria faced by each firm with the weight associated with total cost minimization being set to, in effect, the price the firm is willing to pay per unit of emission. Specifically, for notational convenience and simplicity, we define nonnegative weights associated with the firms $i = A, B$ and links $a \in \mathcal{L}_i$, as follows $\omega_{ia} \equiv 0$ if link $a \notin \mathcal{L}_i$ and $\omega_{ia} = \omega_i$, otherwise, where ω_i is decided upon by the decision-making authority of firm i. Consequently, the pre-merger multicriteria decision-making problem can be expressed as

$$\text{Minimize} \quad \sum_{a \in \mathcal{L}^0} \sum_{i=A,B} \hat{c}_a(f_a) + \omega_{ia} e_a(f_a) \tag{9.9}$$

subject to: constraints (9.1) - (9.3) and (9.7).

Under the above imposed assumptions, the optimization problem is a convex optimization problem. If we further assume that the feasible set underlying the problem represented by the constraints (9.1) through (9.3) and (9.7) is non-empty, then it follows from the standard theory of nonlinear programming that an optimal solution exists.

Let $\mathcal{K}^0 \equiv \{f | \exists x \geq 0, \text{ and } (9.1) - (9.3) \text{ and } (11.7) \text{ hold}\}$. Also, associate the Lagrange multiplier β_a with constraint (9.7) for link a and denote the associated optimal Lagrange multiplier by β_a^*. As was the case in the model in Chapter 8, this term may also be interpreted as the price or value of an additional unit of capacity on link a. We now provide the variational inequality formulation of the problem.

Theorem 9.1: Variational Inequality Formulation of the Pre-Merger Multicriteria Decision-Making Supply Chain Network Problem
The vector of link flows $f^* \in \mathcal{K}^0$ is an optimal solution to the pre-merger multicriteria decision-making problem (9.9) if and only if it satisfies the following variational inequality problem with the vector of optimal nonnegative Lagrange multipliers β^*

$$\sum_{a \in \mathcal{L}^0} \sum_{i=A,B} [\frac{\partial \hat{c}_a(f_a^*)}{\partial f_a} + \omega_{ia} \frac{\partial e_a(f_a^*)}{\partial f_a} + \beta_a^*] \times [f_a - f_a^*] + \sum_{a \in L^0} [u_a - f_a^*] \times [\beta_a - \beta_a^*] \geq 0,$$

$$\forall f \in \mathcal{K}^0, \forall \beta_a \geq 0, \forall a \in \mathcal{L}^0. \qquad (9.10)$$

Proof: See proof of Theorem 8.1.

It is interesting to compare variational inequality (9.10) with variational inequality (8.7). Note that in (9.10) there is an additional term after the marginal total cost for link a, which corresponds to the weighted marginal emissions on that link. One may interpret the sum of these two marginal expressions as the marginal generalized cost on the link. Observe that the feasible set $\mathcal{K}^0$ in this section is identical to the one underlying the pre-merger network model in Section 8.2.1 because the constraints (but not the corresponding objective functions) are one and the same.

9.2.2 The Post-Merger Supply Chain Network Model with Environmental Concerns

We now formulate the post-merger case. Because we are concerned only with the full merger (referred to as Case 3 in Chapter 8), we, for simplicity, denote the full merger case in this chapter as Case 1. Hence the manufacturing facilities produce the product and then ship it to any distribution center, and the retailers can obtain the product from any distribution center. Because the product is assumed to be homogeneous, after the merger, the retail outlets are indifferent as to at which manufacturing plant the product was produced. Figure 8.4 depicts the post-merger supply chain network topology. Note that there is now a supersource node 0, which represents the merger of the firms with additional links joining node 0 to nodes A and B, respectively.

The post-merger optimization problem is concerned with total cost minimization and the minimization of emissions. Specifically, we retain the nodes and links associated with network $\mathcal{G}^0$ depicted in Figure 8.1, but now we add the additional links connecting the manufacturing plants of each firm and the distribution centers and the links connecting the distribution centers and the retailers of the other firm. We refer to the network underlying this merger as $\mathcal{G}^1 = [\mathcal{N}^1, \mathcal{L}^1]$. We associate total cost functions as in (9.4) and emission functions as in (9.5) with the new links.

A path p now (cf. Figure 8.4) originates at the node 0 and is destined for one of the bottom retail nodes. Let x_p now in the post-merger network configuration given in Figure 8.4 denote the flow of the product on path p joining (origin) node 0 with a (destination) retailer node. Then the following conservation of flow equations must hold

$$d_{R_k^i} = \sum_{p \in P^1_{R_k^i}} x_p, \quad i = A, B; k = 1, \dots, n_R^i, \tag{9.11}$$

where $P^1_{R_k^i}$ denotes the set of paths connecting node 0 with retail node R_k^i in Figure 8.4. The set of paths $P^1 \equiv \cup_{i=A,B;k=1,\dots,n_R^i} P^1_{R_k^i}$.

In addition, as before, we let f_a denote the flow of the product on link a. Hence we must also have the following conservation of flow equations satisfied

$$f_a = \sum_{p \in P^1} x_p \delta_{ap}, \quad \forall p \in P^1. \tag{9.12}$$

Of course, we also have that the path flows must be nonnegative, that is

$$x_p \geq 0, \quad \forall p \in P^1. \tag{9.13}$$

As before, the links representing the manufacturing activities, the shipment, and the storage activities possess nonnegative capacities, denoted as u_a, $\forall a \in \mathcal{L}^1$. This can be expressed as

$$f_a \leq u_a, \quad \forall a \in \mathcal{L}^1. \tag{9.14}$$

Post-merger, the weight associated with the environmental emission cost minimization criterion is denoted by α, and this weight is nonnegative. This is reasonable since, unlike in the pre-merger case, the firms are now merged into a single decision-making economic entity, and there is now a single weight associated with the emissions generated.

Hence the following multicriteria decision-making optimization problem must now be solved

$$\text{Minimize} \quad \sum_{a \in \mathcal{L}^1} [\hat{c}_a(f_a) + \omega e_a(f_a)] \tag{9.15}$$

subject to constraints: (9.11) - (9.14). Note that $\mathcal{L}^1$ now represents all links in the post-merger network belonging to Firm A and to Firm B.

There are distinct options for the weight ω and we explore several in Section 9.4, in concrete numerical examples. When the M&A is an environmentally hostile one, then we may set $\omega = 0$; when it is environmentally conscious, then ω may be set to a positive value; and so on, with ω being a function of the firms' pre-merger weights also a possibility.

The solution to the post-merger multicriteria decision-making optimization problem (9.15) subject to constraints (9.11) through (9.14) can also be obtained as a solution to a variational inequality problem akin to (9.10), where $a \in \mathcal{L}^1$, ω is substituted for ω_i, and the vectors f, x, and β have identical definitions, as before, but are redimensioned accordingly. Finally, instead of the feasible set $\mathcal{K}^0$ we now have $\mathcal{K}^1 \equiv \{f | \exists x \geq 0, \text{ and } (9.11) - (9.14) \text{ hold}\}$. We denote the solution to the variational inequality problem governing Case 1 by (f^*, β^*) because the context should be clear. For completeness, we provide the variational inequality formulation of the Case 1 problem. The proof is immediate.

Theorem 9.2: Variational Inequality Formulation of the Post-Merger Multicriteria Decision-Making Supply Chain Network Problem
The vector of link flows $f^* \in \mathcal{K}^1$ is an optimal solution to the post-merger problem if and only if it satisfies the following variational inequality problem with the vector of optimal nonnegative Lagrange multipliers β^*

$$\sum_{a \in \mathcal{L}^1} [\frac{\partial \hat{c}_a(f_a^*)}{\partial f_a} + \alpha \frac{\partial e_a(f_a^*)}{\partial f_a} + \beta_a^*] \times [f_a - f_a^*] + \sum_{a \in \mathcal{L}^1} [u_a - f_a^*] \times [\beta_a - \beta_a^*] \geq 0,$$

$$\forall f \in \mathcal{K}^1, \forall \beta_a \geq 0, \forall a \in \mathcal{L}^1. \quad (9.16)$$

The feasible set $\mathcal{K}^1$ here is identical to the feasible set $\mathcal{K}^3$ in Section 8.2.2.3, because the constraints underlying both problems are identical (but the objective functions are distinct).

Both variational inequalities (9.10) and (9.16) can be put into standard variational inequality form (see Chapter 2), where $X \equiv (f^T, \beta^T)^T$ and f is the vector of link flows for the specific network problem with β being the vector of Lagrange multipliers associated with the link capacity constraints for the specific problem. The function $F(X)$ that enters the variational inequality problem for (9.10) consists of the vector of the terms $\frac{\partial \hat{c}_a(f_a)}{\partial f_a} + \alpha_{ia} \frac{\partial e_a(f_a)}{\partial f_a} + \beta_a$ for each link $a \in \mathcal{L}^0$ and of the vector of terms: $u_a - f_a$ for each link $a \in \mathcal{L}^0$. In the case of variational inequality (9.16), $F(X)$ consists of the vector of terms $\frac{\partial \hat{c}_a(f_a)}{\partial f_a} + \alpha \frac{\partial e_a(f_a)}{\partial f_a} + \beta_a$ for all $a \in \mathcal{L}^1$ and the vector of the terms $u_a - f_a$ for each link $a \in \mathcal{L}^1$. The feasible set $\mathcal{K}$ is then defined as $\mathcal{K} \equiv \{(f, \beta) \text{ such that } f \in \mathcal{K}^0, \beta_a \geq 0, \forall a \in \mathcal{L}^0\}$. In the case of variational inequality (9.16), the corresponding feasible set for the standard form variational inequality is given by $\mathcal{K} \equiv \{(f, \beta) \text{ such that } f \in \mathcal{K}^0, \beta_a \geq 0, \forall a \in \mathcal{L}^1\}$.

Due to the special structure of these feasible sets, which also arises in the feasible sets for the network problems in Chapter 8, variational inequality problems (9.10) and (9.16) can be easily solved by the modified projection method outlined in Chapter 2. Indeed, in each of the two fundamental steps of the modified projection method, we encounter a problem with network structure, which can be solved via "equilibration" as described in Chapter 2, and a problem in the Lagrange multipliers, which can be determined explicitly and in closed form at any given iteration since the multipliers are constrained only to be nonnegative. The numerical examples in Section 9.4 were solved by the modified projection method embedded with the equilibration algorithm for the network subproblems.

9.3 QUANTIFYING SYNERGY ASSOCIATED WITH ENVIRONMENTAL CONCERNS

We define the total generalized cost TGC^0 associated with Case 0 as the value of the objective function in (9.9) evaluated at its optimal solution f^* and the total generalized cost TGC^1 associated with Case 1 as the value of the objective function in (9.15) evaluated at its optimal solution f^*. These flow vectors we obtain from the solutions of variational inequalities (9.10) and (9.16), respectively. In the next section, we discuss how we use these two total generalized costs to determine the strategic advantage or synergy associated with a M&A. In addition, we define TE^0 as the total emissions generated under solution f^* to the Case 0 problem, TE^1 as the total emissions generated under solution f^* to the Case 1 problem, and TC^0 and TC^1 the corresponding total costs.

The synergy associated with the total generalized costs that captures both the total costs and the weighted/priced total emissions is denoted by $\mathcal{S}^{TGC}$ and is defined as follows

$$\mathcal{S}^{TGC} \equiv [\frac{TGC^0 - TGC^1}{TGC^0}] \times 100\%. \tag{9.17}$$

We can also measure the synergy by analyzing the total costs before and after the merger, as was done in Chapter 8 and the changes in emissions. For example, the synergy based on total costs, but not in a multicriteria decision-making context, which we denote here by $\mathcal{S}^{TC}$, can be calculated as the percentage difference between the total cost before versus the total cost after the merger

$$\mathcal{S}^{TC} \equiv [\frac{TC^0 - TC^1}{TC^0}] \times 100\%. \tag{9.18}$$

Formula (9.18) is analogous to $\mathcal{S}^3$ in (8.23).

The environmental impacts related to the relationship between the pre- and post-merger emission levels can also be calculated using a similar measure as that of the total cost. Toward that end, we define the total emissions synergy, denoted by $\mathcal{S}^{TE}$, as

$$\mathcal{S}^{TE} \equiv [\frac{TE^0 - TE^1}{TE^0}] \times 100\%. \tag{9.19}$$

9.4 NUMERICAL EXAMPLES

In this section, we present numerical examples in which we use the synergy measures defined in Section 9.3. We consider Firm A and Firm B, as depicted in Figure 8.5, for the pre-merger case. Each firm owns and operates two manufacturing plants, M_1^i and M_2^i, and one distribution center and provides the product to meet demand at two retail markets R_1^i and R_2^i for $i = A, B$. Figure 9.2 depicts the post-merger supply chain network. The total cost functions were $\hat{c}_a(f_a) = f_a^2 + 2f_a$ for all links a pre- and post-merger in all the numerical examples, except for the links post-merger that join the node 0 with nodes A and B. These merger links had associated total costs

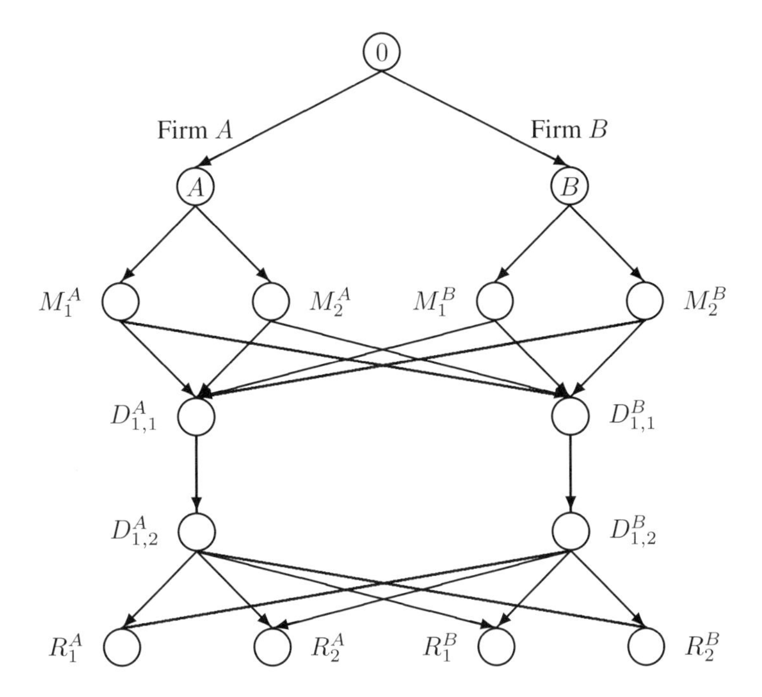

Figure 9.2: Post-merger supply chain network topology for the numerical examples

equal to 0. The links and the associated emission functions for all the examples are given in Table 9.1. The solutions to the numerical examples are given in Table 9.2 for the pre-merger case and in Table 9.3 for the post-merger case. The synergy calculations are presented in Table 9.4.

These numerical examples were solved using the modified projection method embedded with the equilibration algorithm, as discussed in Section 8.4. The same codes may be used as developed for the Case 0 and Case 3 models in Chapter 8, because only the data inputs need to be adapted to reflect the generalized total costs that include multicriteria decision-making associated with environmental concerns. We implemented the modified projection method (and the embedded equilibration algorithm) in Fortran and used a Unix system at the University of Massachusetts for the computations. The algorithm was considered to have converged to the solution when the path flows at the particular and preceding iteration differed by no more than $\epsilon = 10^{-5}$ and the same held for the Lagrange multipliers.

Example 9.1

The demands at the retailers for Firm A and Firm B were set to 5 and the capacity on each link was set to 15 both before and after the merger. The weights $\omega_{ia} = \omega_i$ were set to 1 for both firms $i = A, B$ and for all links $a \in \mathcal{L}^0$. The pre-merger solution f^* for both firms had all components equal to 5 for all links, except for the storage links, which had flows of 10. The associated β^* had all components equal to 0 because the

Table 9.1: The link emission functions for the numerical examples

Link a	From Node	To Node	Ex. 9.1,9.2: $e_a(f_a)$
1	A	M_1^A	$10f_1$
2	A	M_2^A	$10f_2$
3	M_1^A	$D_{1,1}^A$	$10f_3$
4	M_2^A	$D_{1,1}^A$	$10f_4$
5	$D_{1,1}^A$	$D_{1,2}^A$	$10f_5$
6	$D_{1,2}^A$	R_1^A	$10f_6$
7	$D_{1,2}^A$	R_2^A	$10f_7$
8	B	M_1^B	$10f_8$
9	B	M_2^B	$10f_9$
10	M_1^B	$D_{1,1}^B$	$10f_{10}$
11	M_2^B	$D_{1,1}^B$	$10f_{11}$
12	$D_{1,1}^B$	$D_{1,2}^B$	$10f_{12}$
13	$D_{1,2}^B$	R_1^B	$10f_{13}$
14	$D_{1,2}^B$	R_2^B	$10f_{14}$
15	M_1^A	$D_{1,1}^B$	$10f_{15}$
16	M_2^A	$D_{1,1}^B$	$10f_{16}$
17	M_1^B	$D_{1,1}^A$	$10f_{17}$
18	M_2^B	$D_{1,1}^A$	$10f_{18}$
19	$D_{1,2}^A$	R_1^B	$10f_{19}$
20	$D_{1,2}^A$	R_2^B	$10f_{20}$
21	$D_{1,2}^B$	R_1^A	$10f_{21}$
22	$D_{1,2}^B$	R_2^A	$10f_{22}$
23	0	A	.00
24	0	B	.00

flow on any particular link did not meet capacity. The total cost was 660.00, the total emissions generated were 800.00 and the total generalized cost was 1460.00.

Post-merger, for each firm, the cost and emission functions were again set to $\hat{c}_a(f_a) = f_a^2 + 2f_a$ and $e_a(f_a) = 10f_a$, respectively, including those links formed post-merger. The demand at each retail market was kept at 5 and the capacity of each link, including those formed post-merger, was set to 15. The weight ω, post-merger, was set to 1. The solution is as follows (see also Table 9.3). For both firms, the manufacturing link flows were 5; 2.5 was the shipment between each manufacturer and distribution center, 10 was the flow representing storage at each distribution center, and 2.5 was the flow from each distribution/storage center to each demand market. The vector of optimal multipliers β^* post-merger, had all its components equal to .00. The total cost was 560.00, the total emissions generated were 800.00, and the total generalized cost was 1360.00. There were total cost synergistic gains, specifically, $S^{TC} = 15.15\%$, yet no environmental gains, because $S^{TE} = .00\%$. In addition, the total generalized cost synergy was $S^{TGC} = 6.85\%$.

Table 9.2: Pre-merger solutions to the numerical examples

Link a	From Node	To Node	Ex. 9.1 - 9.2: f_a^*
1	A	M_1^A	5.00
2	A	M_2^A	5.00
3	M_1^A	$D_{1,1}^A$	5.00
4	M_2^A	$D_{1,1}^A$	5.00
5	$D_{1,1}^A$	$D_{1,2}^A$	10.00
6	$D_{1,2}^A$	R_1^A	5.00
7	$D_{1,2}^A$	R_2^A	5.00
8	B	M_1^B	5.00
9	B	M_2^B	5.00
10	M_1^B	$D_{1,1}^B$	5.00
11	M_2^B	$D_{1,1}^B$	5.00
12	$D_{1,1}^B$	$D_{1,2}^B$	10.00
13	$D_{1,2}^B$	R_1^B	5.00
14	$D_{1,2}^B$	R_2^B	5.00

Example 9.2

Example 9.2 was constructed from Example 9.1 but with the following modifications. Pre-merger, we assumed that Firm A is environmentally conscious, that is $\omega_{Aa} = 1$ for all links a associated with Firm A; Firm B does not display any concern for the environment, that is, $\omega_{Ba} = 0$ for all its links. In addition, we now assumed that the merger was hostile, with Firm A as the dominant firm, that is, Firm A imposes its environmental concern on Firm B. We assumed that, post-merger, $\omega = 1$. The pre-merger optimal flows are the same as in Example 9.1. The total cost was 660.00, the total emissions generated were 800.00, and the total generalized cost was 1060.00.

The post-merger results were as follows. The total cost was 560.00, the total emissions generated were 800.00, and the total generalized cost was 1360.00. The synergy results were: 15.15% for the total cost; $.00\%$ for the total emissions, and -28.30% for the total generalized cost. When the dominant firm in the proposed merger was more concerned with the environmental impacts, the overall total generalized cost synergy was the lowest. This example illustrates the importance of not only demonstrating concern for the environment but also to take action to reduce the emission functions.

9.4.1 Additional Examples

To explore the impacts of improved technologies associated with distribution we constructed the following variants of the above numerical examples. We assumed that the pre-merger data were as in Examples 9.1 through 9.2 as were the post-merger data except that we assumed that the emission functions associated with the new "merger" links were all identically equal to zero. The post-merger link flow solutions

Table 9.3: Post-merger solutions to the numerical examples

Link a	Ex. 9.1 - 9.2: f_a^*
1	5.00
2	5.00
3	2.50
4	2.50
5	10.00
6	2.50
7	2.50
8	5.00
9	5.00
10	2.50
11	2.50
12	10.00
13	2.50
14	2.50
15	2.50
16	2.50
17	2.50
18	2.50
19	2.50
20	2.50
21	2.50
22	2.50
23	10.00
24	10.00

are given in Table 9.5 and the synergy computations in Table 9.6 for these additional examples, which are labeled as Examples 9.1b and 9.2b, respectively.

The synergies computed for Examples 9.1b through 9.2b suggest an inverse relationship between total cost synergy and emission synergy. It is also interesting to compare the results for Example 9.1b and Example 9.2b in Table 9.6. Despite the fact that they both have identical total cost and total emission synergies, their respective total generalized cost synergies are, nevertheless, distinct. This can be attributed to the difference in concern for the environment pre- and post-merger.

9.5 SUMMARY AND CONCLUDING REMARKS

In this chapter, we presented a multicriteria decision-making framework to evaluate the environmental impacts associated with mergers and acquisitions. The framework is based on a supply chain network perspective, in a system-optimization context,

Table 9.4: Synergy values for the numerical examples

Example	9.1	9.2
TC^0	660.00	660.00
TC^1	560.00	560.00
S^{TC}	15.15%	15.15%
TE^0	800.00	800.00
TE^1	800.00	800.00
S^{TE}	.00%	.00%
TGC^0	1460.00	1060.00
TGC^1	1360.00	1360.00
S^{TGC}	6.85%	−28.30%

Table 9.5: Post-merger solutions to additional numerical examples

Link a	Ex. 9.1b, Ex. 9.2b: f_a^*
1	5.00
2	5.00
3	.00
4	.00
5	10.00
6	.00
7	.00
8	5.00
9	5.00
10	.00
11	.00
12	10.00
13	.00
14	.00
15	5.00
16	5.00
17	5.00
18	5.00
19	5.00
20	5.00
21	5.00
22	5.00
23	10.00
24	10.00

that captures the economic activities of a firm such as manufacturing/production, storage, as well as distribution. We presented the pre-merger and the post-merger network models, derived their variational inequality formulations, and then defined a total generalized cost synergy measure as well as a total cost synergy measure and a total emissions synergy measure. The firms, pre-merger, were assigned a weight

Table 9.6: Synergy values for the additional numerical examples

Example	9.1b	9.2b
TC^0	660.00	660.00
TC^1	660.00	660.00
S^{TC}	.00%	.00%
TE^0	800.00	800.00
TE^1	400.00	400.00
S^{TE}	50.00%	50.00%
TGC^0	1460.00	1060.00
TGC^1	1060.00	1060.00
S^{TGC}	27.40%	.00%

representing their individual environmental concerns; post-merger, the weight was uniform.

Solutions to several numerical examples were computed, which, although stylized, demonstrated the generality of the approach and how the new framework can be used to assess apriori synergy associated with mergers and acquisitions and with an environmental focus. We concluded that the operating economies (resulting from greater economies of scale that improve productivity or cut costs) may have an inverse impact on the environmental effects to society, depending on the level of concern that each firm has for the environment and their joint actions taken to reduce emissions.

With the results in this chapter, one can begin to further explore numerous questions associated with mergers and acquisitions, environmental synergies, and industrial organization. For example, we note that this chapter has focused on horizontal mergers. Additional research is needed to evaluate the possible synergy associated with vertical integrations and the impacts on the environment. We expect related issues will be extremely relevant to different industrial settings.

9.6 SOURCES AND NOTES

This chapter is based on Nagurney and Woolley (2008). However, unlike the models therein, here we do not assume that the total cost functions associated with the top-most merger links need be zero. Hence one can explicitly incorporate cost functions associated not only with the manufacturing, distribution, and storage activities but also with the actual merger itself (coupled with environmental costs). In addition, all the numerical examples in this chapter were solved using an implemented Fortran code rather than MATLAB, as in Nagurney and Woolley (2008).

The weights associated with the environmental emissions in this chapter, as discussed here, have the interpretation of prices. These prices can be self-selected or imposed externally, as in the form of environmental taxes. Nagurney, Liu, and Woolley (2006) discuss carbon taxes and their relationships to self-imposed weights (in the context of electric power supply chains). Woolley, Nagurney, and Stranlund (2009), in turn, formulated environmental policies in the form of tradable pollution permits in the case of multipollutant electric power supply chains.

Additional multicriteria network equilibrium models and applications, along with references, can be found in papers by Nagurney and Dong (2002b, c).

CHAPTER 10

MULTIPRODUCT SUPPLY CHAIN NETWORK INTEGRATION

10.1 INTRODUCTION

In this chapter we turn to multiproduct supply chains and their integration. Although there are numerous articles discussing multi-echelon supply chains, the majority deal with a homogeneous product [see, e.g., Dong, Zhang, and Nagurney (2004), Nagurney (2006a), and Wang, Zhang, and Wang (2007)]. Firms are seeing the need to spread their investment risk by building multiproduct supply facilities, which also may give the advantage of flexibility to meet changing market demands. According to a study of the U.S. supply output at the firm-product level between 1972 and 1997, on the average, two-thirds of U.S. supply firms altered their mix of products every 5 years [Bernard, Redding, and Schott (2006)]. By running a multi-use plant, costs of supply may be divided among different products, which may increase efficiencies.

It is interesting to note the relationships between merger activity to multiproduct output. For example, according to a study of the U.S. supply output at the firm-product level between 1972 and 1997, less than 1% of a firm's product additions occurred due to M&As. In fact, 95% of firms engaging in M&As were found to adjust their product mix, which can be associated with ownership changes [Bernard, Redding, and Schott (2006)]. The importance of the decision as to what to offer (e.g., products and services) and the ability of firms to realize synergistic opportunities of the proposed merger, if any, can add tremendous value. We emphasize again that a successful merger depends on the ability to measure the anticipated synergy of the proposed merger [cf. Chang (1988)].

In this chapter, we focus on the case of horizontal mergers (or acquisitions) but in the more general and richer setting of multiple product supply chains. The modeling framework extends the single product one developed in Chapter 8. From a broader literature perspective, our approach is most closely related to that of Dafermos (1972) who proposed network models with multiple modes/classes of transportation. In particular, we develop a system-optimization approach to the modeling of multiproduct supply chains and their integration and we explicitly introduce capacities on the various economic activity links associated with manufacturing/production, storage, and distribution. Moreover, in this chapter, we analyze the synergy effects associated with horizontal multiproduct supply chain network integration, in terms of the operational synergy, that is, the reduction, if any, in the cost of production, storage, and distribution. Finally, the proposed computational procedure fully exploits the underlying network structure of the supply chain optimization problems both before and after integration.

Nagurney (2006b) proved that supply chain network equilibrium problems, in which there is cooperation between tiers but competition among decision-makers within a tier can be reformulated and solved as transportation network equilibrium problems. Cheng and Wu (2006) proposed a multiproduct, and multicriterion, supply-demand network equilibrium model. Davis and Wilson (2006), in turn, studied differentiated product competition in an equilibrium framework.

This chapter is organized as follows. The pre-integration multiproduct supply chain network model is developed in Section 10.2. Section 10.2 also introduces the horizontally merged (or integrated) multiproduct supply chain model. The method of quantification of the synergistic gains, if any, is provided in Section 10.3, along with new theoretical results. For completeness, we demonstrate, in Section 10.4, how a multiproduct supply chain network problem may be transformed into a single product one but with *extended* total cost functions. We use such a transformation in solving the numerical examples. In Section 10.5 we present numerical examples, which not only illustrate the richness of the framework proposed in this paper but also demonstrate quantitatively how the costs associated with horizontal multiproduct supply chain network integration affect the possible synergies. In Section 10.6, we describe how the framework may be used in the case of a non-corporate application – that of humanitarian logistics.

10.2 THE PRE- AND POST-INTEGRATION MULTIPRODUCT SUPPLY CHAIN NETWORK MODELS

This section develops the pre- and post-integration supply chain network multiproduct models using a system-optimization approach but with the inclusion of explicit capacities on the various links and the incorporation of different product volumes.

Moreover, here, we provide a variational inequality formulation of multiproduct supply chains and their integration, which enables a computational approach which fully exploits the underlying network structure. We also identify the supply chain network structures both before and after the merger and construct a synergy measure.

Section 10.2.1 describes the underlying pre-integration supply chain network associated with an individual firm and its respective economic activities of production/manufacturing, storage, distribution, and retailing. Section 10.2.2 develops the post-integration model. The models are extensions of the models in Chapter 8 to the more complex, and richer, multiproduct domain.

10.2.1 The Pre-Integration Multiproduct Supply Chain Network Model

We first formulate the pre-integration multiproduct decision-making optimization problem faced by Firms A and B and we refer to this model as the multiproduct Case 0. We assume that each firm is represented as a network of its economic activities and the notation for the pre-merger supply chain networks follows that given in Chapter 8 for the single product model. Please refer to Figure 8.1 for the supply chain network before the merger. However, now there are multiple products that are produced, stored, and distributed on these networks. In effect, one envisions now multiple products flowing on the networks in Figure 8.1. As in Section 8.2.1, we let $\mathcal{G}_i = [\mathcal{N}_i, \mathcal{L}_i]$ for $i = A, B$ denote the graph consisting of the nodes and links representing the economic activities of firm i; $i = A, B$. In Section 10.4, we show how the model in this subsection (and, analogously, the model in the next one) can be transformed into a single product network model but over J copies of the particular supply chain network, where J is the number of products that the firm can supply. Unlike the models in Chapters 8 and 9, the total cost functions on the links [cf. (8.4)] are no longer separable, that is, the total cost on a link depends not only on the flow on the particular link.

The demands for the products are assumed as given and are associated with each product, and each firm and retail pair. Let $d^j_{R^i_k}$ denote the demand for product j; $j = 1, \ldots, J$, at retail outlet R^i_k associated with firm i; $i = A, B$; $k = 1, \ldots, n^i_R$. A path (cf. Figure 8.1) consists of a sequence of links originating at a node i; $i = A, B$ and denotes supply chain activities comprising manufacturing, storage, and distribution of the products to the retail nodes. Let x^j_p denote the nonnegative flow of product j on path p. Let $P^0_{R^i_k}$ denote the set of all paths joining an origin node i with (destination) retail node R^i_k as in Figure 8.1. The following conservation of flow equations must hold for each firm i, each product j, and each retail outlet R^i_k

$$d^j_{R^i_k} = \sum_{p \in P^0_{R^i_k}} x^j_p, \quad i = A, B; \quad j = 1, \ldots, J; \quad k = 1, \ldots, n^i_R, \tag{10.1}$$

that is, the demand for each product must be satisfied at each retail outlet.

Links are denoted by a, b, etc. Let f_a^j denote the flow of product j on link a. We must have the following conservation of flow equations satisfied

$$f_a^j = \sum_{p \in P^0} x_p^j \delta_{ap}, \quad j = 1 \ldots, J; \quad \forall a \in \mathcal{L}^0, \tag{10.2}$$

where $\delta_{ap} = 1$ if link a is contained in path p and $\delta_{ap} = 0$, otherwise. Here P^0 denotes the set of *all* paths in Figure 8.1, that is, $P^0 = \cup_{i=A,B;k=1,\ldots,n_R^i} P^0_{R_k^i}$. Expression (10.2) states that the flow on a link of each product is equal to the sum of the flows of the product on paths that contain that link. As usual, the path flows must be nonnegative, that is

$$x_p^j \geq 0, \quad j = 1, \ldots, J; \quad \forall p \in P^0. \tag{10.3}$$

We group the multiproduct path flows into the vector x and the multiproduct link flows into the vector f.

Note that the different products flow on the supply chain networks depicted in Figure 8.1 and share resources with one another. To capture the costs, we proceed as follows. There is a total cost associated with each product j; $j = 1, \ldots, J$, and each link (cf. Figure 8.1) of the network corresponding to each firm i; $i = A, B$. We denote the total cost on a link a associated with product j by $\hat{c}_a^j$. The total cost of a link associated with a product, be it a manufacturing link, a shipment/distribution link, or a storage link is assumed to be, in general, a function of the flow of all the products on the link. Hence we have

$$\hat{c}_a^j = \hat{c}_a^j(f_a^1, \ldots, f_a^J), \quad j = 1, \ldots, J; \quad \forall a \in \mathcal{L}^0. \tag{10.4}$$

The top tier links in Figure 8.1 now have associated total cost functions that capture the manufacturing costs of the products; the second tier links have multiproduct total cost functions associated with them that correspond to the total costs associated with the subsequent shipment to the storage facilities; and the third tier links, because they are the storage links, have associated with them multiproduct total cost functions that correspond to storage. Finally, the bottom-tiered links, since they correspond to the shipment links to the retailers, have total cost functions associated with them that capture the costs of shipment of the products.

We assume that the total cost function for each product on each link is convex, continuously differentiable, and has a bounded third order partial derivative. Because the firms' supply chain networks, pre-integration, have no links in common (cf. Figure 8.1), their individual cost minimization problems can be formulated jointly as follows

$$\text{Minimize} \quad \sum_{j=1}^{J} \sum_{a \in \mathcal{L}^0} \hat{c}_a^j(f_a^1, \ldots, f_a^J) \tag{10.5}$$

subject to: constraints (10.1) – (10.3) and the following capacity constraints

$$\sum_{j=1}^{J} \alpha_j f_a^j \leq u_a, \quad \forall a \in \mathcal{L}^0. \tag{10.6}$$

The term α_j denotes the positive volume taken up by product j, whereas u_a denotes, as before, the positive capacity of link a. In the single product model(s) introduced in Chapters 8 and 9 there was no need to introduce volume coefficients. Of course, in the case of a single product, this model collapses to the model developed in Section 8.2.1.

Under the above imposed assumptions, the optimization problem is a convex optimization problem. If we further assume that the feasible set underlying the problem represented by the constraints (10.1) through (10.3) and (10.6) is non-empty, then it follows from the standard theory of nonlinear programming that an optimal solution exists. Of course, if the total cost functions are strictly convex, then the optimal multiproduct link flow pattern is unique.

Let $\mathcal{K}^0_M$ denote the feasible set for the multiproduct supply chain network problem, where $\mathcal{K}^0_M \equiv \{f|\exists x \text{ such that } (10.1) - (10.3) \text{ and } (10.6) \text{ hold}\}$. We assume that the feasible set $\mathcal{K}^0$ is non-empty. We associate the Lagrange multiplier β_a with constraint (10.6) for each $a \in \mathcal{L}^0$. We denote the associated optimal Lagrange multiplier by β^*_a. We now provide the variational inequality formulation of the problem. For convenience and because we are considering Case 0, we denote the solution of variational inequality (10.7) below as (f^{0*}, β^{0*}), and we refer to the corresponding vectors of variables (f, x, β) with superscripts of 0 to make explicit that this model is a multiproduct one, not a single product one, as in the preceding two chapters.

Theorem 10.1: Variational Inequality Formulation of the Multiproduct Supply Chain Network Pre-Integration Problem

The vector of link flows $f^{0*} \in \mathcal{K}^0_M$ is an optimal solution to the multiproduct supply chain network pre-integration problem if and only if it satisfies the following variational inequality problem with the vector of optimal nonnegative Lagrange multipliers β^{0*}

$$\sum_{j=1}^{J}\sum_{l=1}^{J}\sum_{a\in\mathcal{L}^0}\left[\frac{\partial \hat{c}^l_a(f^{1*}_a,\ldots,f^{J*}_a)}{\partial f^j_a} + \alpha_j\beta^*_a\right] \times [f^j_a - f^{j*}_a]$$

$$+\sum_{a\in\mathcal{L}^0}\left[u_a - \sum_{j=1}^{J}\alpha_j f^{j*}_a\right] \times [\beta_a - \beta^*_a] \geq 0, \quad \forall f^0 \in \mathcal{K}^0_M, \forall \beta^0 \geq 0. \tag{10.7}$$

Proof: See proof of Theorem 8.1.

10.2.2 The Post-Integration Multiproduct Supply Chain Network Model

We now formulate the post-integration case, referred to as Case 1 (because this is the only multiproduct case we consider here). Cases 1 and 2, as in Chapter 8, are, in fact, special instances of that Case 3. Case 1 here denotes the full merger, which in Chapter 8 corresponded to Case 3. Figure 8.4 depicts the post-integration supply chain network topology. Note that there is now a *supersource* node 0 which represents the integration of the firms in terms of their supply chain networks with additional links joining node 0 to nodes A and B, respectively.

As in the pre-integration case, the post-integration optimization problem is also concerned with total cost minimization. Specifically, we retain the nodes and links associated with the network depicted in Figure 8.1 but now we add the additional links connecting the manufacturing facilities of each firm and the distribution centers of the other firm as well as the links connecting the distribution centers of each firm and the retail outlets of the other firm. We refer to the network in Figure 8.4, underlying this integration, as $\mathcal{G}^1 = [\mathcal{N}^1, \mathcal{L}^1]$, where $\mathcal{N}^1 \equiv \mathcal{N}^0 \cup$ node 0 and $\mathcal{L}^1 \equiv \mathcal{L}^0 \cup$ the additional links as in Figure 8.4. We associate total cost functions as in (10.4) with the new links, for each product j. Note that if the total cost functions associated with the integration/merger links connecting node 0 to node A and node 0 to node B are set equal to zero, this means that the supply chain integration is *costless* in terms of the supply chain integration/merger of the two firms. Of course, non-zero total cost functions associated with these links may be used also to capture the risk associated with the integration. We will explore such issues numerically in Section 10.4.

A path p now (cf. Figure 8.4) originates at the node 0 and is destined for one of the bottom retail nodes. Let x_p^j, in the post-integrated network configuration given in Figure 8.4, denote the flow of product j on path p joining (origin) node 0 with a (destination) retail node. Then the following conservation of flow equations must hold

$$d_{R_k^i}^j = \sum_{p \in P_{R_k^i}^1} x_p^j, \quad i = A, B; \quad j = 1, \ldots, J; \quad k = 1, \ldots, n_R^i, \tag{10.8}$$

where $P_{R_k^i}^1$ denotes the set of paths connecting node 0 with retail node R_k^i in Figure 8.4. Due to the integration, the retail outlets can obtain each product j from any manufacturing facility and any distributor. The set of paths $P^1 \equiv \cup_{i=A,B;k=1,\ldots,n_R^i} P_{R_k^i}^1$.

In addition, as before, let f_a^j denote the flow of product j on link a. Hence we must also have the following conservation of flow equations satisfied

$$f_a^j = \sum_{p \in P^1} x_p^j \delta_{ap}, \quad j = 1, \ldots, J; \quad \forall a \in \mathcal{L}^1. \tag{10.9}$$

The path flows must be nonnegative for each product j, that is

$$x_p^j \geq 0, \quad j = 1, \ldots, J; \quad \forall p \in P^1. \tag{10.10}$$

Assume that the supply chain network activities have positive capacities, denoted as u_a, $\forall a \in \mathcal{L}^1$, with α_j representing the volume factor for product j. Thus the following constraints must be satisfied

$$\sum_{j=1}^{J} \alpha_j f_a^j \leq u_a, \quad \forall a \in \mathcal{L}^1. \tag{10.11}$$

Consequently, the optimization problem for the integrated multiproduct supply chain network is

$$\text{Minimize} \quad \sum_{j=1}^{J} \sum_{a \in \mathcal{L}^1} \hat{c}_a^j(f_a^1, \ldots, f_a^J) \tag{10.12}$$

subject to constraints: (10.8) - (10.11).

The solution to the optimization problem (10.12) subject to constraints (10.8) through (10.11) can also be obtained as a solution to a variational inequality problem akin to (10.7) where now $a \in \mathcal{L}^1$. The vectors f, x, and β have identical definitions as before, but are re-dimensioned/expanded accordingly and superscripted with a 1. Finally, instead of the feasible set $\mathcal{K}_M^0$ we now have $\mathcal{K}_M^1 \equiv \{f | \exists x \text{ such that } (10.8) - (10.11) \text{ hold}\}$. We assume that $\mathcal{K}_M^1$ is non-empty. We denote the solution to the variational inequality problem (10.13) below governing Case 1 by (f^{1*}, β^{1*}) and denote the vectors of corresponding variables as (f^1, β^1). We now, for completeness, provide the variational inequality formulation of the multiproduct Case 1 problem. The proof is immediate.

Theorem 10.2: Variational Inequality Formulation of the Multiproduct Supply Chain Network Post-Integration Problem
The vector of link flows $f^{1*} \in \mathcal{K}_M^1$ is an optimal solution to the multiproduct supply chain network post-integration problem if and only if it satisfies the following variational inequality problem with the vector of optimal nonnegative Lagrange multipliers β^{1*}

$$\sum_{j=1}^{J}\sum_{l=1}^{J}\sum_{a\in\mathcal{L}^1}\left[\frac{\partial \hat{c}_a^l(f_a^{1*},\ldots,f_a^{J*})}{\partial f_a^j} + \alpha_j\beta_a^*\right] \times [f_a^j - f_a^{j*}]$$

$$+ \sum_{a\in\mathcal{L}^1}[u_a - \sum_{j=1}^{J}\alpha_j f_a^{j*}] \times [\beta_a - \beta_a^*] \geq 0, \quad \forall f^1 \in \mathcal{K}_M^1, \forall \beta^1 \geq 0. \qquad (10.13)$$

Let MTC^0 denote the (multiproduct) total cost $\sum_{j=1}^{J}\sum_{a\in\mathcal{L}^0}\hat{c}_a^j(f_a^1,\ldots,f_a^J)$ evaluated under the solution f^{0*} to (10.7), and let MTC^1 denote the (multiproduct) total cost $\sum_{j=1}^{J}\sum_{a\in\mathcal{L}^1}\hat{c}_a^j(f_a^1,\ldots,f_a^J)$ evaluated under the solution f^{1*} to (10.13). Due to the similarity of variational inequalities (10.7) and (10.13), the same computational procedure can be used to compute the solutions. Indeed, we use the variational inequality formulations of the respective pre- and post-integration supply chain network problems because we can then exploit the simplicity of the underlying feasible sets $\mathcal{K}^0$ and $\mathcal{K}^1$, which include constraints with a network structure identical to that underlying multimodal system-optimized network problems (cf. Chapter 2).

10.3 QUANTIFYING SYNERGY ASSOCIATED WITH MULTIPRODUCT SUPPLY CHAIN NETWORK INTEGRATION

We measure the synergy by analyzing the multiproduct total costs before and after the supply chain network integration. For example, the synergy based on total costs [cf. (8.23)] discussed in Chapter 8, but now in a multiproduct context, which we denote here by $\mathcal{S}^{MTC}$, can be calculated as the percentage difference between the total cost before and the total cost after the integration:

$$\mathcal{S}^{MTC} \equiv \left[\frac{MTC^0 - MTC^1}{MTC^0}\right] \times 100\%. \qquad (10.14)$$

From (10.14), one can see that the lower the total cost MTC^1, the higher the synergy associated with the multiproduct supply chain network integration. Of course, in specific firm operations, one may wish to evaluate the integration of supply chain networks with only a subset of the links joining the original two supply chain networks. In that case, Figure 8.4 would be modified accordingly and the synergy as in (10.14), computed with MTC^1 corresponding to that new supply chain network topology.

We now provide a theorem that shows that if the total costs associated with the integration of the multiproduct supply chain networks of the two firms are identically equal to zero, then the associated synergy can never be negative.

Theorem 10.3: Multiproduct Synergy
If the total cost functions associated with the integration/merger links from node 0 *to nodes A and B for each product are identically equal to zero, then the associated synergy* $\mathcal{S}^{MTC}$ *can never be negative.*

Proof: We first note that the pre-integration supply chain optimization problem can be defined over the same expanded network as in Figure 10.4 but with the cross-shipment links extracted and with the paths defined from node 0 to the retail nodes. In addition, the total costs from node 0 to nodes A and B must all be equal to zero. Clearly, the total cost minimization solution to this problem yields the same total cost value as obtained for MTC^0. We must now show that $MTC^0 - MTC^1 \geq 0$.

Assume not, that is, that $MTC^0 - MTC^1 < 0$, then clearly we have not obtained an optimal solution to the post-integration problem because the new links need not be used, which would imply that $MTC^0 = MTC^1$, which is a contradiction.

Another interpretation of this theorem is that, in the system-optimization context (assuming that the total cost functions remain the same as do the demands), the addition of new links can never make the total cost increase; this is in contrast to what may occur in the context of user-optimized networks, where the addition of a new link may make everyone worse-off in terms of user cost. This is the well-known Braess paradox (1968), which was discussed in Chapters 3 and 7.

10.4 TRANSFORMATION OF A MULTIPRODUCT SUPPLY CHAIN NETWORK INTO A SINGLE PRODUCT ONE

We now show how a multiproduct supply chain network of a firm i can be transformed into a single product supply chain network by constructing multiple copies of the network. Such an idea is an application of a similar construction, but for transportation networks, by Dafermos (1972) [see also Dafermos (1971)]. Please refer to Figure 10.1. We use such a transformation when we solve numerical examples in the next section.

We construct a single product representation of the multiproduct supply chain network of firm i, which is represented by $\hat{\mathcal{G}}^i = [\hat{\mathcal{N}}_i, \hat{\mathcal{L}}_i]$, in the following way: $\hat{\mathcal{N}}_i = \mathcal{N}_i \times J$; $\hat{\mathcal{L}}_i = \mathcal{L}_i \times J$, with corresponding paths: $\mathcal{P}_i = \cup_{k=1,\ldots,n^i_R} P^0_{R^i_k} \times J$, where $\times$ denotes the Cartesian product. The network $\hat{\mathcal{G}}_i$ is hence the union of J

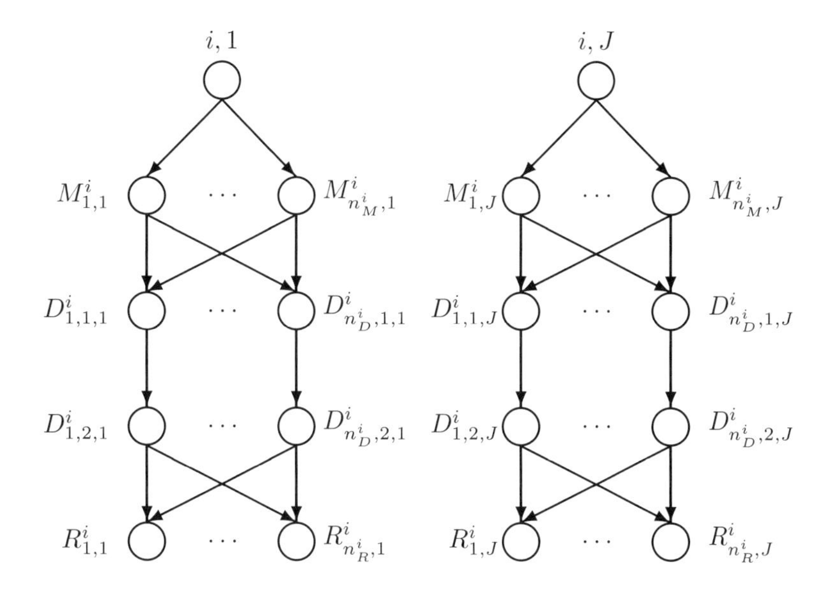

Figure 10.1: J copies of firm i's supply chain network

identical copies of the supply chain network $\mathcal{G}_i$ for firm i, as depicted in Figure 10.1. We then have the indexed set of demands

$$d_{R^i_{k,j}} \equiv d^j_{R^i_k}, \tag{10.16}$$

and the link total cost functions on the replicated network in Figure 10.1 are defined as

$$\hat{c}_{a,j} \equiv \hat{c}^j_a, \quad \forall (a,j) \in \hat{\mathcal{L}}_i, \tag{10.17}$$

where link (a,j) corresponds to link a in the j-th network copy.

We also index the path flows and links as

$$x_{p,j} \equiv x^j_p, \quad \forall (p,j) \in \mathcal{G}_i, \tag{10.18}$$

$$f_{a,j} \equiv f^j_a, \quad \forall (a,j) \in \mathcal{L}_i. \tag{10.19}$$

Finally, we have that constraint (10.5) now becomes

$$\sum_{j=1}^{J} \alpha_j f_{a,j} \leq u_a, \quad \forall (a,j) \in \hat{\mathcal{L}}_i. \tag{10.20}$$

Given any feasible path flow pattern for the multiproduct supply chain network for firm i, which induces a corresponding link flow pattern, the corresponding path flow pattern as defined by (10.18) is feasible for the single product supply chain network problem in Figure 10.1. The converse also holds true.

It follows then that

$$\hat{c}_{a,j} = \hat{c}_{a,j}(f_{a,1}, \ldots, f_{a,J}), \quad \forall (a,j) \in \hat{\mathcal{L}}_i.$$

One sees that the total cost on the single product network in Figure 10.1 corresponds thus to the total cost as in $\sum_{a \in \mathcal{L}_i} \hat{c}_a(f_a^1, \ldots, f_a^J)$. Note that the total cost on a link in the network in Figure 10.1 is determined not only by the flow on that link (as was the case in the models in Chapters 8 and 9) but also by the flows on other links. Thus we have the following.

Proposition 10.1: Multiproduct Supply Chain Network and Multicopy Network Equivalence
A path flow pattern or link flow pattern is system-optimizing for the multiproduct supply chain network of firm i; $i = A, B$, if and only if the flow pattern as defined above is system-optimizing for the single product network in Figure 10.1.

10.5 NUMERICAL EXAMPLES

In this section, we present numerical examples for which we compute the solutions to the supply chains both before and after the integration, along with the associated multiproduct total costs and synergies as defined in Section 10.3. As was also done in Chapters 8 and 9, these examples are solved using the modified projection method embedded with the equilibration algorithm (see Chapter 2). The modified projection method is guaranteed to converge if the function that enters the variational inequality is monotone and Lipschitz continuous (provided that a solution exists). These assumptions are satisfied under the conditions imposed on the multiproduct total cost functions in Section 10.2 as well as by the total cost functions underlying the numerical examples given here. Because we also assume that the feasible sets are non-empty, we are guaranteed that the modified projection method will converge to a solution of variational inequalities (10.7) and (10.13).

We implemented the computational procedure in Fortran and used a Unix system at the University of Massachusetts Amherst for the computations. The algorithm was considered to have converged when the absolute value of the difference between the computed values of the variables (the link flows; respectively, the Lagrange multipliers) at two successive iterations differed by no more than 10^{-5}. To fully exploit the underlying network structure, we first converted the multiproduct supply chain networks, into single-product "extended," ones, as described in Section 10.4, for multimodal/multiclass networks. The link capacity constraints, which do not explicitly appear in the original transportation network models, were adapted accordingly. The modified projection method yielded subproblems at each iteration in flow variables and in price variables. The former were computed using equilibration (see Chapter 2) and the latter were computed explicitly and in closed form (similar to the corresponding approach used to solve the numerical examples in Chapters 8 and 9).

For all the numerical examples, we assumed, as we had in Chapters 8 and 9, that each firm i; $i = A, B$, was involved in the production, storage, and distribution of

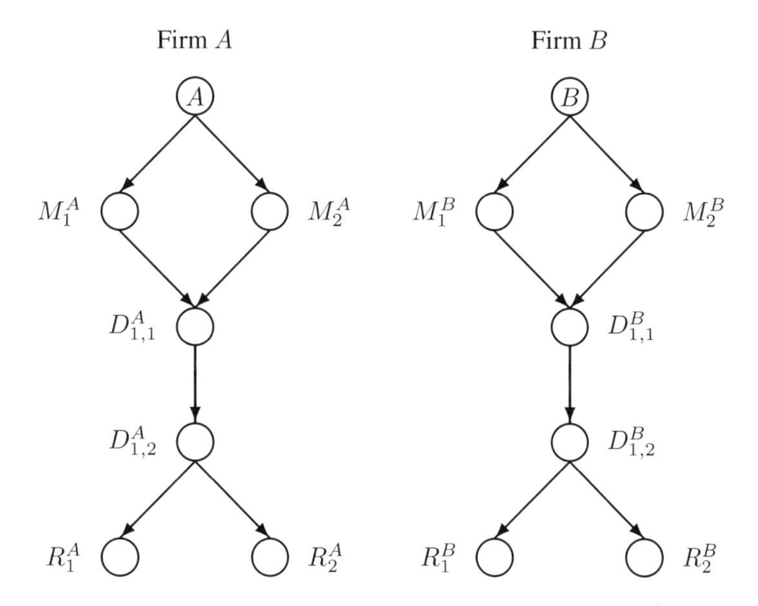

Figure 10.2: Pre-integration supply chain network topology for the numerical examples

two products, and each firm had before the integration/merger two manufacturing plants, and one distribution center, and had supplied the products to two retail outlets.

After the integration of the two firms' supply chain networks, each retailer was indifferent as to which firm supplied the products, and the integrated/merged firms could store the products at any of the two distribution centers and could supply any of the four retailers. For easy reference, Figure 10.2 depicts the pre-integration supply chain network(s); Figure 10.3 depicts the post-integration supply chain network for the numerical examples. Such supply chain network structures were also used in the numerical examples in Chapters 8 and 9.

For all the examples, we assumed that the pre-integration total cost functions and the post-integration total cost functions were nonlinear (quadratic), of the form

$$\hat{c}_a^j(f_a^1, f_a^2) = \sum_{l=1}^{2} g_a^{jl} f_a^j f_a^l + h_a^j f_a^j, \quad \forall a \in \mathcal{L}^0, \forall a \in \mathcal{L}^1; \quad j = 1, 2, \qquad (10.15)$$

with convexity of the total cost functions being satisfied.

Example 10.1

Example 10.1 serves as the baseline for our computations. The links are as defined in Table 8.1. The Example 10.1 data are now described. The pre- and post-integration total cost functions for Products 1 and 2 are listed in Table 10.1. The post-integration links that join the node 0 with nodes A and B had associated total costs equal to zero for each product $j = 1, 2$, for Examples 10.1 through 10.3. The demands at the retail

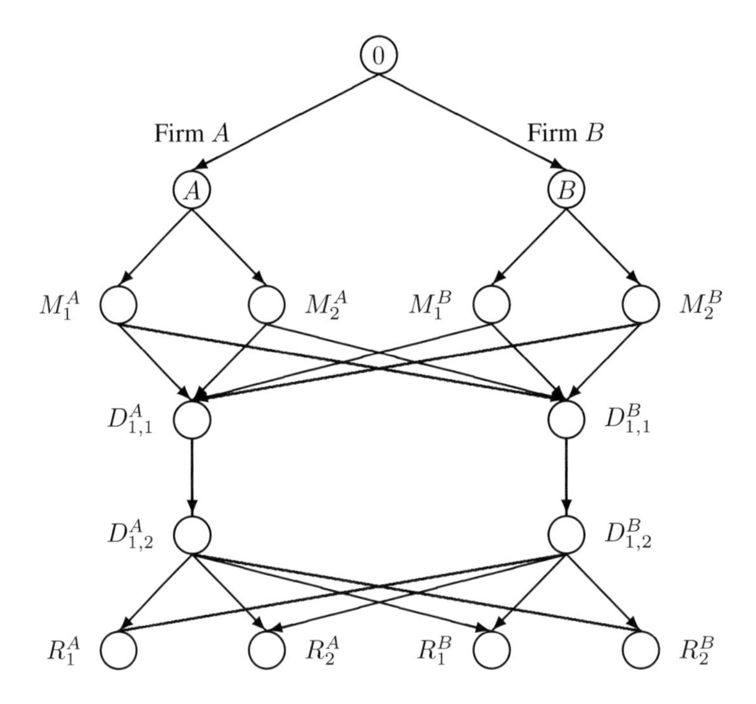

Figure 10.3: Post-integration supply chain network topology for the numerical examples

outlets for Firm A and Firm B were set to 5 for each product. Hence $d^j_{R^i_k} = 5$ for $i = A, B$; $j = 1, 2$, and $k = 1, 2$. The capacity on each link was set to 25 both before and after integration, so that $u_a = 25$ for all links $a \in \mathcal{L}^0$; $a \in \mathcal{L}^1$. The weights $\alpha_j = 1$ were set to 1 for both products $j = 1, 2$, both before and after integration; thus we assumed that the products are equal in volume.

The pre-integration optimal solutions for the product flows for each product for Examples 10.1 through 10.3 are given in Table 10.2. The post-integration optimal solutions are reported in Table 10.3 for Product 1 and in Table 10.4 for Product 2.

Because not one of the link flow capacities was reached, either before or after integration, the vectors β^{0*} and β^{1*} had all their components equal to zero. The multiproduct total cost, pre-merger, $MTC^0 = 5,702.58$. The multiproduct total cost, post-merger, $MTC^1 = 4,240.86$. Please also refer to Table 10.5 for the total cost and synergy values for this example as well as for the next two examples. The synergy $\mathcal{S}^{MTC}$ for the supply chain network integration for Example 10.1 was equal to 25.63%.

It is interesting to note that because the distribution center associated with the original Firm A has total storage costs that are lower for Product 1, whereas Firm B's distribution center has lower costs associated with the storage of Product 2, that Firm A's original distribution center after the integration/merger stores the majority of the volume of Product 1, whereas the majority of the volume of Product 2 is stored,

Table 10.1: Total cost functions for Example 10.1

Link a	$\hat{c}^1_a(f^1_a, f^2_a)$	$\hat{c}^2_a(f^1_a, f^2_a)$
1	$1(f^1_1)^2 + 2f^2_1 f^1_1 + 11f^1_1$	$2(f^2_1)^2 + 2f^1_1 f^2_1 + 8f^2_1$
2	$2(f^1_2)^2 + 2f^2_2 f^1_2 + 8f^1_2$	$1(f^2_2)^2 + 2f^1_2 f^2_2 + 6f^2_2$
3	$3(f^1_3)^2 + 2.5f^2_3 f^1_3 + 7f^1_3$	$4(f^2_3)^2 + 2.5f^1_3 f^2_3 + 7f^2_3$
4	$4(f^1_4)^2 + 1.5f^2_4 f^1_4 + 3f^1_4$	$3(f^2_4)^2 + 1.5f^1_4 f^2_4 + 11f^2_4$
5	$1(f^1_5)^2 + f^2_5 f^1_5 + 6f^1_5$	$4(f^2_5)^2 + f^1_5 f^2_5 + 11f^2_5$
6	$3(f^1_6)^2 + 1.5f^2_6 f^1_6 + 4f^1_6$	$4(f^2_6)^2 + 1.5f^1_6 f^2_6 + 10f^2_6$
7	$4(f^1_7)^2 + 2f^2_7 f^1_7 + 7f^1_7$	$2(f^2_7)^2 + 2f^1_7 f^2_7 + 8f^2_7$
8	$4(f^1_8)^2 + 3f^2_8 f^1_8 + 5f^1_8$	$4(f^2_8)^2 + 3f^1_8 f^2_8 + 6f^2_8$
9	$1(f^1_9)^2 + 1.5f^2_9 f^1_9 + 4f^1_9$	$4(f^2_9)^2 + 1.5f^1_9 f^2_9 + 6f^2_9$
10	$2(f^1_{10})^2 + 3f^2_{10} f^1_{10} + 3.5f^1_{10}$	$3(f^2_{10})^2 + 3f^1_{10} f^2_{10} + 4f^2_{10}$
11	$1(f^1_{11})^2 + 2.5f^2_{11} f^1_{11} + 4f^1_{11}$	$4(f^2_{11})^2 + 2.5f^1_{11} f^2_{11} + 5f^2_{11}$
12	$4(f^1_{12})^2 + 3f^2_{12} f^1_{12} + 6f^1_{12}$	$2(f^2_{12})^2 + 3f^1_{12} f^2_{12} + 5f^2_{12}$
13	$3(f^1_{13})^2 + 3f^2_{13} f^1_{13} + 7f^1_{13}$	$4(f^2_{13})^2 + 3f^1_{13} f^2_{13} + 10f^2_{13}$
14	$4(f^1_{14})^2 + .5f^2_{14} f^1_{14} + 4f^1_{14}$	$4(f^2_{14})^2 + .5f^1_{14} f^2_{14} + 12f^2_{14}$
15	$4(f^1_{15})^2 + 2f^2_{15} f^1_{15} + 6f^1_{15}$	$4(f^2_{15})^2 + 2f^1_{15} f^2_{15} + 7f^2_{15}$
16	$4(f^1_{16})^2 + 2f^2_{16} f^1_{16} + 6f^1_{16}$	$3(f^2_{16})^2 + 2f^1_{16} f^2_{16} + 7f^2_{16}$
17	$1(f^1_{17})^2 + 3.5f^2_{17} f^1_{17} + 4f^1_{17}$	$4(f^2_{17})^2 + 3.5f^1_{17} f^2_{17} + 5f^2_{17}$
18	$4(f^1_{18})^2 + 3f^2_{18} f^1_{18} + 9f^1_{18}$	$4(f^2_{18})^2 + 3f^1_{18} f^2_{18} + 7f^2_{18}$
19	$4(f^1_{19})^2 + 3.5f^2_{19} f^1_{19} + 7f^1_{19}$	$1(f^2_{19})^2 + 3.5f^1_{19} f^2_{19} + 9f^2_{19}$
20	$2(f^1_{20})^2 + 3f^2_{20} f^1_{20} + 5f^1_{20}$	$4(f^2_{20})^2 + 3f^1_{20} f^2_{20} + 6f^2_{20}$
21	$4(f^1_{21})^2 + 2.5f^2_{21} f^1_{21} + 3f^1_{21}$	$3(f^2_{21})^2 + 2.5f^1_{21} f^2_{21} + 9f^2_{21}$
22	$3(f^1_{22})^2 + 2f^2_{22} f^1_{22} + 4f^1_{22}$	$4(f^2_{22})^2 + 2f^1_{22} f^2_{22} + 3f^2_{22}$
23	.00	.00
24	.00	.00

post-integration, at Firm B's original distribution center. It is also interesting to note that, post-integration, the majority of the production of Product 1 takes place in Firm B's original manufacturing plants, whereas the converse holds true for Product 2. This example hence vividly illustrates the types of supply chain cost gains that can be achieved in the integration of multiproduct supply chains.

Example 10.2

Example 10.2 was constructed from Example 10.1 but with the following modifications. We now considered an idealized situation in which we assume that the total costs associated with the new integration links [see Table 10.1 (links 15 through 22)] for each product were identically equal to zero.

Post-integration, the optimal flow for each product, for each firm, has now changed; see Table 10.3 and Table 10.4. It is interesting to note that now the second manufacturing plant associated with the original Firm B produces the majority of Product 1 but the majority of Product 1 is still stored at the original distribution center of Firm A. Indeed, the zero costs associated with distribution between the original supply

Table 10.2: Pre-integration optimal flow solutions to Examples 10.1 through 10.3

Link a	From Node	To Node	f_a^{1*}	f_a^{2*}
1	A	M_1^A	8.50	.80
2	A	M_2^A	1.50	9.20
3	M_1^A	$D_{1,1}^A$	8.50	.80
4	M_2^A	$D_{1,1}^A$	1.50	9.20
5	$D_{1,1}^A$	$D_{1,2}^A$	10.00	10.00
6	$D_{1,2}^A$	R_1^A	5.00	5.00
7	$D_{1,2}^A$	R_2^A	5.00	5.00
8	B	M_1^B	.00	8.03
9	B	M_2^B	10.00	1.97
10	M_1^B	$D_{1,1}^B$	.00	8.03
11	M_2^B	$D_{1,1}^B$	10.00	1.97
12	$D_{1,1}^B$	$D_{1,2}^B$	10.00	10.00
13	$D_{1,2}^B$	R_1^B	5.00	5.00
14	$D_{1,2}^B$	R_2^B	5.00	5.00

chain networks lead to further synergies as compared to those obtained for Example 10.1.

Since, again, none of the link flow capacities were reached, either pre- or post-integration, the vectors β^{0*} and β^{1*} had all their components equal to zero. The total cost, post-merger, $MTC^1 = 2,570.27$. The synergy $\mathcal{S}^{MTC}$ for the supply chain network integration for Example 10.2 was equal to 54.93%. Observe that this obtained synergy is, in a sense, the maximum possible for this example since the total costs for both products on all the new links are all equal to zero.

Example 10.3

Example 10.3 was constructed from Example 10.2 but with the following modifications. We now assumed that the capacities associated with the links that had zero costs between the two original firms had their capacities reduced from 25 to 5. The computed optimal flow solutions are given in Table 10.3 for Product 1 and in Table 10.4 for Product 2.

We now also provide the computed vector of Lagrange multipliers β^{1*}. All terms of β^{1*} were equal to zero except those for links 15 through 20 because the sum of the corresponding product flows on each of these links was equal to the imposed capacity of 5. In particular, we now had $\beta_{15}^* = 40.82$, $\beta_{16}^* = 59.79$, $\beta_{17}^* = 14.35$, $\beta_{18}^* = 53.59$, $\beta_{19}^* = 79.95$, and $\beta_{20}^* = 68.39$.

The multiproduct total cost, after the merger, was $MTC^1 = 3,452.34$. The synergy $\mathcal{S}^{MTC}$ for the supply chain network integration for Example 10.3 was equal to 39.46%. Hence even with substantially lower capacities on the new links, given the zero costs, the synergy associated with the supply chain network integration in Example 10.3 was quite high, although not as high as obtained in Example 10.2.

Table 10.3: Post-integration optimal solutions to Examples 10.1 through 10.3 for Product 1

Link a	From Node	To Node	Ex. 12.1 f_a^{1*}	Ex. 12.2 f_a^{1*}	Ex. 12.3 f_a^{1*}
1	A	M_1^A	5.94	.76	5.36
2	A	M_2^A	.53	.00	1.98
3	M_1^A	$D_{1,1}^A$	5.94	.00	5.36
4	M_2^A	$D_{1,1}^A$	.53	.00	1.98
5	$D_{1,1}^A$	$D_{1,2}^A$	18.27	19.24	17.34
6	$D_{1,2}^A$	R_1^A	5.00	5.00	5.00
7	$D_{1,2}^A$	R_2^A	3.27	4.24	4.27
8	B	M_1^B	6.25	1.67	5.00
9	B	M_2^B	7.29	17.57	7.66
10	M_1^B	$D_{1,1}^B$	.00	.00	.00
11	M_2^B	$D_{1,1}^B$	1.73	.00	2.66
12	$D_{1,1}^B$	$D_{1,2}^B$	1.73	.76	2.66
13	$D_{1,2}^B$	R_1^B	.00	.00	.00
14	$D_{1,2}^B$	R_2^B	.00	.00	1.93
15	M_1^A	$D_{1,1}^B$	.00	.76	.00
16	M_2^A	$D_{1,1}^B$	.00	.00	.00
17	M_1^B	$D_{1,1}^A$	6.25	1.67	5.00
18	M_2^B	$D_{1,1}^A$	5.55	17.57	5.00
19	$D_{1,2}^A$	R_1^B	5.00	5.00	5.00
20	$D_{1,2}^A$	R_2^B	5.00	5.00	3.07
21	$D_{1,2}^B$	R_1^A	.00	.00	.00
22	$D_{1,2}^B$	R_2^A	1.73	.76	.73
23	O	A	6.47	.76	7.34
24	O	B	13.54	19.24	12.66

Firm B's original distribution center now stores more of Products 1 and 2 than it did in Example 10.2 (after the integration). Also, because of capacity reductions associated with the cross-shipment links there is a notable reduction in the volume of shipment of Product 1 from the second manufacturing plant of Firm B to Firm A's original distribution center and in the shipment of Product 2 from Firm A's original second manufacturing plant to Firm B's original distribution center.

10.5.1 Additional Examples

We then proceeded to ask the following question: assuming that the post-merger links joining node 0 to nodes A and B no longer had zero associated total cost for each product but, rather, reflected a cost associated with merging the two firms, what would be the effect on synergy? We further assumed that the cost [cf. (10.15)] was linear and of the specific form given by

$$\hat{c}_a^j = h_a^j f_a^j = h f_a^j, \quad j = 1, 2,$$

Table 10.4: Post-integration optimal solutions to Examples 10.1 through 10.3 for Product 2

Link a	From Node	To Node	Ex. 10.1 f_a^{2*}	Ex. 10.2 f_a^{2*}	Ex. 10.3 f_a^{2*}
1	A	M_1^A	3.44	4.66	5.00
2	A	M_2^A	11.81	11.88	8.74
3	M_1^A	$D_{1,1}^A$	.00	.88	.00
4	M_2^A	$D_{1,1}^A$	4.91	.48	3.74
5	$D_{1,1}^A$	$D_{1,2}^A$	4.91	4.82	3.74
6	$D_{1,2}^A$	R_1^A	1.52	.00	.61
7	$D_{1,2}^A$	R_2^A	2.58	.00	1.20
8	B	M_1^B	2.34	3.46	3.58
9	B	M_2^B	2.42	.00	2.68
10	M_1^B	$D_{1,1}^B$	2.34	.00	3.58
11	M_2^B	$D_{1,1}^B$	2.42	.00	2.68
12	$D_{1,1}^B$	$D_{1,2}^B$	15.09	15.18	16.26
13	$D_{1,2}^B$	R_1^B	4.88	2.72	5.00
14	$D_{1,2}^B$	R_2^B	4.30	2.46	3.07
15	M_1^A	$D_{1,1}^B$	3.44	3.78	5.00
16	M_2^A	$D_{1,1}^B$	6.89	11.40	5.00
17	M_1^B	$D_{1,1}^A$	.00	3.46	.00
18	M_2^B	$D_{1,1}^A$	.00	.00	.00
19	$D_{1,2}^A$	R_1^B	.12	2.28	.00
20	$D_{1,2}^A$	R_2^B	.70	2.54	1.93
21	$D_{1,2}^B$	R_1^A	3.48	5.00	4.39
22	$D_{1,2}^B$	R_2^A	2.42	5.00	3.80
23	0	A	15.25	16.54	13.74
24	0	B	4.76	3.46	6.26

Table 10.5: Total costs and synergy values for the examples

Measure	Example 10.1	Example 10.2	Example 10.3
Pre-Integration MTC^0	5,702.58	5,702.58	5,702.58
Post-Integration MTC^1	4,240.86	2,570.27	3,452.34
Synergy $\mathcal{S}^{MTC}$	25.63%	54.93%	39.46%

for the upper-most links (cf. Figure 10.3). Hence we assumed that all the h_a^j terms were identical and equal to an h. At what value would the synergy then for Examples 10.1, 10.2, and 10.3 become negative? Through computational experiments we were able to determine these values. In the case of Example 10.1, if $h = 36.52$, then the synergy value would be approximately equal to zero since the new total cost would be approximately equal to $MTC^0 = 5,702.58$. For any value larger than this h, one would obtain negative synergy. This has clear implications for mergers in terms of supply chain network integration and demonstrates that the total costs associated with the integration/merger itself have to be carefully weighed against the cost benefits associated with the integrated supply chain activities. In the case of Example 10.2, the h value was approximately equal to 78.3. A higher value than this h for each such merger link would result in the total cost exceeding MTC^0 and hence negative synergy would result.

Finally, for completeness, we also determined the corresponding h in the case of Example 10.3 and found the value to be $h = 78.3$, as in Example 10.2.

10.6 APPLICATION TO HUMANITARIAN LOGISTICS

In this section, we discuss how the above framework may be used and applied in the context of humanitarian operations and logistics. The supply chain is a critical component not only of corporations but also of humanitarian organizations and their logistical operations. Today, humanitarian supply chains are more extended, fragile, and time-sensitive than ever before. Moreover, the need to deliver vital goods (and services) to populations in times of crises is ever more pressing. The current humanitarian logistics environment requires that organizations mitigate risks and operate efficiently, which has spurred interest as to how to use supply chains for humanitarian logistics most efficiently and effectively.

For example, as noted in Balcik and Beamon (2008), the number of disasters is increasing as well as the people affected by them. The period between 2000 and 2004 experienced an average annual number of disasters that was 55% higher than the period 1995 and 1999, with 33% more people affected in the more recent period. According to International Strategy for Disaster Reduction (2006) 157 million people required immediate assistance due to disasters in 2005, with approximately 150 million requiring assistance the year before.

Moreover, during the rescue operations for the 2004 Asian tsunami, United Nations Secretary General Kofi Annan pointed out that the biggest challenge was the logistics of getting the essential supplies to those who needed them the most [UN News Center (2004)] especially because the infrastructure of roads, ports, and communications was completely devastated by the tsunami. The lack of infrastructure resulted in a longer window for the delivery of essential supplies. It normally takes about 2 weeks for a well-functioning aid operation to begin running smoothly but without adequate infrastructure, it can take weeks more [Perry (2005)]. There is an urgent call for effectively allocating resources in disasters relief operations due to a large amount of unsatisfied demands spread over different humanitarian service sectors. Figure 10.4 shows the average percentage of needs met by different humanitarian service sectors from 2000 to 2005 [Development Initiatives (2006)]. We can see that there exists a large portion of unsatisfied demand in such important sectors as health, economic recovery, and infrastructure as well as in water and sanitation. Therefore, how to organize and operate an effective multiproduct humanitarian logistics network is of great value to practitioners and also of interest to researchers.

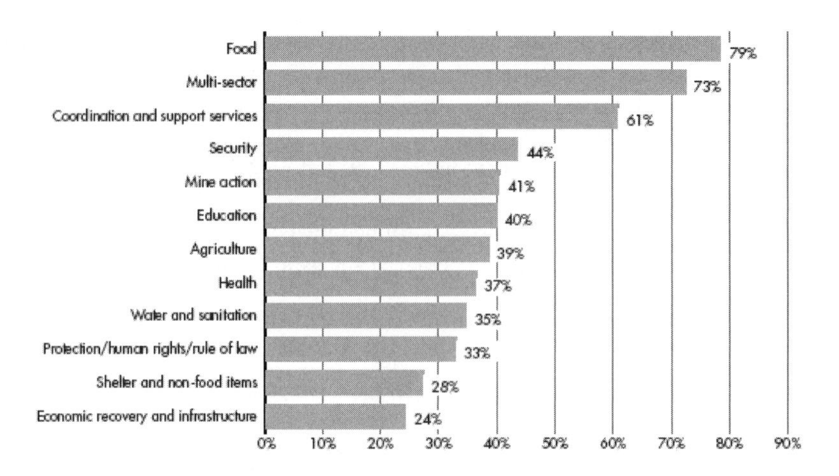

Figure 10.4: Average percentage of needs met by different sectors, 2000-2005 [Source: Financial Tracking System, Office for the Coordination of Humanitarian Affairs; obtained from Development Initiatives (2006)]

In addition, in terms of humanitarian logistics, it is clear that single product supply chains are not sufficient because populations may require not only food but clothing, fuel, medical supplies, shelter, etc. Furthermore, by having humanitarian organizations use multiproduct supply plants and distribution centers, the costs of provision may be divided among different products, which may increase efficiencies and enhance the organizations' operational effectiveness. We note that Haghani and Oh (1996) developed a multicommodity, multimodal network model for disaster relief operations, which is considered the first such general model. The model, however, assumed linear costs and the authors emphasized the need for nonlinear costs. This chapter proposes supply chain network models with nonlinear costs that can also capture the reality of congestion, which may occur in humanitarian disaster relief operations.

Organizations, in general, and humanitarian ones, in particular, may benefit not only from multiproduct supply chains [cf. Perea-Lopez, Ydtsie, and Grossman (2003)] but also from the integrated management and control of the entire supply chain [Thomas and Griffin (1996)]. Coordination enables the sharing of information, which, according to Cachon and Fisher (2000), can reduce supply chain costs by approximately 2.2%. In addition, the traditional competitive barriers between supply chain members may be mitigated thus reducing uncertainty and allowing goods to flow through the supply chain quicker and more evenly, which creates a more effective supply chain.

Furthermore, it is imperative to recognize the unique characteristics of humanitarian logistics operations in a supply chain context and how they differ from commercial supply chains. For example, Beamon (2004), Thomas and Kopczak (2005), and Van Wassenhove (2006) delineate the differences between the environment surrounding disaster relief versus the commercial environment with major implications for the underlying supply chains. In addition, due to the different nature of humanitarian logistics networks, they should be evaluated by using different performance metrics from that of commercial supply chains. Davidson (2006) argued that there are four key performance indicators in assessing humanitarian logistics networks, namely, appeal coverage, donation-to-delivery time, financial efficiency, and assessment accuracy. Appeal coverage includes two metrics: percent of appeal coverage and percent of items delivered. Donation-to-delivery time examines how long it takes for an item to be delivered to the destination country after a donor has pledged to donate it. Financial efficiency assesses how cost effective a humanitarian logistics network transports goods to beneficiaries. Assessment accuracy deals with how quickly and accurately donations are pledged and delivered to beneficiaries.

The tools proposed in this book are also of direct relevance to humanitarian logistical operations. Clearly, the importance of the decisions as to what to offer in terms of products and services as well as the ability of humanitarian organizations to realize synergistic opportunities of integrated supply chain networks, can add tremendous value and, perhaps, even save lives.

In the context of humanitarian logistics, one would consider, in terms of the models in Section 10.2, the specific humanitarian organizations (rather than firms) and their integration for a particular humanitarian disaster relief operation. Rather than manufacturing links, one would simply have supply links with associated total costs of procurement. Because the public who donates funds expects a decent utilization of the financial resources, the framework proposed here could quantify the total costs and the synergy obtained through cooperation. Of course, one could also conduct sensitivity analysis to evaluate a spectrum of possible demands as well as a range of link capacities to make sure that the demand can be satisfied. Moreover, one might incorporate also ideas from multicriteria decision-making, as discussed in Chapter 9. For example, risk may be a very important issue is disaster relief operations and the total cost functions could be generalized costs to capture risk.

Of course, in a humanitarian relief operation requiring multiple products, it may be the case that a given organization supplies distinct products from that of another organization, but to the same points of demand. The supply chain network configuration would then be as given in Figure 10.5, in the case of two organizations A and B. The paths would then be defined from nodes A to the demand markets: R_1, R_2, and, so on, through R_{n_R}, and the conservation of flow equations would correspond to distinct products on each organization's network. If these two organizations were to merge for a given humanitarian operation, then the network representing the merger, given that each organization specializes in particular products, would have additional links added from the supply points to the other organization's distribution centers. The origin points in the resulting paths could remain as before, that is, nodes A and B, and the paths would have the same destination points, that is, the demand points: $R_1, \ldots, R_{n_R}$. If one wished to assign costs to such a merger/integration then one could append two supersource nodes, with the first supersource node joined with a single link to node A and the second one joined by a link to node B. The origin points on the redefined paths would then correspond to the supersource nodes; the same

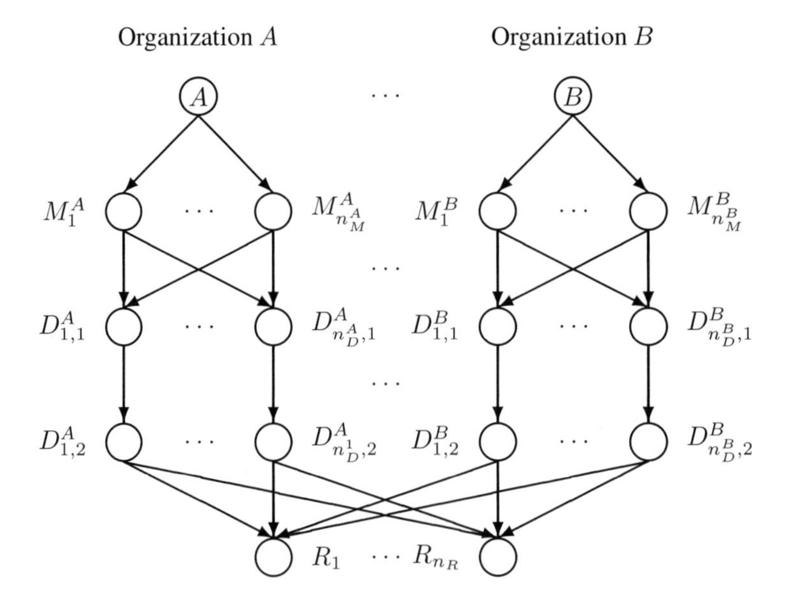

Figure 10.5: Network structure of supply chain networks of two organizations with the same demand markets

destination points would remain. Obviously, the constraints on the new links joining the supply points to the other distribution centers would have to include appropriate capacities and volume constraints as in (10.6). Clearly, such a network representation could also be applied to the merger of firms, each of which produces a distinct set of products but the two firms supply the same retail points with known and fixed demands for the products.

10.7 SUMMARY AND CONCLUSIONS

In this chapter, we developed multiproduct supply chain network models, which allow one to evaluate the total costs associated with production/supply, storage, and distribution in firms' supply chains both before and after integration. The model(s) use a system-optimization perspective and allow for explicit upper bounds on the various links associated with manufacturing, storage, and distribution. The models are formulated and solved as variational inequality problems.

In addition, we used a proposed multiproduct synergy measure to identify the potential cost gains associated with such horizontal supply chain network integrations, which can occur in a variety of application domains, including humanitarian operation ones. We proved that, in the case of zero merging costs, the associated synergy can never be negative. We computed solutions to several numerical examples for which we determined the optimal multiproduct flows and Lagrange multipliers/shadow prices associated with the capacity constraints both before and after the integration. The computational approach allows one to explore many issues regarding multiproduct supply chain network integration and to effectively ascertain the synergies prior to any implementation of a potential merger. In addition, we determined, computationally, for several examples, which identical linear costs would yield zero synergy, with higher values resulting in negative synergy.

There are numerous questions that remain and that will be considered for future research. For example, it would be very interesting to explicitly incorporate the risks associated with supply chain network integration within our framework. We have taken a step in this direction with this chapter by including total costs associated with the top-most merger links.

10.8 SOURCES AND NOTES

This chapter is based on the paper by Nagurney, Woolley, and Qiang (2008) except that here we provide additional information regarding how to transform a multiproduct supply chain network into a single product supply chain network by making as many copies of the network as there are products and by defining the flows, costs, and demands on the new network accordingly. In addition, here we discuss how the multiproduct supply chain network models and their integration can be used in humanitarian logistics to deliver supplies to human populations at their most vulnerable times of need.

In the next chapter, we formulate oligopolistic supply chain network problems. In those models the demand markets are shared by the producing firms and the firms compete.

CHAPTER 11

NETWORK OLIGOPOLIES AND THE MERGER PARADOX

11.1 INTRODUCTION

In this chapter, we formulate and solve network oligopoly problems in which several firms compete and consumers are indifferent as to who supplies the product because the product is assumed to be homogeneous. Examples of some oligopolies include oil, beer, and automobile manufacturing companies in the United States, supermarket chains in the United Kingdom; and media outlets in Australia. Examples of recent mergers of oligopolistic firms include the merger of Wells Fargo & Co. and Wachovia Corp. in financial services [cf. Phoenix Business Journal (2009)], the merger of the airlines Delta and Northwest [Global News Wire (2008)], the merger of Anheuser-Busch and InBev in the beer/beverage industry [TradingMarkets.com (2008)] and that of Molson and Coors [Beverage World (2007)], and the merger of Exxon and Mobil in the oil industry [see CNNMoney.com (1999)].

In particular, in this chapter we formulate coalitions and the associated mergers. Unlike the models in Chapters 8 through 10, here we allow for the merger of more than two firms, if appropriate, and relevant. In addition, in this chapter the demands for the product are no longer assumed to be fixed but rather are elastic and the price for the product at a particular demand market may be a function of the demands for the product at all the demand markets.

The formulation, analysis, and solution of oligopoly problems is of wide theoretical and application-based interest in economics and in operations research/management science [cf. Gabay and Moulin (1980), Murphy, Sherali, and Soyster (1982), Dafermos and Nagurney (1987), Flam and Ben-Israel (1990), Okuguchi and Szidarovsky (1990), Nagurney (1999, 2006a), Nagurney, Dupuis, and Zhang (1994), and references therein]. The topic is also garnering interest from the engineering and computer science communities, who are investigating decentralized resource allocation among competing users in communication networks, along with incentives and pricing issues [cf. Ozdaglar and Srikant (2008), Shakkottai et al. (2008), and the references therein].

Specifically, in this chapter, we consider the modeling of network oligopolies, consisting of firms that compete in a Nash (1950, 1951) and Cournot (1838) framework as well as the modeling of mergers that are formed through coalitions. The topic of mergers in an oligopolistic setting has been a major issue in economics and a subject of much discussion [cf. Salant, Switzer, and Reynolds (1983), Perry and Porter (1985), Fershtman and Judd (1987), and Farrell and Shapiro (1990)].

Much of the economics literature on this topic, however, is limited to linear cost and demand functions. Realistic oligopolistic models may not be amenable to the derivation of analytical expressions for their equilibria and hence classical techniques from industrial organization [cf. Tirole (1988)] may no longer be sufficient. Moreover, numerical methods, when applied to compute equilibria of merger problems associated with oligopolies for different cases of demand and cost parameters, may yield deeper insights and information. See Meschi (1997) for a survey of analytical perspectives for mergers and acquisitions.

In this chapter, we focus on horizontal mergers of firms in the same industry, that are engaged in oligopolistic competition, but we offer a general, network perspective, which provides a powerful graphical means by which to visualize and study different mergers formed through coalitions among firms. Furthermore, it allows the exploration, through computations, of what is known as the merger paradox. According to Pepall, Richards, and Norman (1999), "What may be surprising to you is that it is, in fact, quite difficult to construct a simple economic model in which there are sizable profitability gains for the firms participating in a horizontal merger that is not a merger to monopoly." Earlier, Salant, Switzer, and Reynolds (1983) pointed out that, in quantity-setting games, as we consider here, it is not usually advantageous for the merging firms unless the merger includes the vast majority of the firms, in particular, 80% or more [see also Creane and Davidson (2004)].

As in Chapters 8 through 10, we depict each firm as a network of its economic activities of production, storage, and distribution to the demand markets. We assume that the firms both before and after the merger(s) produce a homogeneous good in a noncooperative manner. We identify the network structure of the horizontally merged firms, which allows one to associate costs with the new links. Both models are formulated as variational inequality problems. The network structure both before and after the merger(s) differs from those associated with the merger of two firms with fixed product demands (see the models in Chapters 8 through 10). We extend the previous models substantively by allowing for competition on the production side, in distribution and storage, and across the demand markets. In addition, here we formulate any possible merger or mergers among firms and not just a partial merger of a number of firms or a complete merger to a monopoly. In other words, we allow subsets of the firms to form coalitions that result in distinct mergers. In this chapter, we utilize the change in total costs and total profits as measures to identify whether or not the merger would make economic sense. For background on game theory as related to supply chains, see Cachon and Netessine (2004).

This chapter is organized as follows. In Section 11.2, we develop the network oligopoly model in which the firms own multiple manufacturing plants, where the product is produced, and then shipped to distribution centers where it is stored before being distributed to the demand markets. We consider competition in production, distribution, and storage in that we allow the underlying functions to depend on the flows not only of the particular firm but, in general, on the flows of all the firms. We provide the game theoretic formulation of the problem, state the governing Nash-Cournot equilibrium conditions, and present alternative variational inequality formulations. In addition, we identify several well-known oligopoly models in the literature that are special cases of the new model. Specifically, we prove that both the spatial oligopoly model of Dafermos and Nagurney (1987) and the classical Cournot (1838) oligopoly model are special cases of the network oligopoly model developed in this chapter.

In Section 11.3, we propose a network economics framework for the formulation of coalitions among the firms that result in mergers. The novelty of the approach is that through a network formalism, one can graphically depict the various coalitions and mergers through the addition of new nodes and links with the subsequent additions of associated costs, if any. The governing equilibrium/optimality conditions are also formulated as variational inequality problems.

In Section 11.4, we consider the solution of the network oligopoly problems, both before and after the mergers. We propose the Euler method (cf. Section 2.4.4), which is a special case of the general iterative scheme introduced by Dupuis and Nagurney (1993) for determining the stationary points of projected dynamical systems; equivalently, solutions of variational inequality problems. We demonstrate that in the context of our models the Euler method resolves the network problems into subproblems that can be solved at each iteration explicitly and in closed form. These explicit formulae may be interpreted as discrete-time adjustment or tatonnement processes. A variety of economic equilibrium problems (and the associated tatonnement/adjustment processes) have been modeled and solved to-date as projected dynamical systems, including dynamic spatial price problems [see Nagurney, Takayama, and Zhang (1995) and Nagurney and Zhang (1996a)] and dynamic network oligopolies [cf. Nagurney, Dupuis, and Zhang (1994)].

In Section 11.5, we present numerical examples for a spectrum of supply chain network structures of oligopolistic firms and evaluate various mergers formed through coalitions. We also explore questions regarding the merger paradox computationally, an approach that allows one to gain insights through numerical experimentation. We

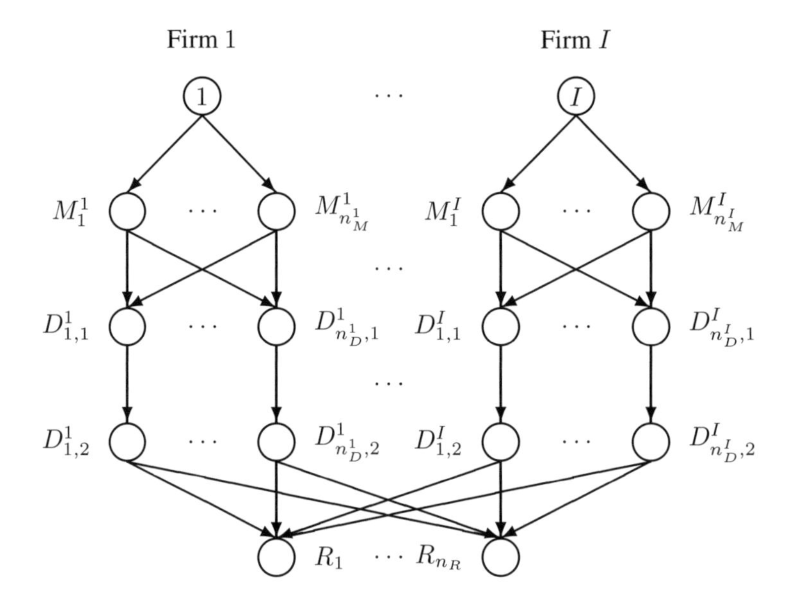

Figure 11.1: Network structure of the oligopoly

compute the total costs, the revenues, and the profits for the supply chain networks before and after the mergers.

In Section 11.6, we summarize the results in this chapter and give suggestions for future research.

11.2 THE NETWORK OLIGOPOLY MODEL

In this section, we develop the network oligopoly model and identify several special cases. We consider a finite number of I firms, with a typical firm denoted by i, that are involved in the production, storage, and distribution of a homogeneous product and that compete noncooperatively in an oligopolistic manner.

We assume that each firm is represented as a network of its economic activities (cf. Figure 11.1). Each firm i; $i = 1, \ldots, I$ has n^i_M manufacturing facilities/plants; n^i_D distribution centers, and serves the same n_R retail outlets/demand markets. Let $\mathcal{L}^0_i$ denote the set of directed links representing the economic activities associated with Firm i; $i = 1, \ldots, I$. Let $\mathcal{L}^0 \equiv \cup_{i=1,I}\mathcal{L}^0_i$. Let $\mathcal{G}^0 = [\mathcal{N}^0, \mathcal{L}^0]$ denote the graph consisting of the set of nodes $\mathcal{N}^0$ and the set of links $\mathcal{L}^0$ in Figure 11.1.

The links from the top-tiered nodes i; $i = 1, \ldots, I$ in Figure 11.1 are connected to the manufacturing nodes of the respective firm i, which are denoted, respectively, by: $M^i_1, \ldots, M^i_{n^i_M}$, and these links represent the manufacturing links. The links from the manufacturing nodes, in turn, are connected to the distribution center nodes of each firm i; $i = 1, \ldots, I$, which are denoted by $D^i_{1,1}, \ldots, D^i_{n_D{}^i,1}$. These links correspond to the shipment links between the manufacturing plants and the distribution centers where the product is stored. The links joining nodes $D^i_{1,1}, \ldots, D^i_{n^i_D,1}$ with nodes $D^i_{1,2}, \ldots, D^i_{n^i_D,2}$ for $i = 1, \ldots, I$ correspond to the storage links. Finally, there are shipment links joining the nodes $D^i_{1,2}, \ldots, D^i_{n^i_D,2}$ for $i = 1, \ldots, I$ with the demand market nodes $R_1, \ldots, R_{n_R}$. Note that the competing firms share the same demand markets (unlike in the models in Chapters 8 through 10), and hence the links from the distribution centers terminate in the same destination nodes in Figure 11.1.

Assume that associated with each link (cf. Figure 11.1) of the network corresponding to each firm i; $i = 1, \ldots, I$ is a total cost. We denote, without any loss in generality, the links by a, b, etc., and the total cost on a link a by $\hat{c}_a$. Let d_{R_k} denote the demand for the product at demand market R_k; $k = 1, \ldots, n_R$. Let x_p denote the nonnegative flow of the product on path p joining (origin) node i; $i = 1, \ldots, I$ with a (destination) demand market node. Then the following conservation of flow equations must hold

$$d_{R_k} = \sum_{p \in P^0_{R_k}} x_p, \quad k = 1, \ldots, n_R, \tag{11.1}$$

where $P^0_{R_k}$ denotes the set of paths connecting the (origin) nodes i; $i = 1, \ldots, I$ with (destination) demand market R_k. In particular, we have that $P^0_{R_k} = \cup_{i=1,\ldots,I} P^0_{R^i_k}$, where $P^0_{R^i_k}$ now denotes the set of paths from origin node i to demand market k as in Figure 11.1. According to (11.1), the demand at each demand market must be equal to the sum of the product flows from all firms to that demand market. The product is assumed to be homogeneous, and thus, at least in principle, the product at a particular demand market can be obtained from any of the manufacturing firms.

Assume that there is a demand price function associated with the product at each demand market. We denote the demand price at demand market R_k by ρ_{R_k} and assume, as given, the demand price functions

$$\rho_{R_k} = \rho_{R_k}(d), \quad k = 1, \ldots, n_R, \tag{11.2}$$

where d is the n_R-dimensional vector of demands at the demand markets. Note that we consider the general situation in which the demand price for the product at a particular demand market may depend on the demand for the product at the other demand markets. Assume that the demand price functions are continuous, continuously differentiable, and monotone decreasing. Note that the consumers at each demand market are indifferent as to which firm produced the homogeneous product.

In addition, let f_a denote the flow of the product on link a. Hence the following conservation of flow equations must be satisfied

$$f_a = \sum_{p \in P^0} x_p \delta_{ap}, \quad \forall a \in \mathcal{L}^0, \tag{11.3}$$

where $\delta_{ap} = 1$ if link a is contained in path p and $\delta_{ap} = 0$, otherwise. Here P^0 denotes the set of *all* paths in Figure 11.1, that is, $P^0 = \cup_{k=1,\ldots,n_R} P^0_{R_k}$. There are

n_{P^0} paths in the network in Figure 11.1. Obviously, before any coalition formation and merger the paths associated with a given firm have no links in common with paths of any other firm. This changes when the coalitions are formed, in which case the number of paths and the sets of paths also change, as do the number of links and the sets of links, as we demonstrate in Section 11.3. We use P_i^0 to denote the set of all paths from firm i to all the demand markets for $i = 1, \ldots, I$. There are $n_{P_i^0}$ paths from the firm i node to the demand markets.

Of course, the path flows must be nonnegative, that is

$$x_p \geq 0, \quad \forall p \in P^0. \tag{11.4}$$

The total cost on a link, be it a manufacturing link, a distribution link, or a storage link is assumed, in general, to be a function of the flows of the product on all the links. We denote the total cost on link a by $\hat{c}_a$ and assume that

$$\hat{c}_a = \hat{c}_a(f), \quad \forall a \in \mathcal{L}^0, \tag{11.5}$$

where f is the vector of all the link flows. The total cost on each link is assumed to be convex and continuously differentiable. The same is assumed for all links that are added to form the mergers. Observe that the link total cost functions (11.5) are more general than those in the network models in Chapters 8 and 9 in that the total cost on a link no linger depends solely on the flow of the particular link. This more general function can capture competition associated with the particular economic link activities. For example, there may be competition to secure resources needed for producing the product. Of course, we are interested in competition between or among the firms and not within a firm (although that may, of course, occur in certain situations as between manufacturing plants).

The profit function u_i of firm i; $i = 1, \ldots, I$, is the difference between the firm's revenue and its total costs, that is

$$u_i = \sum_{k=1}^{n_R} \rho_{R_k}(d) \sum_{p \in P^0_{R_k^i}} x_p - \sum_{a \in \mathcal{L}_i^0} \hat{c}_a(f). \tag{11.6}$$

In view of (11.1), (11.2), (11.3), and (11.5), we may write

$$u = u(x), \tag{11.7}$$

where x is the vector of all the path flows $\{x_p, p \in P^0\}$, and u is the I-dimensional vector of the firms' profits.

We now consider the oligopolistic market mechanism in which the I firms produce and distribute the product in a noncooperative manner, each one trying to maximize its own profit. We seek to determine a nonnegative path flow pattern x^* for which the I firms will be in a state of equilibrium as defined next.

Definition 11.1: Network Cournot-Nash Equilibrium
A product flow pattern $x^* \in R_+^{n_{P^0}}$ is said to constitute a network Cournot-Nash equilibrium if for each firm i; $i = 1, \ldots, I$

$$u_i(x_i^*, \hat{x}_i^*) \geq u_i(x_i, \hat{x}_i^*), \quad \forall x_i \in R_+^{n_{P_i^0}}, \tag{11.8}$$

where $x_i \equiv \{\{x_p\}|p \in P_i^0\}$ and $\hat{x}_i^* \equiv (x_1^*, \ldots, x_{i-1}^*, x_{i+1}^*, \ldots, x_I^*)$.

Note that, according to (11.8), a Cournot-Nash equilibrium has been established if no firm can increase its profits unilaterally.

The variational inequality formulation of the Cournot-Nash [Cournot (1838), Nash (1950, 1951)] network equilibrium satisfying Definition 11.1 is given in the following theorem.

Theorem 11.1: Variational Inequality Formulation of Network Cournot-Nash Equilibrium
Assume that for each firm i; $i = 1, \ldots, I$, the profit function $u_i(x)$ is concave with respect to the variables x_p; $p \in P_i^0$ and is continuously differentiable. Then $x^* \in R_+^{n_{P^0}}$ is a network Cournot-Nash equilibrium according to Definition 11.1 if and only if it satisfies the variational inequality

$$-\sum_{i=1}^{I} \sum_{p \in P_i^0} \frac{\partial u_i(x^*)}{\partial x_p} \times (x_p - x_p^*) \geq 0, \quad \forall x \in R_+^{n_{P^0}}, \tag{11.9}$$

or, equivalently, due to (11.1), (11.2), and (11.3): determine $x^* \in \mathcal{K}^0$ satisfying:

$$\sum_{i=1}^{I} \sum_{k=1}^{n_R} \sum_{p \in P^0_{R_k^i}} [\frac{\partial \hat{C}_p(x^*)}{\partial x_p} - \rho_{R_k}(x^*) - \sum_{l=1}^{n_R} \frac{\partial \rho_{R_l}(x^*)}{\partial d_{R_k}} \sum_{p \in P^0_{R_k^i}} x_p^*] \times [x_p - x_p^*] \geq 0,$$

$$\forall x \in \mathcal{K}^0, \tag{11.10}$$

where $\mathcal{K}^0 \equiv \{x | x \in R_+^{n_{P^0}}\}$ and $\frac{\partial \hat{C}_p(x)}{\partial x_p} \equiv \sum_{b \in L_i^0} \sum_{a \in \mathcal{L}_i^0} \frac{\partial \hat{c}_b(f)}{\partial f_a} \delta_{ap}$ for paths $p \in P_i^0$.

Proof: Follows directly from Gabay and Moulin (1980) and Dafermos and Nagurney (1987). Here we have also used the fact that the demand price functions (11.2) can be reexpressed, in light of (11.1), directly as functions of path flows.

It is interesting to relate this network oligopoly model to the spatial oligopoly model proposed by Dafermos and Nagurney (1987), which assumed a finite number of competing manufacturing firms that supplied spatially separated demand markets. The costs faced by the manufacturing firms included both production costs and transportation costs.

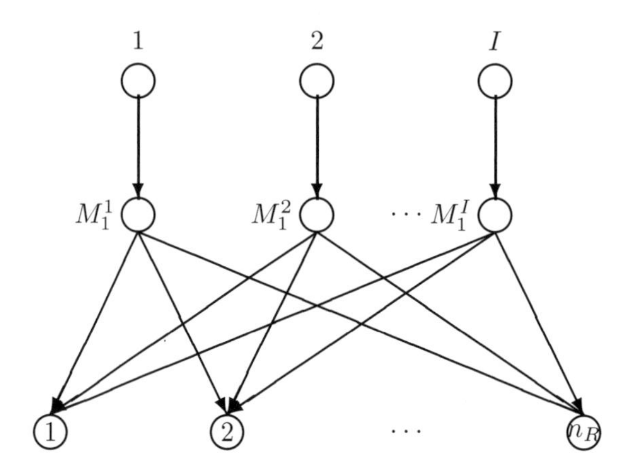

Figure 11.2: Network structure of the spatial oligopoly

Corollary 11.1: Relationship between the Network Oligopoly Model and a Spatial Oligopoly
Assume that there are I firms in the network oligopoly model and that each firm has a single manufacturing plant and a single distribution center. Assume also that the distribution costs from each manufacturing plant to the distribution center and the storage costs are all equal to zero. Then the resulting model is isomorphic to the spatial oligopoly model of Dafermos and Nagurney (1987) whose underlying network structure is given in Figure 11.2.

Proof: Follows from Dafermos and Nagurney (1987) and Nagurney (1999).

Hence the network oligopoly model, which captures the full supply chain activities, is an extension of the spatial oligopoly model of Dafermos and Nagurney (1987). It is also interesting to note that in the new network oligopoly model there is competition on the supply and distribution sides as well as on the demand side, because the cost functions associated with production, with distribution, and with storage may, in general, depend on not only the flows of the particular firm but also on the flows of all the firms.

In Dafermos and Nagurney (1987) and in Nagurney (1999) the network structure of the spatial oligopoly problem is depicted as a bipartite graph with the manufacturing assumed to take place at the manufacturing/production nodes. In the network formalism here, we associate the production with links, and the outputs are hence the link flows, and Figure 11.2 makes this explicit. Dafermos and Nagurney (1987) also establish the relationships between spatial Cournot oligopolies and perfectly competitive spatial price equilibrium problems. Nagurney, Dupuis, and Zhang (1994), in turn, developed a dynamic version of the model of Dafermos and Nagurney (1987) using projected dynamical systems theory [see also Nagurney and Zhang (1996)].

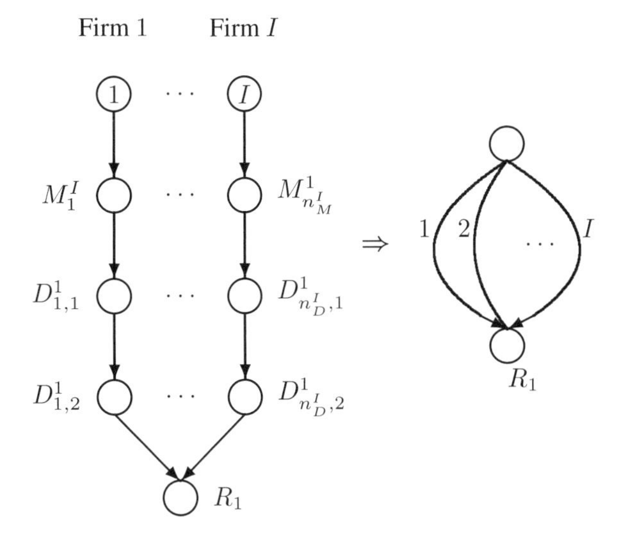

Figure 11.3: Network structure of the classical oligopoly

It is also relevant to demonstrate the relationship between the new network oligopoly model and the classical Cournot (1838) oligopoly model in quantity variables, which has been formulated as a variational inequality problem by Gabay and Moulin (1980) and has been studied by both economists and operations researchers [cf. Murphy, Sherali, and Soyster (1982), Harker (1984), Flam and Ben-Israel (1990), Nagurney (1993), and references therein]. Indeed, we have the following corollary, the proof of which is immediate.

Corollary 11.2: Relationship between the Network Oligopoly Model and the Classical Oligopoly
Assume that there is a single manufacturing plant associated with each firm in the above model and a single distribution center. Assume also that there is a single demand market and that the manufacturing cost of each manufacturing firm depends upon only its own output. Then, if the storage and distribution cost functions are all identically equal to zero, the model collapses to the classical oligopoly model in quantity variables. Furthermore, if $I = 2$, one then obtains the classical duopoly model.

The network structure for the classical oligopoly problem is depicted in Figure 11.3. The classical model is an *aspatial* model because there are no explicit transportation/transaction costs between the firms/manufacturers and the demand market. In the case of a linear demand price function and quadratic manufacturing total cost functions (which are separable, that is, the total cost associated with a firm depends only on that firm's output), then the equilibrium production pattern can be computed using a special purpose algorithm which is similar to the exact equilibration algorithm for elastic demand transportation network equilibrium problems with the special network structure given in Figure 11.3 [see Nagurney (1999)].

Existence results for both spatial and aspatial oligopoly problems can be found in Nagurney and Zhang (1996) and the references therein. Typically, either strong monotonicity or coercivity conditions (cf. Chapter 2) are imposed on the relevant functions to guarantee existence of a Cournot-Nash equilibrium in such applications.

11.3 MERGERS THROUGH COALITION FORMATION

We now consider mergers through coalition formation. In the literature, usually, one considers mergers between/among a subset of firms in an oligopoly and then explores the effect on demands, prices, profits, etc. Here we would like to exhaust all possibilities in terms of possible mergers through different coalition formations. Obviously, an extreme situation would occur if all the firms in the industry merge, leading to the much-studied monopoly case.

Specifically, we assume that coalitions are formed among the I firms as follows. The first $n_{1'}$ firms join to form new Firm $1'$, the second group of $n_{2'}$ firms join to form Firm $2'$, and so on, through the remaining $n_{I'}$ firms joining to form the I'th firm. Associated with a coalition formation in the form of a merger, we construct a new *supersource* node to represent the new firm, and we construct new links from each such supersource node, which now becomes an origin node, to the respective top-most original firm nodes. If firms do not enter into any merger/coalition we simply retain the original nodes for that firm and retain their top-most nodes as the origin nodes. In addition, because the newly merged firms now share resources, including their distribution centers, we now add new links from their original manufacturing nodes to the other firms' in the merger distribution center nodes and associate total cost functions with these new links. For example, in Figure 11.4, we depict a coalition formation resulting in a merger among the first $n_{1'}$ firms yielding new Firm $1'$ with the remainder of the firms merging with one another to form Firm $n_{2'}$.

Associated with the coalition formation resulting in a particular set of mergers is a new graph denoted by $\mathcal{G}^1$, which consists of the original nodes and links as in $\mathcal{G}^0$ but with the new nodes and links $[\mathcal{N}^1, \mathcal{L}^1]$ to represent the formation of the new firms.

This model is interesting and relevant, because not all firms in an industry necessarily need to merge when a merger occurs. Such a model also allows one to evaluate the effect of the merger on total costs and profits of firms not associated with the merger. Moreover, it allows one to explore questions regarding the merger paradox.

We define $P^1_{R^i_k}$ as the set of paths joining origin node i with demand market R_k, where $i = 1', 2', \ldots, I'$ with the proviso that we relabel the origin nodes of the unmerged firms accordingly.

Let x_p now denote (cf. Figure 11.4) the nonnegative flow on a path joining an (origin) node i with a demand market node. Then the following conservation of flow equations must now be satisfied

$$d_{R_k} = \sum_{p \in P^1_{R_k}} x_p, \quad k = 1, \ldots, n_R, \tag{11.11}$$

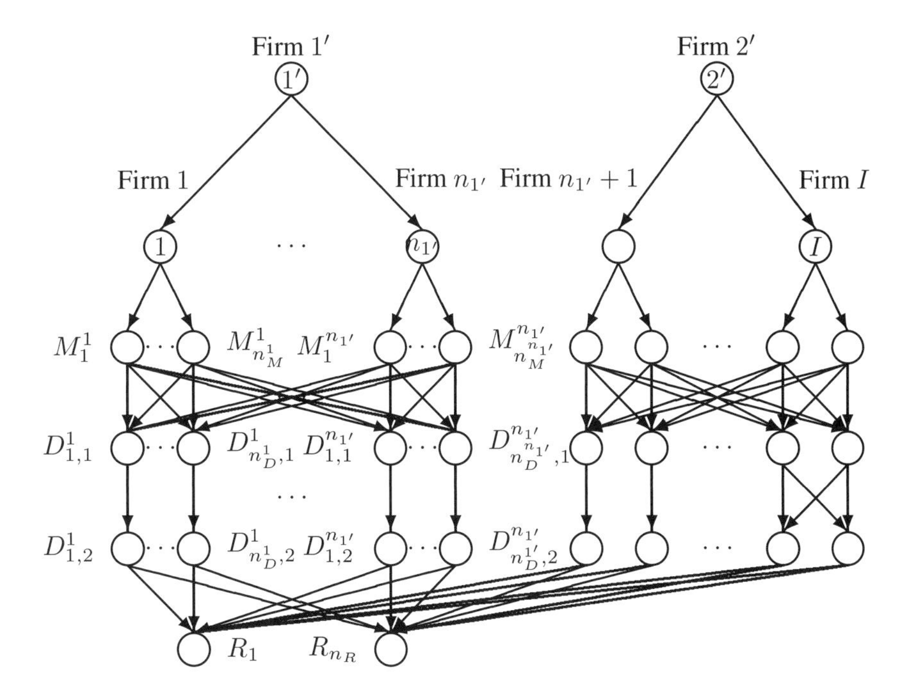

Figure 11.4: The network for the mergers of the first $n_{1'}$ firms and the next $n_{2'}$ firms

where $P_{R_k}^1 = \cup_{i=1',\dots,I'} P_{R_k^i}^1$. We let P_i^1 denote the set of all paths emanating from node i to the demand markets for $i = 1', \dots, I'$.

The demand price functions remain as in (11.2). The link conservation of flow equations now take the form

$$f_a = \sum_{p \in P^1} x_p \delta_{ap}, \quad \forall a \in \mathcal{L}^1, \tag{11.12}$$

where $P^1 = \cup_{k=1,\dots,n_R} P_{R_k}^1$.

Of course, as in the preceding model, we must have that the path flows are nonnegative, that is

$$x_p \geq 0, \quad \forall p \in P^1. \tag{11.13}$$

Again, we assume that the new links that correspond to the merger have total cost functions associated with them; hence

$$\hat{c}_a = \hat{c}_a(f), \quad \forall a \in \mathcal{L}^1. \tag{11.14}$$

The new total cost functions on the new links have properties corresponding to those as in the original links. We retain the original total cost functions on the original links.

The profit now for the firms, with the firms renumbered as $i = 1', \ldots, I'$ can be expressed as

$$u_i = \sum_{k=1}^{n_R} \rho_{R_k}(d) \sum_{p \in P^1_{R^i_k}} x_p - \sum_{a \in \mathcal{L}^1_i} \hat{c}_a(f), \tag{11.15}$$

where $\mathcal{L}^1_i$ denotes that subset of links in $\mathcal{L}^1$ corresponding to firm i; $i = 1', \ldots, I'$.

We can now adapt Definition 11.1 to the merger/coalition setting, in which firms $1', \ldots, I'$ compete with one another in a Cournot-Nash setting until the equilibrium is attained. We impose the same assumptions on the utility functions here as were imposed on the utility functions in Theorem 11.1.

Definition 11.2: Network Merger Cournot-Nash Equilibrium
A product flow pattern $x^* \in R^{n_{P^1}}_+$ is said to constitute a network Cournot-Nash equilibrium for the particular merger, due to coalition formation, if for firm i; $i = 1', \ldots, I'$

$$u_i(x^*_i, \hat{x}^*_i) \geq u_i(x_i, \hat{x}^*_i), \quad \forall x_i \in R^{n_{P^1_i}}_+, \tag{11.16}$$

where now, w.l.o.g. $x_i \equiv \{\{x_p\} | p \in P^1_i\}$ and $\hat{x}^*_i \equiv (x^*_{1'}, \ldots, x^*_{i-1}, x^*_{i+1}, \ldots, x_{I'}{}^*)$.

Theorem 11.2
Assume that for each firm i; $i = 1', \ldots, I'$, the profit function $u_i(x)$ given by (11.15) is concave with respect to the variables x_p; $p \in P^1_i$ and is continuously differentiable. Then $x^* \in R^{n_{P^1}}_+$ is a network merger Cournot-Nash equilibrium according to Definition 11.2 if and only if it satisfies the variational inequality

$$-\sum_{i=1}^{I'} \sum_{p \in P^1_i} \frac{\partial u_i(x^*)}{\partial x_p} \times (x_p - x^*_p) \geq 0, \quad \forall x \in R^{n_{P^1}}_+, \tag{11.17}$$

or, equivalently, in view of (11.11), (11.12), and (11.2), determine $x^* \in \mathcal{K}^1$, satisfying

$$\sum_{i=1}^{I'} \sum_{k=1}^{n_R} \sum_{p \in P^1_{R^i_k}} [\frac{\partial \hat{C}_p(x^*)}{\partial x_p} - \rho_{R_k}(x^*) - \frac{\partial \rho_{R_l}(x^*)}{\partial d_{R_k}} \sum_{p \in P^1_{R^i_k}} x^*_p] \times [x_p - x^*_p] \geq 0,$$

$$\forall x \in \mathcal{K}^1, \tag{11.18}$$

where $\mathcal{K}^1 \equiv \{x | x \in R^{n_{P^1}}_+\}$ and here $\frac{\partial \hat{C}_p(x)}{\partial x_p} \equiv \sum_{b \in \mathcal{L}^1_i} \sum_{a \in \mathcal{L}^1_i} \frac{\partial \hat{c}_b(f)}{\partial f_a} \delta_{ap}$ for all paths $p \in P^1_i$.

For existence results for both classical and spatial oligopoly problems, see Nagurney and Zhang (1996). The results therein can be adapted to this more general oligopoly network setting as well.

11.4 THE ALGORITHM

In this section, we recall the Euler method [see also (2.54)] for the solution of both network equilibrium problems governed by variational inequalities (11.10) and (11.18). Specifically, at an iteration $\mathcal{T}$ of the Euler method, one computes

$$X^{\mathcal{T}} = P_{\mathcal{K}}(X^{\mathcal{T}-1} - \alpha_{\mathcal{T}} F(X^{\mathcal{T}-1})), \tag{11.19}$$

where $P_{\mathcal{K}}$ is the projection [cf. (2.56)] on the feasible set $\mathcal{K}$ and F is the function that enters the variational inequality problem: determine $X^* \in \mathcal{K}$ such that

$$\langle F(X^*)^T, X - X^* \rangle \geq 0, \quad \forall X \in \mathcal{K}, \tag{11.20}$$

where $\langle \cdot, \cdot \rangle$ is the inner product in N-dimensional Euclidean space, $X \in R^N$, and $F(X)$ is an N-dimensional function from $\mathcal{K}$ to R^N, with $F(X)$ being continuous.

Both variational inequality problems (11.10) and (11.18) can be put into the standard form (2.19) [or (11.20)]. Recall that for convergence of the Euler method the sequence $\{\alpha_{\mathcal{T}}\}$ must satisfy $\sum_{\mathcal{T}=1}^{\infty} \alpha_{\mathcal{T}} = \infty$, $\alpha_{\mathcal{T}} > 0$, $\alpha_{\mathcal{T}} \to 0$, as $\alpha_{\mathcal{T}} \to \infty$. Specific conditions for convergence of this scheme to the solution of variational inequality (11.20) can be found for a variety of network-based problems, similar to those constructed here, in Nagurney and Zhang (1996) and the references therein.

Explicit Formulae for the Euler Method Applied to the Network Variational Inequality (11.10)

The elegance of this procedure for the computation of solutions to the network problems modeled in Section 11.2 can be seen in the following explicit formulae. Indeed, (11.19) for the network oligopoly problem governed by variational inequality problem (11.10) yields the following: for all i, k, and paths $p \in P^0_{R^i_k}$

$$x_p^{\mathcal{T}} = \max\{0, x_p^{\mathcal{T}-1} + \alpha_{\mathcal{T}}(\rho_{R_k}(x^{\mathcal{T}-1}) - \sum_{l=1}^{n_R} \frac{\partial \rho_l(x^{\mathcal{T}-1})}{\partial d_{R_k}} \sum_{p \in P^0_{R^i_k}} x_p^{\mathcal{T}-1} - \frac{\partial \hat{C}_p(x^{\mathcal{T}-1})}{\partial x_p})\}. \tag{11.21}$$

Explicit Formulae for the Euler Method Applied to the Network Merger Variational Inequality (11.18)

In the case of the merger supply chain network problem, the iterative step (11.19) for the corresponding variational inequality problem (11.18) yields the following expression: for all i, k, and paths $p \in P^1_{R^i_k}$

$$x_p^{\mathcal{T}} = \max\{0, x_p^{\mathcal{T}-1} + \alpha_{\mathcal{T}}(\rho_{R_k}(x^{\mathcal{T}-1}) - \sum_{l=1}^{n_R} \frac{\partial \rho_l(x^{\mathcal{T}-1})}{\partial d_{R_k}} \sum_{p \in P^1_{R^i_k}} x_p^{\mathcal{T}-1} - \frac{\partial \hat{C}_p(x^{\mathcal{T}-1})}{\partial x_p})\}. \tag{11.22}$$

In particular, both (11.21) and (11.22) are similar to the iterative step of the Euler method for elastic demand network equilibrium problems [cf. (2.60)]. Note that variational inequality problems (11.10) and (11.18) can also be reformulated in link flow variables. However, we have provided formulations in path flow variables, because, computationally, these lead to the above simple and explicit formulae. Bertsekas and Gafni (1982) also proposed projection methods in path flow variables for the traffic assignment (network equilibrium) problem, along with convergence results. It is also worth noting that both (11.21) and (11.22) can be implemented on parallel architectures, if available [see also Bertsekas and Tsitsiklis (1989) and Nagurney (1996)].

11.5 NUMERICAL EXAMPLES

In this section, we present three sets of numerical network oligopoly examples of increasing complexity. In Set 1, reported in Section 11.5.1, we present mergers associated with oligopoly examples consisting of four firms where each firm has a single manufacturing plant and a single distribution center and there is a single retailer/demand market that each of the firms competes in. In Set 2, reported in Section 11.5.2, we again consider such oligopoly problems but, unlike the problems in Set 1, the total cost functions on the new merger links are no longer equal to zero. In Section 11.5.3, we report Set 3 numerical examples, in which we compute solutions to the mergers of more complex networks with multiple demand markets.

We implemented the Euler method, as discussed in Section 11.4, for each of the particular supply chain network problems, before and after the mergers. The codes were implemented in Fortran and the computer system used for the computations was a Unix system at the University of Massachusetts Amherst. The convergence criterion and tolerance were $|X^{\mathcal{T}} - X^{\mathcal{T}-1}| \leq .001$ for all the examples. Specifically, we assumed that the algorithm had converged of the path flows between two successive iterations differed by no more than $\epsilon = .001$. The sequence $\{\alpha_{\mathcal{T}}\}$ used [cf. (11.19)] was $.1\{1, \frac{1}{2}, \frac{1}{2}, \frac{1}{3}, \frac{1}{3}, \frac{1}{3}, \ldots, \}$. This sequence satisfies the condition required of it for convergence of the algorithm.

11.5.1 Problem Set 1

In this set, we solved oligopolistic network problems both before and after various mergers.

Example 11.1.1

As described earlier, the original/baseline problem, Example 11.1.1, consisted of four oligopolistic firms (cf. Figure 11.5). For simplicity, we let all the total cost functions on the links representing this baseline problem be equal and given by

$$\hat{c}_a(f) = 2f_a^2 + f_a, \quad \forall a \in \mathcal{L}_i^0; i = 1, 2, 3, 4. \tag{11.23}$$

The demand price function at the single demand market was given by

$$\rho_{R_1}(d) = -d_{R_1} + 200. \tag{11.24}$$

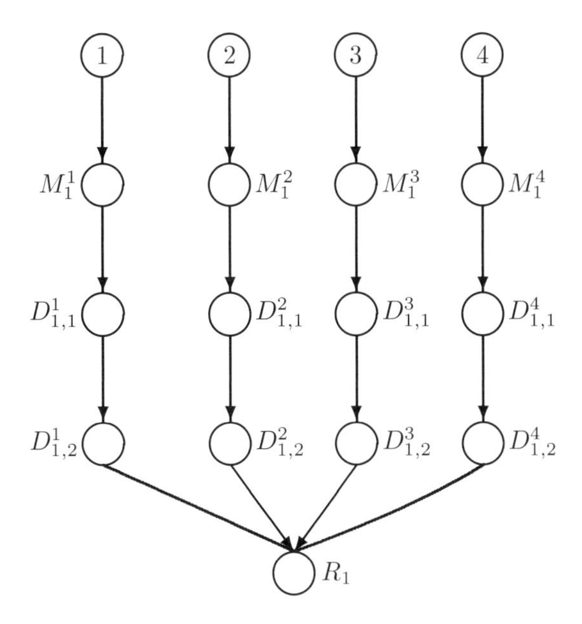

Figure 11.5: Network structure of the four firm oligopoly for Problem Sets 1 and 2

We denoted the paths by p_1, p_2, p_3, and p_4 corresponding to Firm 1 through Firm 4, respectively, with each path originating in its top-most firm node and ending in the demand market node (cf. Figure 11.5).

The Euler method converged to the equilibrium solution

$$x_{p_1}^* = x_{p_2}^* = x_{p_3}^* = x_{p_4}^* = 9.33, \quad d_{R_1}^* = 37.33.$$

The demand market price was $\rho_{R_1}(d^*) = 162.67$. The total cost was $2,936.50$; the total revenue was $6,072.79$, and the total profit was $3,136.29$. Each firm in this four firm oligopoly hence earned an individual profit of 784.07.

Example 11.1.1a

We then considered the case of the first two firms in the four-firm oligopoly Example 11.1.1 merging. Recall that, according to Salant, Switzer, and Reynolds (1983), in a Cournot oligopoly, it is not usually advantageous for quantity-setting firms to merge unless almost all of them merge. In investigating the merger of two firms out of four, we clearly do not have the majority merging. We assumed in this merger example, as well as in the remainder of the examples in Problem Set 1, that the total costs on the new links associated with the particular merger were all identically equal to zero. Obviously, this represents an ideal type of merger, in a sense.

Please refer to Figure 11.6 for the network topology associated with this example. We let path p_1 now originate in node $1'$ but follow then the same sequence of nodes as path p_1 in Example 11.1.1; the same for path p_2. Paths p_3 and p_4 remained as in

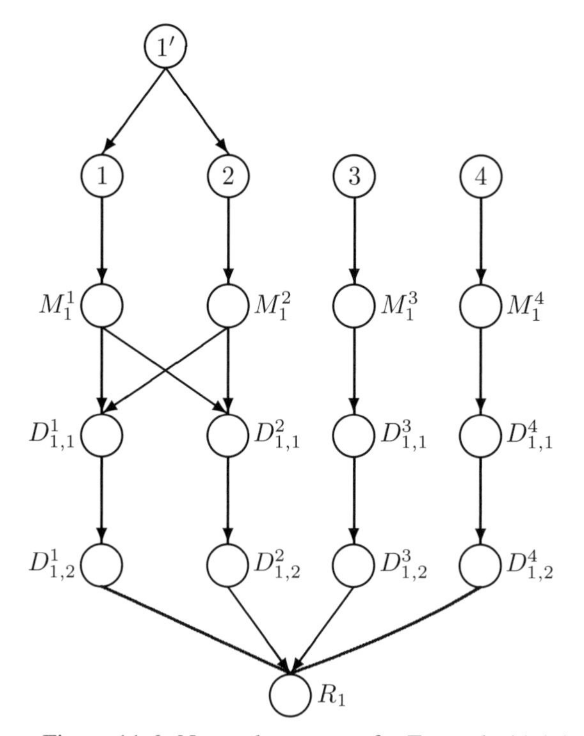

Figure 11.6: Network structure for Example 11.1.1a

Example 11.1.1. There were two additional paths associated with new firm $1'$ and we denote these cross-hauling paths, respectively, by paths p_5 and p_6.

The computed equilibrium solution was now

$$x_{p_1}^* = x_{p_2}^* = 0.00, \quad x_{p_3}^* = x_{p_4}^* = 9.14, \quad x_{p_5}^* = x_{p_6}^* = 11.16$$

with an equilibrium demand $d_{R_1}^* = 40.60$. The demand market price was $\rho_{R_1}(d^*) = 159.40$. The total cost was $2,971.56$. The total revenue was $6,472.11$, and the total profit was $3,500.54$. Each firm in the merged firm earned a profit of 998.23, whereas each of the two unmerged firms earned a profit of 752.04. Thus, each of the "insiders" gained considerably, whereas the firms that did not merge (the "outsiders") now had lower profits than in Example 11.1.1.

Hence, through computations, we were able to construct a simple counterexample to the ideas set forth in Salant, Switzer, and Reynolds (1983) but in the more general framework of supply chain network oligopolies.

Example 11.1.1b

We next considered the merger of the first three firms in the oligopoly in Example 11.1.1. The resulting network structure after the merger is given in Figure 11.7.

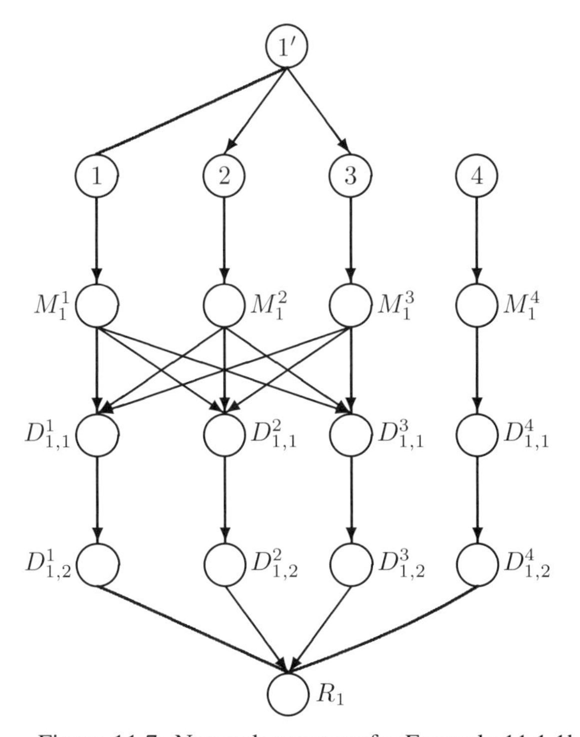

Figure 11.7: Network structure for Example 11.1.1b

Again, as in Example 11.1.1a, we assumed that the total cost functions on all the new links establishing the merger were identically equal to zero.

There are now nine paths joining node $1'$ to demand market node R_1 in Figure 11.7. The computed equilibrium solution was as follows. Each of the original paths associated with the original first three firms before the merger (but extended to include node $1'$) had flow equal to zero, whereas the flow on each of the new paths resulting from the merger was 5.22. The flows on the path for the fourth firm, which did not enter into the merger was 9.15. The demand $d_{R_1}^* = 40.44$ and the demand market price $\rho_{R_1}(d^*) = 159.56$. The total cost was now $2,758.85$. The total revenue was now $6,453.71$, and the total profit was $3,694.86$. Each of the firms in the three-firm merger now earned a profit of 980.45, whereas the unmerged firm earned a profit of 753.49. Hence for the fourth firm, from a profit perspective, it was better when three firms, rather than only two, merged. However, the individual profit for the first two firms was higher when they did not merge with the third firm but merged only with one another as in Example 11.2.1a.

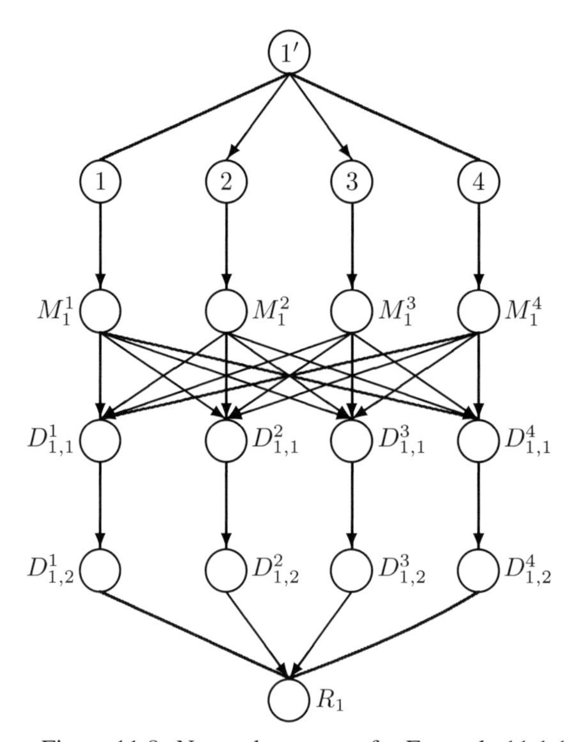

Figure 11.8: Network structure for Example 11.1.1c

Example 11.1.1c

This example consisted of all the four firms in Example 11.1.1 merging to form a monopoly. The total link cost functions on the new links were equal to zero; the original total cost functions were retained as was the demand market price function (as we had also done in Examples 11.1.1a through 11.1.1c). The network topology for this merger is given in Figure 11.8. There are now 16 paths joining node $1'$ to node R_1. The computed equilibrium solution is as follows. The equilibrium path flows on all the original firm paths (cf. Figure 11.5 but extended to node $1'$ as in Figure 11.8) were equal to zero. The flow on each of the cross-hauling paths, of which there were 12 such paths, was equal to 3.28. The equilibrium demand was 39.38 and the demand market price was 160.62. The total cost was now $2,444.58$. The total revenue was $6,326.91$, and the profit was $3,882.33$. Each firm in the monopoly earned an individual profit of 970.58. Hence as predicted by economic theory, the total profit in the monopoly was the highest of all the examples reported here.

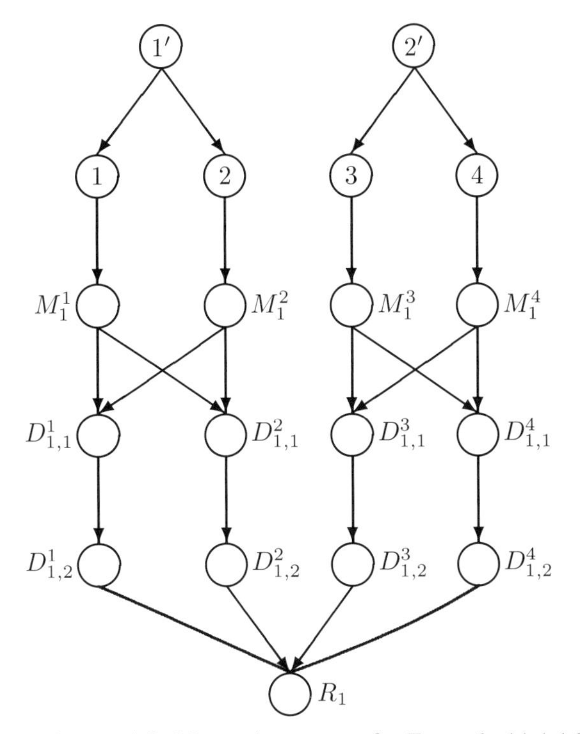

Figure 11.9: Network structure for Example 11.1.1d

Example 11.1.1d

For completeness, we also investigated the merger in the case of a merger of the first two firms and the merger of the next two firms in Example 11.1.1 yielding the supply chain network topology in Figure 11.9. We retained the original functions as in Example 11.1.1 and assigned zero total costs to all the new links.

The computed equilibrium solution was now as follows. The flows on the cross-hauling paths for each new firms were all equal to 10.94 with the other path flows all equal to zero. The equilibrium demand was $d_{R_1}^* = 43.74$ and the demand market price was 156.26. The total cost was 3,000.82, the total revenue was 6,835.27, and the total profit was 3,834.45. Each firm individually earned a profit of 958.61, which is lower than that earned in the case of a merger to a monopoly as in Example 11.1.1c.

In Table 11.1, we present a summary of the results for Examples 11.1.1, 11.1.1a through 11.1.1d.

Table 11.1: Summary of results for Problem Set 1

Measure	Ex. 11.1.1	Ex. 11.1.1a	Ex. 11.1.1b	Ex. 11.1.1c	Ex. 11.1.1d
Total Profit	3,136.29	3,500.54	3,694.86	3,882.33	3,834.45
Total Cost	2,936.50	2,971.56	2,758.85	2,444.58	3,000.82
Total Revenue	6,072.79	6,472.11	6,453.71	6,326.91	6,835.27
Equ. Demand	37.33	40.60	40.44	39.38	43.74
Equ. Price	162.67	159.40	159.56	160.62	156.26

11.5.2 Problem Set 2

In the second problem set, we, again, used Example 11.1.1 as a baseline, and we constructed Examples 11.2.1a through 11.2.1d to mimic Examples 11.1.1a through 11.1.1d, respectively, with the proviso that the new links for the particular mergers no longer had associated zero total costs, but rather *all* now had associated total cost functions as in (11.23).

Next we report the computed path flows, etc., and in Table 11.2 we provide a summary of the results for the examples in Problem Set 2.

Example 11.2.1a
This merger corresponded to the merger of the four oligopolistic firms depicted in Figure 11.6. The computed equilibrium path flows were now as follows. The flows on all the paths corresponding to the new Firm $1'$ were equal to 4.00 whereas the flow on the path of each unmerged firm was 9.47. The equilibrium demand was 34.94 and the demand price was 165.05. The total cost was $2,743.91$. The total revenue was $5,768.47$, and the total profit was $3,024.55$. The profit for each merged firm was now 704.57 and the profit for each unmerged firm was 807.70. It is quite interesting to compare these results with those obtained in Example 11.1.1a, in which the firms in the merger profited substantially, whereas those who were not in the merger lost out as compared to the individual profits in the four firm oligopoly Example 11.1.1 before the the mergers. Note that in this merger example, in contrast to Example 11.1.1a, there were now non-zero total cost functions associated with the merger links. In contrast to the results obtained for Example 11.1.1a, the firms now in the merger had the individual profits reduced from 998.23 to 704.57, whereas the two firms who did not enter into the merger each had its profits grow from 752.04 to 807.70.

Example 11.2.1b
This merger corresponds to the network in Figure 11.7. The equilibrium path flows associated with the newly merged firm were all equal to 2.65. There were nine such path flows. The path flow for the unmerged firm was 9.56. The equilibrium demand was 33.41 and the equilibrium demand price was now 166.59. The total cost was $2,530.71$. The total revenue was $5,565.32$, and the total profit was $3,034.61$. The profit for the unmerged firm was 823.24. The profit for each of the three merged firms was 737.12. In this example, the outsider clearly gained by not entering into the merger.

Table 11.2: Summary of results for Problem Set 2

Measure	Ex. 13.1.1	Ex. 13.2.1a	Ex. 13.2.1b	Ex. 13.2.1c	Ex. 13.2.1d
Profit	3,136.29	3,024.55	3,034.61	3,040.47	2,937.62
Total Cost	2,936.50	2,743.91	2,530.71	2,225.56	2,506.12
Revenue	6,072.79	5,768.47	5,565.32	5,266.03	5,443.74
Equ. Demand	37.33	34.94	33.41	31.20	32.50
Equ. Price	162.67	165.05	166.59	168.80	167.50

Example 11.2.1c

This merger corresponds to the network in Figure 11.8 and represents a merger of the four firms in Example 11.1.1 to a monopoly. The equilibrium path flows were now all equal to: 1.95 and the equilibrium demand was 31.20 with an equilibrium price of 168.80. The total cost was $2,225.56$. The total revenue was $5,266.03$, and the total profit was $3,040.47$. Hence the profit of each of the original firms was 760.10, which is lower than before the merger but, understandable, because we now, unlike in Example 11.1.1c, have non-zero total cost functions associated with the merger links.

Example 11.2.1d

This merger corresponds to the network in Figure 11.9. Each of the path flows was now 4.06 with a demand of 32.50 and a demand price of 167.50. The total cost was $2,506.12$. The revenue was $5,443.74$, and the profit was $2,937.62$. The individual profit was, thus, 734.40, which is lower than in Example 11.1.1d, in which the merger links have zero associated total cost functions.

In Table 11.2, we present a summary of the results for Examples 11.1.1, 11.2.1a through 11.2.1d.

The examples in Problem Sets 1 and 2, although stylized, nevertheless, demonstrate the power of being able to solve a spectrum of mergers associated with oligopolistic firms and to explore the effects of such mergers on individual firm profits. Also, the availability of appropriate algorithms that are easy to implement allows one to investigate questions associated with the effects of different costs associated with merger links.

Obviously, the depiction of these problems as network problems greatly facilitates both the possible computational procedures and the ease of the investigation of different mergers that are associated with different network topologies and different costs associated with the new merger links.

11.5.3 Problem Set 3

In this subsection, we continued our computational/numerical experiments. In this set, we solved pre- and post-merger problems consisting of four firms and two demand markets. These examples were more complex than those reported earlier.

Example 11.3.1

In this example, we, again, considered four firms competing in an oligopolistic manner. The total cost functions on each of the links was given by

$$\hat{c}_a(f) = 2f_a^2 + f_a, \quad \forall a \in \mathcal{L}^0.$$

However, rather than a single demand market, we now had two demand markets. The demand price function associated with the first demand market was as in (11.24). The demand price function for the second demand market was

$$\rho_{R_2}(d) = -d_{R_2} + 100. \tag{11.25}$$

The Cournot-Nash equilibrium solution consisted of each firm in the oligopoly supplying demand market R_1 an amount 9.33 of the product and each supplying demand market R_2 an amount: .00 of the product. The total cost was 2, 935.94, the revenue was 6, 072.54, and the profit hence was 3, 136.59 with each firm earning a profit of 784.15. The demand market price at R_1 was 162.67 and at R_2: 100.00.

Example 11.3.1.1

We then assumed that the demand for the product at the second demand market increased, so that (11.25) was now

$$\rho_{R_2}(d) = -d_{R_2} + 300, \tag{11.26}$$

with the remainder of the data unchanged.

The new computed equilibrium solution was: each firm in the oligopoly produced and shipped an amount 1.91 of the product to demand market R_1 and an amount: 13.00 to demand market R_2. The total cost was now 6, 954.69. The total revenue was 14, 365.24, and the total profit was 7, 410.55, with each individual firm earning a profit of 1, 852.64. The demand price at demand market R_1 was: 192.37 and the demand price at R_2 was: 248.00.

Example 11.3.2

Example 11.3.2 was a partial merger problem. We assumed that the first two firms in Example 11.3.1.1 formed a coalition and merged. Also, we assumed that the new links associated with the merger (the top-most links and the added distribution links between the first two original firms) had total cost functions that were all equal to zero. The total cost was now 7, 425.73, the total revenue was 15, 902.39, and the total profit was 8, 476.66. The newly merged firm supplied demand market R_1 at an amount of 4.90 on each of its two cross-hauling (new) paths, and supplied demand market R_2 at a level of 14.47 on the same two paths. Each of the unmerged two firms supplied R_1 at a level of 1.53 and R_2 at an amount 13.09 of the product. The demand market price at R_1 was 187.47 and was 244.89 at R_2. Note that the profit now was substantially higher than that in Example 11.3.1.1.

The profit of each of the two unmerged firms was now 1, 804.13 whereas the profit of each merged firm was 2, 434.20, a value significantly higher than before the merger.

Again, we have constructed a relatively simple example for which the "insiders" in the merger gain. We were able to accomplish this through the powerful tool of computational methods and numerical experimentation.

11.6 SUMMARY AND CONCLUSIONS

In this chapter, we developed a network framework in an oligopolistic setting that allows one to explore mergers formed through coalitions among the competing firms. We formulated all the network problems uniformly as variational inequality problems and then proposed an algorithm, which fully exploits the underlying structure of these problems and yields closed form expressions at each iterative step. We conducted extensive computational experiments to explore issues surrounding what has been termed in the literature as the merger paradox.

The network formalism proposed here, which captures competition on the production, distribution, and demand market dimensions, enables the investigation of economic issues surrounding mergers and the associated costs, revenues, and profits. In addition, it allows for the identification of special cases of oligopolistic market equilibrium problems, spatial, and aspatial, through the underlying network structure, that have appeared in the literature. Furthermore, the network structure allows one to visualize graphically the problems both before and after the particular mergers and to assign costs as relevant. It is important that this chapter illustrates the power of computational methodologies for exploring issues regarding competing firms and possible mergers before any actualization of mergers.

The research in this chapter can be extended in several directions. One can construct multiproduct versions of the oligopolistic supply chain network models developed here, and one can also consider international issues concerning mergers and acquisitions. Obviously, further computational experimentation as well as theoretical developments would also be of value, but we leave such research questions/problems for the future.

11.7 SOURCES AND NOTES

Nagurney (2008a) developed an oligopolistic supply chain network framework for mergers and acquisitions, that was subsequently extended in Nagurney (2008b) to more general total cost and demand price functions. This chapter is based on the latter paper. The former paper also contains additional theoretical results that enable one to determine whether a merger in the case of a duopoly will be profitable, based on the total costs associated with the new merger links and the demand price function at the demand market.

Note that Part III of this book deals with synergies associated with network integration through mergers and acquisitions in terms of cost savings. On the other hand, Part II of this book deals with the opposite: the efficiency lost and the associated cost increments after network components are eliminated or partially damaged. Although seemingly different, we believe that they are intimately related in that both concepts provide valuable insights into the protection and enhancement of network systems on which our societies and economies critically depend.

Appendix

In this appendix, we review the fundamentals of optimization theory at a level used in this book. The goal of this appendix is to provide the reader who may not be as familiar with the fundamentals of optimization theory with the basic mathematical background needed for an adequate understanding of the models in this book. The reader may skip this appendix at first reading and refer back to it as needed.

We first recall some basic definitions and results from convexity theory and mathematical optimization at a level that is relevant to the material presented in this book. The results are standard. For a more detailed presentation of the aforementioned areas, we refer the reader to Bradley, Hax, and Magnanti (1977), Bazaraa, Jarvis, and Sherali (1990), Bazaraa, Sherali, and Shetty (1993), and Nagurney (1999).

Definition A.1: A Convex Set
A set $\mathcal{K} \subset R^N$ is called a convex set if, given any two vectors X_1 and X_2 in $\mathcal{K}$, $\lambda X_1 + (1-\lambda)X_2 \in \mathcal{K}$, for every $\lambda \in [0,1]$.

In the case of two dimensions, that is, when one is dealing with a set $\mathcal{K} \in R^2$, a set is convex if the line that connects any two points in that set also lies in the interior of that set.

Example A.1: A Convex Set
For example, consider the set $\mathcal{K}$ consisting of x_1, x_2, such that $x_1 \geq 0$ and $x_2 \geq 0$. Moreover, assume that x_1 and x_2 must also satisfy the constraint

$$x_1 + x_2 \leq 5.$$

Clearly, the set $\mathcal{K}$ is convex.

Certain functions that arise often in the models developed in this book are now defined.

Definition A.2: A Continuous Function
A function $f : \mathcal{K} \mapsto R$ is said to be continuous at $X \in \mathcal{K}$, if for any given $\epsilon > 0$ there is a $\delta > 0$, such that $\psi \in \mathcal{K}$ and $\|\psi - X\| < \delta$, imply that $|f(\psi) - f(X)| < \epsilon$.

Definition A.3: A Differentiable Function
Consider a nonempty set $\mathcal{K}$ such that $\mathcal{K} \subset R^N$, a column vector X lying in the interior of $\mathcal{K}$, and let f be a function such that $f : \mathcal{K} \mapsto R$. Then f is said to be differentiable, if there exists a column vector $\nabla f(X) \in R^N$, called the gradient of f at X, and defined as

$$\nabla f(X)^T = \left(\frac{\partial f(X)}{\partial X_1}, \frac{\partial f(X)}{\partial X_2}, \ldots, \frac{\partial f(X)}{\partial X_N} \right), \tag{A.1}$$

and a function $\beta(X; y) \to 0$ as $y \to x$ such that

$$f(y) = f(X) + \langle \nabla f(X)^T, (y - X) \rangle + \|y - X\| \beta(X; y), \quad \forall y \in \mathcal{K}, \tag{A.2}$$

where $\langle \cdot, \cdot \rangle$ denotes the inner product in the N-dimensional Euclidean space.

Definition A.4: A Twice Differentiable Function
The function f is said to be twice differentiable at X if, in addition to the gradient vector, there exists an $N \times N$ matrix $H(X)$, called the Hessian matrix of the function f at X, defined as

$$H = \begin{pmatrix} \frac{\partial^2 f}{\partial X_1 \partial X_1} & \frac{\partial^2 f}{\partial X_1 \partial X_2} & \cdots & \frac{\partial^2 f}{\partial X_1 \partial X_N} \\ \frac{\partial^2 f}{\partial X_2 \partial X_1} & \frac{\partial^2 f}{\partial X_2 \partial X_2} & \cdots & \frac{\partial^2 f}{\partial X_2 \partial X_N} \\ \vdots & \ddots & \ddots & \vdots \\ \frac{\partial^2 f}{\partial X_N \partial X_1} & \cdots & \cdots & \frac{\partial^2 f}{\partial X_N \partial X_N} \end{pmatrix}, \tag{A.3}$$

and a function $\beta(X; y) \to 0$ as $y \to X$ such that

$$f(y) = f(X) + \langle \nabla f(X)^T, (y - X) \rangle$$

$$+ \frac{1}{2} (y - X)^T H(X) (y - X) + ||y - X||^2 \beta(X; y), \quad \forall y \in \mathcal{K}. \tag{A.4}$$

Example A.2
An example of a twice differentiable function in two dimensions is $f(x) = 8x_1^2 + 5x_2^2$.

Definition A.5: Convex and Concave Functions
Let $\mathcal{K}$ be a nonempty convex set and consider a function $f : \mathcal{K} \mapsto R$. Then the function $f(X)$ is said to be a convex function on $\mathcal{K}$ if for any two distinct points $X_1, X_2 \in \mathcal{K}$, and for all $\lambda \in [0,1]$, the following holds

$$f\left[\lambda X_2 + (1-\lambda) X_1\right] \leq \lambda f(X_2) + (1-\lambda) f(X_1). \qquad (A.5)$$

The function f is said to be strictly convex on $\mathcal{K}$ if the above inequality holds as a strict inequality. A function $f(X)$ is said to be concave (strictly concave) if $-f(X)$ is convex (strictly convex).

In two dimensions, a function $f(X)$ is convex (concave) if a line segment joining any two points $[X_1, f(X_1)], [X_2, f(X_2)]$ on the surface of $f(X)$ lies on or above (below) that surface.

The previous definition can be used to determine whether a function is convex or not. For example, if one assumes that the function $f(X)$ is continuous and that it has second-order partial derivatives over $\mathcal{K}$, then an alternative way to determine whether a function is convex or not is to evaluate whether the Hessian of the function is positive semidefinite or not. In particular, if the Hessian matrix H of second-order partial derivatives is positive semidefinite, then $f(X)$ is convex; if H is negative semidefinite, then $f(x)$ is concave. Recall that a matrix is positive (negative) definite if all its eigenvalues are positive (negative) or if all of its principal determinants have positive (negative) value. A matrix is positive semidefinite if all its eigenvalues are nonnegative as are all of its principal determinants.

Example A.3: A Convex Function
Consider the function $f(x_1, x_2) = 9x_1^2 + 7x_2^2$, whose Hessian matrix is

$$H = \begin{pmatrix} 18 & 0 \\ 0 & 14 \end{pmatrix}.$$

This Hessian matrix is a diagonal matrix with positive elements on the diagonal and hence it is positive definite, so f is clearly convex and, in fact, it is strictly convex.

Definition A.6: Quasiconvex and Quasiconcave Functions
Let $\mathcal{K}$ be a nonempty convex set and consider a function $f : \mathcal{K} \mapsto R$. The function $f(X)$ is said to be a quasiconvex function on $\mathcal{K}$, if for any two distinct points $X_1, X_2 \in \mathcal{K}$, and $\forall \lambda \in [0,1]$, we have

$$f\left[\lambda X_2 + (1-\lambda) X_1\right] \leq \textit{maximum} \ \left(f(X_1), f(X_2)\right).$$

The function $f(X)$ is said to be quasiconcave on $\mathcal{K}$ if $-f(X)$ is quasiconvex.

Definition A.7: Pseudoconvex and Pseudoconcave Functions
Let $\mathcal{K}$ be a nonempty convex set and consider a function $f : \mathcal{K} \mapsto R$, which is differentiable on $\mathcal{K}$. Then the function $f(X)$ is said to be a pseudoconvex function

on $\mathcal{K}$, if for any two distinct points $X_1, X_2 \in \mathcal{K}$, with $\langle \nabla f(X_1)^T, X_2 - X_1 \rangle \geq 0$ we have $f(X_2) \geq f(X_1)$.

The function $f(X)$ is said to be pseudoconcave on $\mathcal{K}$ if $-f(X)$ is pseudoconvex.

We now present some of the fundamentals of optimization theory in addition to some of the concepts and ideas that are used throughout this book.

An optimization problem in R^N is a problem in which one seeks to optimize a function f, which is said to be the *objective function*, subject to some constraints. For example, an objective function may represent profits to be maximized or costs or risk to be minimized (or a combination thereof, as in the case of multicriteria decision-making problems). Constraints, in turn, can reflect budget or other resource constraints as well as nonnegativity constraints that occur frequently in network problems. For example, in network problems, flows tend to be nonnegative because they correspond to vehicles, telecommunication messages, products, prices, etc. A point $X^0 \in R^N$ is called a *feasible solution* to the optimization problem if it satisfies all the constraints of the problem. Moreover, a point $X^* \in R^N$ is called an *optimal solution* to the optimization problem if it is a feasible solution and it provides the best possible value for the objective function.

There are different classes of optimization problems, depending on the structure of the objective function and the constraints. If the objective function is linear, as are the constraints, and the variables are continuous (rather than discrete), then the problem is a *linear* programming problem, whose principal method of solution is the simplex method [see Bazaraa, Jarvis, and Sherali (1990)]. On the other hand, if either the objective function or the constraints are nonlinear expressions of the variables, then the problem is a *nonlinear* programming problem. For further details, see Bazaraa, Sherali, and Shetty (1993). Clearly, the classic system-optimization problems as well as the optimization reformulation of the symmetric user-optimization problem (see, e.g., Chapters 2 and 8 through 10) are nonlinear programming problems.

On the other hand, if one or more of the variables in a problem are constrained to be discrete, that is, to take on integer values, then one has an *integer* programming problem at hand. We do not consider such problems in this book.

We discuss important conditions for optimality, but we first present some basic definitions.

Definition A.8: Global Maximum and Minimum
The function $f : \mathcal{K} \mapsto R^N$ is said to take its global maximum at point X^* if

$$f(X) \leq f(X^*), \quad \forall X \in \mathcal{K}. \tag{A.6}$$

Similarly, the function $f : \mathcal{K} \mapsto R^N$ is said to take its global minimum at point ψ^* if

$$f(\psi) \geq f(\psi^*), \quad \forall \psi \in \mathcal{K}. \tag{A.7}$$

Definition A.9: Local Maximum and Minimum
The function $f : \mathcal{K} \mapsto R^N$ is said to take its local maximum at point X^* if there exists a $\delta > 0$ such that for every $X \neq X^*$ that belongs to $\mathcal{K}$ and is in a δ-neighborhood

of X^*, the following holds

$$f(X) \leq f(X^*), \quad \forall X \in (\mathcal{K} \cap B(X^*, \delta)), \tag{A.8}$$

where $B(X^*, \delta)$ denotes the ball with center X^* and radius δ.

Similarly, the function $f : \mathcal{K} \mapsto R^N$ is said to take its local minimum at point ψ^* if there exists a $\delta > 0$ such that for every $\psi \neq \psi^*$ that belongs to $\mathcal{K}$ and is in a δ-neighborhood of ψ^*, the following holds

$$f(\psi) \geq f(\psi^*), \quad \forall \psi \in (\mathcal{K} \cap B(\psi^*, \delta)). \tag{A.9}$$

Karush-Kuhn-Tucker (KKT) Conditions

Karush (1939) and Kuhn and Tucker (1951) independently proposed a set of necessary and sufficient conditions for an optimal solution of a general mathematical optimization problem. Their work provided the mathematical framework on which the qualitative theory as well as the computational methods in optimization have been based. A compact presentation of these conditions is given for reference purposes.

Karush-Kuhn-Tucker Necessary Conditions

Let $\mathcal{K}$ be a nonempty open set such that $\mathcal{K} \subset R^N$, and let $\{f : R^N \mapsto R\}$, $\{g_i : R^N \mapsto R\}$ for $i = 1, 2, \ldots, m$, and $\{h_j : R^N \mapsto R\}$ for $j = 1, 2, \ldots, t$. Consider the general optimization problem of the following form

$$\text{Minimize} \quad f(X)$$

subject to:

$$g_i(X) \leq 0, \quad \text{for} \quad i = 1, 2, \ldots, m,$$

$$h_j(X) = 0, \quad \text{for} \quad j = 1, 2, \ldots, t,$$

$$X \in \mathcal{K}.$$

Let X^* be a feasible solution and let $I = \{i : g_i(X^*) = 0\}$. Furthermore, assume that f and g_i are differentiable at X^* for $i \in I$ and that the g_i are continuous at X^* for $i \in I$. Finally, assume that the h_j are continuously differentiable at X^* for all $j = 1, 2, \ldots, t$. Further, suppose that $\nabla g_i(X^*)$ for $i \in I$ and $\nabla h_j(X^*)$ for $j = 1, 2, \ldots, t$ are linearly independent. If X^* locally solves the minimization problem then there exist unique scalars v_i^* for $i \in I$, and γ_j^* for $j = 1, 2, \ldots, t$, such that

$$\nabla f(X^*) + \sum_{i=1}^{m} v_i^* \nabla g_i(X^*) + \sum_{j=1}^{t} \gamma_j^* \nabla h_j(X^*) = 0, \tag{A.10}$$

$$v_i^* g_i(X^*) = 0, \text{ for } i = 1, 2, \ldots, m, \tag{A.11}$$

$$v_i^* \geq 0, \text{ for } i = 1, 2, \ldots, m. \tag{A.12}$$

The scalars v_i^* and γ_j^* are called *Lagrange* multipliers. Note that there is one Lagrange multiplier associated with each constraint and that they represent the marginal rate of change in the objective function f with respect to each per unit change in the right-hand side of the corresponding constraint. Lagrange multipliers have important economic interpretations, as revealed in different supply chain network models that are presented in Chapters 8 through 10 in this book.

Any point that satisfies the KKT conditions is called a *KKT point*.

As mentioned earlier, in the case of network optimization problems, nonnegativity constraints are imposed on the variables, that is, $X \geq 0$. Clearly, the KKT conditions that were just presented will still hold. Many times, however, for reasons of convenience and simplicity, the Lagrange multipliers associated with the nonnegativity constraints are eliminated, and the conditions are reduced to

$$\nabla f(X^*) + \sum_{i=1}^{m} v_i^* \nabla g_i(X^*) + \sum_{j=1}^{t} \gamma_j^* \nabla h_j(X^*) \geq 0, \tag{A.13}$$

$$\left[\nabla f(X^*) + \sum_{i=1}^{m} v_i^* \nabla g_i(X^*) + \sum_{j=1}^{t} \gamma_j^* \nabla h_j(X^*) \right]^T X^* = 0, \tag{A.14}$$

$$v_i^* g_i(X^*) = 0, \text{ for } i = 1, \ldots, m, \tag{A.15}$$

$$v_i^* \geq 0, \text{ for } i = 1, \ldots, m. \tag{A.16}$$

A geometric interpretation of the KKT conditions is that a vector X^* is a KKT point if and only if $-\nabla f(X^*)$ lies in the cone spanned by the gradients of the binding constraints, that is, those constraints that hold as equalities.

Karush-Kuhn-Tucker Sufficient Conditions

Let $\mathcal{K}$ be a nonempty open set such that $\mathcal{K} \subset R^N$, and let $\{f : R^N \mapsto R\}$, $\{g_i : R^N \mapsto R\}$ for $i = 1, 2, \ldots, m$, and $\{h_j : R^N \mapsto R\}$ for $j = 1, 2, \ldots, t$. Moreover, consider, again, the general optimization problem of the following form

$$\text{Minimize} \quad f(X)$$

subject to:

$$g_i(X) \leq 0, \quad \text{for} \quad i = 1, 2, \ldots, m,$$
$$h_j(X) = 0, \quad \text{for} \quad j = 1, 2, \ldots, t,$$
$$X \in \mathcal{K}.$$

Let X^* be a feasible solution, and let $I = \{i : g_i(X^*) = 0\}$. Assume that the KKT conditions hold at X^*. In other words, assume that there exist scalars $\overline{v}_i \geq 0$ with $i \in I$ and $\overline{\gamma}_j$ with $j = 1, 2, \ldots, t$, such that

$$\nabla f(X^*) + \sum_{i \in I} \overline{v}_i \nabla g_i(X^*) + \sum_{j=1}^{t} \overline{\gamma}_j \nabla h_j(X^*) = 0. \tag{A.17}$$

Let $J = \{j : \bar{\gamma}_j > 0\}$ and $L = \{j : \bar{\gamma}_j < 0\}$. Further, suppose that f is pseudoconvex at X^*, the constraints g_i are quasiconvex at X^* for $i \in I$, h_j is quasiconvex for the

$j \in J$ and quasiconcave for the $j \in L$. Then X^* is a global optimal solution to the general minimization problem.

For a maximization problem, the KKT conditions are similar, where now the function f has to be pseudoconcave.

Definition A.10: Lagrangian Function

Consider, once again, the general minimization problem described previously. Then the function such that

$$\phi(X, v, \gamma) = f(X) + \sum_{i=1}^{m} v_i g_i(X) + \sum_{j=1}^{t} \gamma_j h_j(X) \tag{A.18}$$

is said to be the Lagrangian function of the general optimization problem.

If we let X^* be a KKT point to the general optimization problem with v^*, γ^* being the Lagrange multipliers that correspond to the constraints of the problem, then the function

$$L(X) \equiv \phi(X, v^*, \gamma^*) = f(X) + \sum_{i \in I} v_i^* g_i(X) + \sum_{j=1}^{t} \gamma_j^* h_j(X) \tag{A.19}$$

is said to be the restricted Lagrangian function.

Let $\nabla^2 L$ denote the Hessian of (A.19). Then if $\nabla^2 L$ is

- positive semidefinite for all X in the feasible set, X^* is a global minimum;
- positive semidefinite for all X in the feasible set and in a δ-neighborhood $B(X^*, \delta)$, for a $\delta > 0$, X^* is a local minimum;
- positive definite, X^* is a strict local minimum.

Bibliography

Aashtiani, H. Z., Magnanti, T. L., 1981. Equilibrium on a congested transportation network. *SIAM Journal on Algebraic and Discrete Methods* 2, 213-226.

Ahuja, R. K., Magnanti, T. L., Orlin, J. B., 1993. *Network Flows: Theory, Algorithms, and Applications*. Prentice-Hall, Upper Saddle River, New Jersey.

Akcelik, R., Besley, M., 2003. Operating cost, fuel consumption, and emission models in aaSIDRA and aaMOTION. Paper presented at the 25th Conference of Australian Institutes of Transport Research (CAITR 2003), University of South Australia, Adelaide, Australia, December 3-5.

Albert, R., Albert, I., Nakarado, G. L., 2004. Structural vulnerability of the North American power grid. *Physical Review E* 69, Article no. 025103.

Albert, R., Jeong, H., Barabási, A. L., 2000. Attack and error tolerance of complex networks. *Nature* 401, 130-131.

Alexopoulos, A., Assimacopoulos, D., 1993. Model for traffic emissions estimation. *Atmospheric Environment B* 27, 435-466.

Ali, S., Maciejewski, A. A., Siegel, H. J., Kim, J., 2003. Definition of a robustness metric for resource allocation. In *Proceedings of the 17th International Parallel and Distributed Processing Symposium*, IEEE CS Press.

Allen, F., Gale, D., 1998. Optimal financial crises. *Journal of Finance* 53, 1245-1284.

Allen, F., Gale, D., 2000. Financial contagion. *Journal of Political Economy* 108, 1-33.

Alptekinoglu, A., Tang, C. S., 2005. A model for analyzing multi-channel distribution systems. *European Journal of Operational Research* 163, 802-824.

Altay, N., Green III, W. G., 2006. OR/MS research in disaster operations management. *European Journal of Operational Research* 175, 475-493.

Amaral, L. A. N., Scala, A., Barthélèmy, M., Stanley, H. E., 2000. Classes of small-world networks. *Proceedings of the National Academy of Sciences USA* 97, 11149-11152.

American Society of Civil Engineers, 2005. Report card for America's infrastructure. http://www.asce.org/reportcard/2005/index.cfm Accessed on November 15, 2007.

Arkell, B. P., Darch, G. J. C., 2006. Impact of climate change on London's transport network. *Municipal Engineer* 4, 231-237.

Arora, S., Cason, T. N., 1996. Why do firms volunteer to exceed environmental regulations? Understanding participation in EPA's 33/50 Program. *Land Economics* 72, 413-432.

Bagchi, K. K., Tang, Z., 2005. Network size, deterrence effects, and Internet attack incident growth. *Journal of Information Technology Theory and Application* 6, 117-137.

Balcik, B., Beamon, B., 2008. Facility location in humanitarian relief. *International Journal of Logistics: Research and Applications* 11, 101-121.

Barabási, A. L., 2003. *Linked: How Everything Is Connected to Everything Else and What It Means*. Plume, New York.

Barabási, A. L., Albert, R., 1999. Emergence of scaling in random networks. *Science* 286, 509-512.

Barbagallo, A., 2007. Regularity results for time-dependent variational and quasi-variational inequalities and application to the calculation of dynamic traffic networks. *Mathematical Models and Methods in Applied Sciences* 17, 277-304.

Bar-Gera, H., 2002. Origin-based algorithm for the traffic assignment problem. *Transportation Science* 36, 398-417.

Bar-Gera, H., 2008. Transportation network test problems.
http://www.bgu.ac.il/~bargera/tntp/ Accessed on November 15, 2008.

Barling, R., 2005. Profit lies in improved outsourcing: Current supply chain situation in China is 10 years behind Europe, says report. *South China Morning Post*, April 3.

Barrat, A., Barthélémy, M., Pastor-Satorras, R., Vespignani, A., 2004. The architecture of complex weighted networks. *Proceedings of the National Academy of Sciences USA* 101, 3747-3752.

Barrat, A., Barthélèmy, M., Vespignani, A., 2005. The effects of spatial constraints on the evolution of weighted complex networks. *Journal of Statistical Mechanics: Theory and Experiment*, Article no. P05003.

Bazaraa, M. S., Jarvis, J. J., Sherali, H. D., 1993. *Linear Programming and Network Flows*. John Wiley & Sons, New York.

Bazaraa, M. S., Sherali, H. D., Shetty, C. M., 1993. *Nonlinear Programming: Theory and Algorithms*. John Wiley & Sons, New York.

BBC News, 2008a. China freeze has cost billions. February 1. http://news.bbc.co.uk/2/hi/asia-pacific/7221456.stm Accessed on February 15, 2008.

BBC News, 2008b. Food warnings amid China freeze. January 31. http://news.bbc.co.uk/2/hi/asia-pacific/7219092.stm Accessed on February 15, 2008.

Beamon, B. M., 1998. Supply chain design and analysis: Models and methods. *International Journal of Production Economics* 55, 281-294.

Beamon, B. M., 1999. Measuring supply chain performance. *International Journal of Operations and Production Management* 19, 275-292.

Beamon, B. M., 2004. Humanitarian relief chains: Issues and challenges. In *Proceedings of the 34th International Conference on Computers and Industrial Engineering*, San Francisco, California.

Beckmann, M. J., 1967. On the theory of traffic flows in networks. *Traffic Quarterly* 21, 109-116.

Beckmann, M. J., McGuire, C. B., Winsten, C. B., 1956. *Studies in the Economics of Transportation*. Yale University Press, New Haven, Connecticut.

Bell, M. G. H., 2000. A game theory approach to measuring the performance reliability of transport networks. *Transportation Research* 34, 533-545.

Ben-Tal, A., Nemirovski, A., 1998. Robust convex programming. *Mathematics of Operations Research* 23, 769-805.

Ben-Tal, A., Nemirovski, A., 1999. Robust solutions of uncertain linear programs. *Operations Research Letters* 25, 1-13.

Bernard, A. B., Redding, S. J., Schott, P. K., 2006. Multi-product firms and product switching. Working paper, Dartmouth College, Hanover, New Hampshire. http://mba.tuck.dartmouth.edu/pages/faculty/andrew.bernard/pswitch.pdf Accessed on October 12, 2008.

Bertsekas, D. P., Gafni, E. M., 1982. Projection methods for variational inequalities with application to the traffic assignment problem. *Mathematical Programming* 17, 139-159.

Bertsekas, D. P., Gallager, R. G., 1987. *Data Networks*. Prentice-Hall, Englewood Cliffs, New Jersey.

Bertsekas, D. P., Tsitsiklis, J. N., 1989. *Parallel and Distributed Computation - Numerical Methods*. Prentice Hall, Englewood Cliffs, New Jersey.

Beverage World, 2007. Molson Coors Brewing Co. http://www.beverageworld.com/content/view/33296/ Accessed on January 15, 2009.

Bienenstock, E. J., Bonacich, P., 2003. Balancing efficiency and vulnerability in social networks. In *Dynamic Social Network Modeling and Analysis: Workshop Summary and Papers*. The National Academy of Sciences, pp. 253-264.

Boginski, V., Butenko, S., Pardalos, P. M., 2003. On structural properties of the market graph. In *Innovations in Financial and Economic Networks*. Nagurney, A., Editor, Edward Elgar Publishing, Cheltenham, England, pp. 29-42.

Bonacich, P., 1972. Factoring and weighting approaches to status scores and clique identification. *Journal of Mathematical Sociology* 2, 13-20.

Boss, M., Elsinger, H., Summer, M., Thurner, S., 2004. The network topology of the interbank market. *Quantitative Finance* 4, 677-684.

Boyce, D. E., Mahmassani, H. S., Nagurney, A., 2005. A retrospective on Beckmann, McGuire, and Winsten's studies in the economics of transportation. *Papers in Regional Science* 84, 85-103.

Boyce, D. E., Nagurney, A., 2005. Preface to "On a paradox of traffic planning." *Transportation Science* 39, 443-445.

Bradley, S. P., Hax, A. C., Magnanti, T. L., 1977. *Applied Mathematical Programming.* Addison-Wesley, Reading, Massachusetts.

Braess, D., 1968. Uber ein paradoxon aus der verkehrsplanung. *Unternehmensforschung* 12, 258-268.

Braess, D., Nagurney, A., Wakolbinger, T., 2005. On a paradox of traffic planning, translation of the 1968 article by Braess. *Transportation Science* 39, 446-450.

Braine, T., 2006. Was 2005 the year of natural disasters? *Bulletin of the World Health Organization* 84, 1-80.

Brown, G., Keegan, J., Vigus, B., Wood, K., 2001. The Kellogg company optimizes production, inventory and distribution. *Interfaces* 31, 1-15.

Buchanan, J. M., Musgrave, R. A., 1999. *Public Finance and Public Choice – Two Contrasting Visions of the State.* MIT Press, Cambridge, Massachusetts.

Bundschuh, M., Klabjan, D., Thurston, D. L., 2003. Modeling robust and reliable supply chains. Optimization Online e-print. http://www.optimization-online.org Accessed on June 4, 2008.

Bureau of Public Roads, 1964. *Traffic Assignment Manual.* U.S. Department of Commerce, Urban Planning Division, Washington, DC.

Cachon, G. P., Fisher, M., 2000. Supply chain inventory management and the value of shared information. *Management Science* 46, 1032-1048.

Cachon, G., Netessine, S., 2004. Game theory in supply chain analysis. In *Handbook of Quantitative Supply Chain Analysis Modeling in the eBusiness Era.* Simchi-Levi, D., Wu, S. D., Shen, Z. J., Editors, Kluwer Academic Publishers, Norwell, Massachusetts, pp. 13-66.

Caldarelli, G., Battiston, S., Garlaschelli, D., Catanzaro, M., 2004. Emergence of complexity in financial networks, *Lecture Notes in Physics* 650, 399-423.

California Air Resource Board, 2005. Methods to find the cost-effectiveness of funding air quality projects.
http://www.arb.ca.gov/planning/tsaq/eval/mv_fees_cost-effectiveness_methods_may05.pdf Accessed on January 15, 2008.

Callaway, D. S., Newman, M. E. J., Strogatz, S. H., Watts, D. J., 2000. Network robustness and fragility: percolation on random graphs. *Physical Review Letters* 85, 5468-5471.

Canadian Competition Bureau, 2006. Competition bureau concludes examination into gasoline price spike following hurricane Katrina. http://www.competitionbureau.gc.ca Accessed on January 4, 2009.

Cantor, D. G., Gerla, M., 1974. Optimal routing in a packet-switched computer network. *IEEE Transactions on Computers* 23, 1062-1069.

Catto, S., 2003. Travel advisory: Toronto contends with SARS outbreak. *New York Times*, April 27.

Chang, P. C., 1988. A measure of the synergy in mergers under a competitive market for corporate control. *Atlantic Economic Journal* 16, 59-62.

Chankong, V., Haimes, Y. Y., 1983. *Multiobjective Decision Making: Theory and Methodology*. North-Holland, New York.

Chari, V. V., Christiano, L. J., Kehoe, P. J., 2008. Facts and myths about the financial crisis of 2008. Working paper 666, The Federal Reserve Bank of Minneapolis, Minneapolis, Minnesota.

Charnes, A., Cooper, W. W., 1967. Some network characterizations for mathematical programming and accounting approaches to planning and control. *The Accounting Review* 42, 24-52.

Chassin, D. P., Posse, C., 2005. Evaluating North American electric grid reliability using the Barabási-Albert network model. *Physica A* 355, 667-677.

Chatterjee, S., 1986. Types of synergy and economic value: The impact of acquisitions on merging and rival firms. *Strategic Management Journal* 7, 119-139.

Cheng, T. C. E, Wu, Y. N., 2006. A multiproduct, multicriterion supply-demand network equilibrium model. *Operations Research* 54, 544-554.

Chipman, J. S., 1987. Compensation principle. In *The New Palgrave: A Dictionary of Economics, Vol. 1*. Eatwell, J., Milgate, M., Newman, P., Editors, The Stockton Press, New York, pp. 524-531.

Christofides, N., Hewins, R. D., Salkin, G. R., 1979. Graph theoretic approaches to foreign exchange operations. *Journal of Financial and Quantitative Analysis* 14, 481-500.

CNNMoney.com, 1999. Exxon-Mobil merger done. November 30.

Coffman, K. G., Odlyzko, A. M., 2002. Internet growth: Is there a "Moore's law" for data traffic? In *Handbook of Massive Datasets*. Abello, J., Pardalos, P. M., Resende, M. G. C., Editors, Kluwer Academic Publishers, Norwell, Massachusetts, pp. 47-93.

Cohen, R., Erez, K., ben-Avraham, D., Havlin, S., 2000a. Resilience of the Internet to random breakdowns. *Physical Review Letters* 85, 4626-4628.

Cohen, R., Erez, K., ben-Avraham, D., Havlin, S., 2000b. Breakdown of the Internet under intentional attack. *Physical Review Letters* 86, 3682-3685.

Cojocaru, M. G., Daniele, P., Nagurney, A., 2005. Projected dynamical systems and evolutionary variational inequalities via Hilbert spaces with applications. *Journal of Optimization Theory and Applications* 27, 1-15.

Cojocaru, M. G., Daniele, P., Nagurney, A., 2006. Double-layered dynamics: A unified theory of projected dynamical systems and evolutionary variational inequalities. *European Journal of Operational Research* 175, 494-507.

Cojocaru, M. G., Daniele, P., Nagurney, A., 2007. Projected dynamical systems, evolutionary variational inequalities, applications, and a computational procedure. In *Pareto Optimality, Game Theory and Equilibria*. Migdalas, A., Pardalos, P. M., Pitsoulis, L., Editors, Springer Verlag, New York, pp. 169-188.

Committee on Climate Change and U.S. Transportation Research Board, 2006. Potential impacts of climate change on U.S. transportation. http://onlinepubs.trb.org/onlinepubs/sr/sr290.pdf Accessed on December 30, 2008.

Committee on Disaster Research in the Social Sciences: Future Challenges and Opportunities, 2006. *Facing Hazards and Disasters: Understanding Human Dimensions*. National Academies Press, Washington, DC.

Copeland, M. A., 1952. *A Study of Moneyflows in the United States*, National Bureau of Economic Research, New York.

Cournot, A. A., 1838. *Researches into the Mathematical Principles of the Theory of Wealth*. English translation. Macmillan, London, England, 1897.

Craighead, C. W., Blackhurst, J., Rungtusanatham, M. J., Handfield, R. B., 2007. The severity of supply chain disruptions: Design characteristics and mitigation capabilities. *Decision Sciences* 38, 131-156.

Creane, A., Davidson, C., 2004. Multidivisional firms, internal competition and the merger paradox. *Canadian Journal of Economics* 37, 951-977.

Crum, R. L., Nye, D. J., 1981. A network model of insurance company cash flow management. *Mathematical Programming Study* 15, 86-101.

Cruz, J. M., Nagurney, A., Wakolbinger, T., 2006. Financial engineering of the integration of global supply chain networks and social networks with risk management. *Naval Research Logistics* 53, 674-696.

Dafermos, S. C., 1971. An extended traffic assignment problem with applications to two-way traffic. *Transportation Science* 5, 366-389.

Dafermos, S. C., 1972. The traffic assignment problem for multiclass-user transportation networks. *Transportation Science* 6, 73-87.

Dafermos, S. C., 1973. Toll patterns for multiclass-user transportation networks. *Transportation Science* 7, 211-223.

Dafermos, S., 1980. Traffic equilibrium and variational inequalities. *Transportation Science* 14, 42-54.

Dafermos, S., 1982. The general multimodal network equilibrium problem with elastic demand. *Networks* 12, 57-72.

Dafermos, S., 1983. An iterative scheme for variational inequalities. *Mathematical Programming* 28, 57-72.

Dafermos, S., Nagurney, A., 1984. Stability and sensitivity analysis for the general network equilibrium - travel choice model. In *Proceedings of the Ninth International*

Symposium on Transportation and Traffic Theory. Volmuller, J., Hamerslag, R., Editors, VNU Science Press, Utrecht, The Netherlands, pp. 217-232.

Dafermos, S., Nagurney, A. 1987. Oligopolistic and competitive behavior of spatially separated markets. *Regional Science and Urban Economics* 17, 245-254.

Dafermos, S. C., Sparrow, F. T., 1969. The traffic assignment problem for a general network. *Journal of Research of the National Bureau of Standards 73B*, 91-118.

Daganzo, C. F., 1977a. On the traffic assignment problem with flow dependent costs-I. *Transportation Research* 11, 433-437.

Daganzo, C. F., 1977b. On the traffic assignment problem with flow dependent costs-II. *Transportation Research* 11, 439-441.

Dall'Asta, L., Barrat, A., Barthélemy, M., Vespignani, A., 2006. Vulnerability of weighted networks. *Journal of Statistical Mechanics*, Article no. P04006.

Daniele, P., 2003a. Evolutionary variational inequalities and economic models for demand supply markets. *Mathematical Models and Methods in Applied Sciences* 4, 471-489.

Daniele, P., 2003b. Variational inequalities for evolutionary financial equilibrium. In *Innovations in Financial and Economic Networks.* Nagurney, A., Editor, Edward Elgar Publishing, Cheltenham, England, pp. 84-108.

Daniele, P., 2004. Time-dependent spatial price equilibrium problem: existence and stability results for the quantity formulation model. *Journal of Global Optimization* 28, 283-295.

Daniele, P., 2006. *Dynamic Networks and Evolutionary Variational Inequalities.* Edward Elgar Publishing, Cheltenham, England.

Daniele, P., Maugeri, A., Oettli, W., 1999. Time-dependent traffic equilibria. *Journal of Optimization Theory and its Applications* 103, 543-555.

Davidson, A. L., 2006. *Key Performance Indicators in Humanitarian Logistics.* Master of Engineering Dissertation, Massachusetts Institute of Technology, Cambridge, Massachusetts.

Davidson, K. B., 1966. The theoretical basis of a flow-travel time relationship for use in transportation planning. *Australian Road Research* 8, 32-35.

Davis, D. D., Wilson, B. J., 2006. Equilibrium price dispersion, mergers and synergies: An experimental investigation of differentiated product competition. *International Journal of the Economics of Business* 13, 169-194.

deLisle, J., 2003. SARS, greater China, and the pathologies of globalization and transition. *Orbis* 47, 587-604.

DesRoches, G. J., Rix, D., 2006. Hurricane Katrina's impact on Louisiana's transportation infrastructure. School of Civil and Environmental Engineering, Georgia Institute of Technology, Atlanta, Georgia.

Development Initiatives, 2006. Global humanitarian assistance 2006 report. http://www.globalhumanitarianassistance.org/pdfdownloads/GHA%202006.pdf Accessed on January 5, 2009.

Devero, A. J., 2004. Look beyond the deal. *American Management Association MWORLD*, 30-32.

Dhanda, K. K., Nagurney, A., Ramanujam, P., 1999. *Environmental Networks: A Framework for Economic Decision-Making and Policy Analysis*. Edward Elgar Publishing, Cheltenham, England.

Doherty, N. A., 1997. Financial innovation in the management of catastrophe risk. *Journal of Applied Corporate Finance* 10, 84-95.

Dong, J., Zhang, D., Nagurney, A., 2002. Supply chain networks with multicriteria decision-makers. In *Transportation and Traffic Theory in the 21st Century*. Taylor, M. A. P., Editor, Pergamon-Elsevier, Oxford, England, pp. 179-196.

Dong, J., Zhang, D., Nagurney, A., 2004. A supply chain network equilibrium model with random demands. *European Journal of Operational Research* 156, 194-212.

Dong, J., Zhang, D., Yan, H., Nagurney, A., 2005. Multitiered supply chain networks: Multicriteria decision-making under uncertainty. *Annals of Operations Research* 135, 155-178.

Dow Theory Forecasts, 2006. HP makes printers more profitable. November 13.

Doyle, J. C., Alderson, D. L., Lun, L., Low, S., Roughan, M., Shalunov, S., Tanaka, R., Willinger, W., 2005. The "robust yet fragile" nature of the Internet. *Proceedings of the National Academy of Sciences USA* 102, 14497-14502.

Drug Week, 2005. Business Update: Healthcare products provider creates $70 billion supply chain services unit. October 7.

Dueñas-Osorio, L. A., Craig, J. I., Goodno, B. G., 2005. Optimal flow approach to quantify performance of networked systems. Civil and Environmental Engineering Department, Georgia Institute of Technology, Atlanta, Georgia.

Dupuis, P., Nagurney, A., 1993. Dynamical systems and variational inequalities. *Annals of Operations Research* 44, 9-42.

Eccles, R. G., Lanes, K. L., Wilson, T. C., 1999. Are you paying too much for that acquisition? *Harvard Business Review* 77, 136-146.

Emergency Events Database, 2008. Center for Research on the Epidemiology of Disasters, Catholic University of Louvain, Belgium. http://www.emdat.be/ Accessed on December 15, 2008.

Environment News Service, January 22, 2008. Governors, mayor form coalition to rebuild ailing U.S. infrastructure. http://www.ens-newswire.com/ens/jan2008/2008-01-22-02.asp Accessed on November 15, 2008.

Fairtrade Foundation, 2002. Spilling the bean. http://www.fairtrade.org.uk Accessed on June 5, 2008.

Faloutsos, M., Faloutsos, P., Faloutsos, C., 1999. On power-law relationships of the Internet topology. In *SIGCOMM '99: Proceedings of the Conference on Applications, Technologies, Architectures, and Protocols for Computer Communication*, pp. 251-262.

Farrell, J., Shapiro, C., 1990. Horizontal mergers: An equilibrium analysis. *American Economic Review* 80, 107-126.

Federal Emergency Management Agency, 1992. Federal response plan. FEMA Publication 229.

Fei, J. C. H., 1960. The study of the credit system by the method of the linear graph. *The Review of Economics and Statistic* 42, 417-428.

Feldman, S. J., Soyka, P. A., Ameer, P. G., 1997. Does improving a firm's environmental management system and environmental performance result in a higher stock price? *Journal of Investing* 6, 87-97.

Fershtman, C., Judd, K., 1987. Equilibrium incentives in oligopoly. *American Economic Review* 77, 927-940.

Fishburn, P. C., 1970. *Utility Theory for Decision Making*. John Wiley & Sons, New York.

Fisk, C., Boyce, D. E., 1983. Alternative variational inequality formulation of the network equilibrium-travel choice problem. *Transportation Science* 17, 454-463.

Flam, S. P., Ben-Israel, A., 1990. A continuous approach to oligopolistic market equilibrium. *Operations Research* 38, 1045-1051.

Ford, L. R., Fulkerson, D. R., 1962. *Flows in Networks*. Princeton University Press, Princeton, New Jersey.

Freeman, L. C., 1979. Centrality in social networks: Conceptual clarification. *Social Networks* 1, 215-239.

Freeman, L. C., Borgatti, S. P., White, D. R., 1991. Centrality in valued graphs: A measure of betweenness based on network flow. *Social Networks* 13, 141-154.

Gabay, D., Moulin, H., 1980. On the uniqueness and stability of Nash equilibria in noncooperative games. In *Applied Stochastic Control of Econometrics and Management Science*. Bensoussan, A., Kleindorfer, P., Tapiero, C. S., Editors, North-Holland, Amsterdam, The Netherlands, pp. 271-294.

Gallager, R. G., 1977. A minimum delay routing algorithm using distributed computation. *IEEE Transaction on Communications* 25, 73-85.

Garbin, D. A., Shortle, J. F., 2007. Measuring resilience in network-based infrastructures. In *Critical Thinking: Moving from Infrastructure Protection to Infrastructure Resilience*. Critical Infrastructure Discussion Paper Series, George Mason University School of Law, Fairfax, Virginia, pp. 72-86.

Garten, J. E., 1999. Lessons for the next financial crisis. *Foreign Affairs* 78, 76-92.

Gartner, N. H., 1980a. Optimal traffic assignment with elastic demands: A review; Part I. Analysis framework. *Transportation Science* 14, 174-191.

Gartner, N. H., 1980b. Optimal traffic assignment with elastic demands: A review; Part II. Algorithmic approaches. *Transportation Science* 14, 192-208.

Geunes, J., Pardalos, P. M., 2003. Network optimization in supply chain management and financial engineering: An annotated bibliography. *Networks* 42, 66-84.

Gilli, M., Këllezi, E., 2006. An application of extreme value theory for measuring financial risk. *Computational Economics* 27, 207-228.

Girvan, M., Newman, M. E. J., 2002. Community structure in social and biological networks. *Proceedings of the National Academy of Sciences USA* 99, 8271-8276.

Global News Wire, 2008. Delta and Northwest merge, creating premier global airline. Press release, October 29.

Gribble, S. D., 2001. Robustness in complex systems. In *Proceedings of the 8th Workshop on Hot Topics in Operating Systems (HotOS-VIII)*, pp. 21-26.

Gupta, D., 1996. The (Q, r) inventory system with an unreliable supplier. *INFOR* 34, 59-76.

Gupta, D., Gerchak, Y., 2002. Quantifying operational synergies in a Merger/ Acquisition. *Management Science* 48, 517-534.

Haghani, A., Oh, S., 1996. Formulation and solution of a multi-commodity, multi-modal network flow model for disaster relief operations. *Transportation Research A*, 30, 231-250.

Haimes, Y. Y., Longstaff, T., 2002. The role of risk analysis in the protection of critical infrastructures against terrorism. *Risk Analysis* 22, 439-444.

Hakkinen, L., Norrman, A., Hilmola, O.-P., Ojala, L., 2004. Logistics integration in horizontal mergers and acquisitions. *The International Journal of Logistics Management* 15, 27-42.

Handfield, R., 2007. Avoid supply chain risk.
http://www.sas.com/news/sascom/2007q2/feature_supplychain.html Accessed on November 12, 2008.

Hanneman, R. A., 2001. Introduction to social network models. Mimeo, University of California, Riverside, California.

Hansson, S. O., Helgesson, G., 2003. What is stability? *Synthese* 136, 219-235.

Harker, P. T., 1984. A variational inequality approach for the determination of oligopolistic market equilibrium. *Mathematical Programming* 30, 105-111.

Hart, S., Ahuja, G., 1996. Does it pay to be green? An empirical examination of the relationship between emission reduction and firm performance. *Business Strategy and the Environment* 5, 30-37.

Hendricks, K. B., Singhal, V. R., 2005. An empirical analysis of the effect of supply chain disruptions on long-term stock price performance and risk of the firm. *Production and Operations Management* 14, 35-52.

Herd, T., Saksena, A. K., Steger, T. W., 2005. Delivering merger synergy: A supply chain perspective on achieving high performance. *Outlook – Point of View*, Accenture, May.

Hizir, A. E., 2006. *Using Emission Functions in Mathematical Programming Models for Sustainable Urban Transportation: An Application in Bilevel Optimization.* Master of Science Dissertation, Sabanci University, Turkey.

Holme, P., Kim, B. J., 2002. Vertex overload breakdown in evolving networks. *Physical Review E* 65, Article no. 066109.

Holmgren, A. J., 2007. A framework for vulnerability assessment of electric power systems. In *Reliability and Vulnerability in Critical Infrastructure: A Quantitative*

Geographic Perspective. Murray, A., Grubesic, T., Editors, Springer, New York, pp. 31-55.

International Strategy for Disaster Reduction, 2006. Press Release. January 30, 2006. http://www.unisdr.org/eng/media-room/press-release/2006/ PR-2006-02-Disasters -increase-18-per-cent-2005-but-death-rates-drop.pdf Accessed on January 5, 2009.

Internet World Stats, 2009. http://www.internetworldstats.com Accessed on January 6, 2009.

Institute of Electrical and Electronics Engineers, 1990. *IEEE Standard Computer Dictionary: a Compilation of IEEE Standard Computer Glossaries*. New York.

Insurance Journal, 2007. Major Internet disruption would cost $250 billion in economic damages. September 25.

Izquierdo, L.R., Hanneman, R. A., 2006. *Introduction to the Formal Analysis of Social Networks Using Mathematica*. University of California, Riverside, California.

Jackson, M. O., Wolinsky, A., 1996. A strategic model of social and economic networks. *Journal of Economic Theory* 71, 44 -74.

Jeanneret, M., 2006. Federal government freight bottleneck report highlights America's growing transportation infrastructure crisis.
http://www.artba.net/news/press_releases/2006/02-03-06.htm Accessed on November 12, 2007.

Jenelius, E., Petersen, T., Mattsson, L. G., 2006. Road network vulnerability: Identifying important links and exposed regions. *Transportation Research A* 20, 537-560.

Judge, G. G., Takayama, T., Editors, 1973. *Studies in Economic Planning Over Space and Time*. North-Holland, Amsterdam, The Netherlands.

Juga, J., 1996. Organizing for network synergy in logistics: A case study. *International Journal of Physical Distribution & Logistics* 26, 51-66.

Kali, R., Reyes, J., 2005. Financial contagion in the international trade network. To appear in *Economic Inquiry*.

Karush, W., 1939. *Minima of Functions of Several Variables with Inequalities as Side Conditions*. Master's Thesis, Department of Mathematics, University of Chicago, Chicago, Illinois.

Keeney, R. L., Raiffa, H., 1992. *Decisions with Multiple Objectives: Preferences and Value Tradeoffs*. Cambridge University Press, Cambridge, England.

Kembel, R., 2000. *The Fibre Channel Consultant: A Comprehensive Introduction*. Northwest Learning Associates, Tucson, Arizona.

Kim, D. H., Jeong, H., 2005. Systematic analysis of group identification in stock markets. *Physical Review E* 72, Article no. 046133.

Kinderlehrer, D., Stampacchia, G., 1980. *An Introduction to Variational Inequalities and Their Applications*. Academic Press, New York.

Klassen, R. D., McLaughlin, C. P., 1996. The impact of environmental management on firm performance. *Management Science* 42, 1199-1214.

Klein, B., Leffler, K. B., 1981. The role of market forces in assuring contractual performance. *Journal of Political Economy* 89, 615-641.

Kleindorfer, P. R., Saad, G. H., 2005. Managing disruption risks in supply chains. *Production and Operations Management* 14, 53-68.

Knight, F. H., 1924. Some fallacies in the interpretation of social cost. *Quarterly Journal of Economics* 38, 582-606.

Knowledge@Wharton, 2005. Sears-Kmart merger: Is it a tough sell? http://knowledge.wharton.upenn.edu/article.cfm?articleis=1081 Accessed on September 3, 2007.

Knudsen, T., Bang, B., 2007. Environmental consequences of better roads. SINTEF Technology and Society, Road and Transport Studies, Trondheim, Norway.

Kohl, J. E., 1841. Der verkehr und die ansiedelungen der menschen in ihrer abhangigkeit von den gestaltung der erdorberflache. Dresden, Leipzig, Germany.

Kontoghiorghes, E. J., Rustem, B., Siokos, S., Editors, 2002. *Computational Methods in Decision-Making, Economics and Finance, Optimization Models.* Kluwer Academic Publishers, Norwell, Massachusetts.

Korilis, Y. A., Lazar, A. A., Orda, A., 1999. Avoiding the Braess paradox in noncooperative networks. *Journal of Applied Probability* 36, 211-222.

Korpelevich, G. M., 1977. The extragradient method for finding saddle points and other problems. *Matekon* 13, 35-49.

Koschützki, D., Lehmann, K. A., Petters, L., Richter, S., Tenfelde-Podehl, D., Zlotowski, O., 2005. Centrality indices. In *Network Analysis: Methodological Foundations.* Brandes, U., Erlebach, T., Editors, Springer-Verlag, Berlin, Germany, pp. 16-61.

Koutsoupias, K., Papadimitrou, C. H., 1999. Worst-case equilibria. In *Proceedings of the 16th Annual Symposium on Theoretical Aspects of Computer Science (STACS), Lecture Series in Computer Science* 1563, pp. 404-413.

Krauss, C., 2003. Toronto mayor calls for understanding from business and consumers. *New York Times*, April 25.

Kuhn, H. W., Tucker, A. W., 1951. Nonlinear programming. In *Proceedings of 2nd Berkeley Symposium*, pp. 481-492.

Kusstatscher, V., Cooper, C. L., 2005. *Managing Emotions in Mergers and Acquisitions.* Edward Elgar Publishing, Cheltenham, England.

Lai, K. H., Ngai, E. W. T., Cheng, T. C. E., 2002. Measures for evaluating supply chain performance in transport logistics. *Transportation Research E* 38, 439-456.

Lambert, D. M., Pohlen, T. L., 2001. Supply chain metrics. *International Journal of Logistics Management* 12, 1-19.

Lambertini, L., Mantovani, A., 2007. Collusion helps abate environmental pollution: A dynamic approach. Department of Economics, University of Bologna, Italy.

Langabeer, J., 2003. An investigation of post-merger supply chain performance. *Journal of Academy of Business and Economics* 2, 14-25.

Langabeer, J., Seifert, D., 2003. Supply chain integration: The key to merger success (synergy). *Supply Chain Management Review* 7, 58-64.

Latora, V., Marchiori, M., 2001. Efficient behavior of small-world networks. *Physical Review Letters* 87, Article no. 198701.

Latora, V., Marchiori, M., 2002. Is the Boston subway a small-world network? *Physica A* 314, 109-113.

Latora, V., Marchiori, M., 2003. Economic small-world behavior in weighted networks. *The European Physical Journal B* 32, 249-263.

Latora, V., Marchiori, M., 2004. How the science of complex networks can help developing strategies against terrorism. *Chaos, Solitons and Fractals* 20, 69-75.

Latour, A., 2001. Trial by fire: A blaze in Albuquerque sets off major crisis for cell-phone giants. *Wall Street Journal*, January 29.

Lazaro, M., 2002. West coast port lockout leaves suppliers dry. http://hometextilestoday.com/article/CA250892.html Accessed on October 15, 2008.

LeBlanc, L. J., Morlok, E. K., Pierskalla, W. P. 1975. An efficient approach to solving the road network equilibrium traffic assignment problem. *Transportation Research* 9, 309-318.

Lee, H., Whang, S., 1999. Decentralized multi-echelon supply chains: Incentives and information. *Management Science* 45, 633-640.

Lemon, S., 2006. Internet access recovers in Asia after quake. *NetworkWorld*, December 28.

Leventhal, T., Nemhauser, G., Trotter, L., 1973. A column generation algorithm for optimal traffic assignment. *Transportation Science* 7, 168-176.

Lewis, T. G., 2006. *Critical Infrastructure Protection in Homeland Security: Defending a Networked Nation*. John Wiley & Sons, New York.

Liu, Z., Nagurney, A., 2007. Financial networks with intermediation and transportation network equilibria: A supernetwork equivalence and computational management reinterpretation of the equilibrium conditions with computations. *Computational Management Science* 4, 243- 281.

Liu, Z., Nagurney, A., 2009. An integrated electric power supply chain and fuel market network framework: theoretical modeling with empirical analysis for New England. *Naval Research Logistics*, in press.

Loubergé, H., Këllezi, E., Gilli, M., 1999. Using catastrophe-linked securities to diversify insurance risk: A financial analysis of cat-bonds. *The Journal of Insurance Issues* 2, 125-146.

Louisiana Department of Health and Hospitals, 2006. Reports of missing and deceased. http://www.dhh.louisiana.gov/offices/page.asp?ID=192&Detail=5248 Accessed on September 12, 2008.

Lynn, B. C., 2006. *End of the Line: The Rise and Coming Fall of the Global Corporation*. Doubleday, New York.

Lyon, T. P., 2003. Green firms bearing gifts. *Regulation* 26, 36-40.

Marks, M. L., Mirvis, P. H., 2001. Making mergers and acquisitions work: Strategic and psychological preparation. *Academy of Management Executive* 15, 80-92.

Marquez Diez-Canedo, J., Martinez-Jaramillo, S., 2007. Financial contagion: A network model for estimating the distribution of losses for the financial system. Paper presented at the 13th International Conference on Computing in Economics and Finance, Montreal, Canada.

Mas-Colell, A., Whinston, M., and Green, J. R., 1995. *Microeconomic Theory*. Oxford University Press, New York.

McCarthy, J. A., 2007. Introduction: From protection to resilience: Injecting 'moxie' into the infrastructure security continuum. In *Critical Thinking: Moving from Infrastructure Protection to Infrastructure Resilience*, Critical Infrastructure Discussion Paper Series, George Mason University School of Law, Fairfax, Virginia, pp. 1-8.

Meschi, M., 1997. Analytical perspectives on mergers and acquisitions: A survey. Paper number 5-97, ISSN number 1366-6290, Centre for International Business Studies, South Bank University, London, England.

Messmer, E., 2003. Blaster worm racks up victims. *PC World*, August 15.

Min, H., Zhou, G. 2002. Supply chain modeling: Past, present, future. *Computers and Industrial Engineering* 43, 231-249.

Monahan, S., Nardone, R., 2007. How Unilever aligned its supply chain and business strategies. *Supply Chain Management Review* November, 44-50.

Mulvey, J. M., 1987. Nonlinear networks in finance. *Advances in Mathematical Programming and Financial Planning* 20, 187-217.

Murphy, F. H., Sherali, H. D., Soyster, A. L., 1982. A mathematical programming approach for determining oligopolistic market equilibrium. *Mathematical Programming* 24, 92-106.

Murray, A. T., Grubesic, T. H., Editors, 2007. *Critical Infrastructure: Reliability and Vulnerability*. Springer, New York.

Murray-Tuite, P. M., Mahmassani, H. S., 2004. Methodology for determining vulnerable links in a transportation network. *Transportation Research Record* 1882, 88-96.

Nagurney, A., 1984. Comparative tests of multimodal traffic equilibrium methods. *Transportation Research B* 18, 469-485.

Nagurney, A., 1999. *Network Economics: A Variational Inequality Approach*, second and revised edition. Kluwer Academic Publishers, Dordrecht, The Netherlands.

Nagurney, A., 2000. *Sustainable Transportation Networks*. Edward Elgar Publishing, Cheltenham, England.

Nagurney, A., Editor, 2003. *Innovations in Financial and Economic Networks*. Edward Elgar Publishing, Cheltenham, England.

Nagurney, A., 2006a. *Supply Chain Network Economics: Dynamics of Prices, Flows, and Profits*. Edward Elgar Publishing, Cheltenham, England.

Nagurney, A., 2006b. On the relationship between supply chain and transportation network equilibria: A supernetwork equivalence with computations. *Transportation Research E* 42, 293-316.

Nagurney, A., 2008a. Formulation and analysis of horizontal mergers among oligopolistic firms with insights into the merger paradox: A supply chain network perspective. To appear in *Computational Management Science*.

Nagurney, A., 2008b. To merge or not to merge: Multimarket supply chain network oligopolies, coalitions, and the merger paradox. Isenberg School of Management, University of Massachusetts, Amherst, Massachusetts.

Nagurney, A., 2009. A system-optimization perspective for supply chain network integration: The horizontal merger case. *Transportation Research E* 45, 1-15.

Nagurney, A., Cruz, J., 2003a. International financial networks with electronic transactions. In *Innovations in Financial and Economic Networks*. Nagurney, A., Editor, Edward Elgar Publishing, Cheltenham, England, pp. 136-168.

Nagurney, A., Cruz, J., 2003b. International financial networks with intermediation: modeling, analysis, and computations. *Computational Management Science* 1, 31-58.

Nagurney, A., Cruz, J., Dong, J., Zhang, D., 2005. Supply chain networks, electronic commerce, and supply side and demand side risk. *European Journal of Operational Research* 164, 120-142.

Nagurney, A., Dong, J., 2002a. *Supernetworks: Decision-Making for the Information Age*. Edward Elgar Publishers, Cheltenham, England.

Nagurney, A., Dong, J., 2002b. A multiclass, multicriteria network equilibrium model with elastic demand. *Transportation Research B* 36, 445-469.

Nagurney, A., Dong, J., 2002c. Urban location and transportation in the information age: A multiclass, multicriteria network equilibrium perspective. *Environment & Planning B* 29, 53-74.

Nagurney, A., Dong, J., Hughes, M., 1992. Formulation and computation of general financial equilibrium. *Optimization* 26, 339-354.

Nagurney, A., Dong, J., Mokhtarian, P. L., 2002a. Traffic network equilibrium and the environment: A multicriteria decision-making perspective. In *Computational Methods in Decision-Making, Economics and Finance*, Kontoghiorges, E. J., Rustem, B., Siokos, S., Editors, Kluwer Academic Publishers, Dordrecht, The Netherlands, pp. 501-523.

Nagurney, A., Dong, J., Mokhtarian, P. L., 2002b. Teleshopping versus shopping: A multicriteria network equilibrium framework. *Mathematical and Computer Modelling* 34, 783-798.

Nagurney, A., Dong, J., Mokhtarian, P. L., 2002c. Multicriteria network equilibrium modeling with variable weights for decision-making in the information age with applications to telecommuting and teleshopping. *Journal of Economic Dynamics and Control* 26, 1629-1650.

Nagurney, A., Dong, J., Zhang, D., 2002. A supply chain network equilibrium model. *Transportation Research E* 8, 281-303.

Nagurney, A., Dupuis, P., Zhang, D., 1994. A dynamical systems approach for network oligopolies and variational inequalities. *Annals of Operations Research* 28, 263-293.

Nagurney, A., Hughes, M., 1992. Financial flow of funds networks. *Networks* 22, 145-161.

Nagurney, A., Ke, K., 2001. Financial networks with intermediation. *Quantitative Finance* 1, 309-317.

Nagurney, A., Ke, K., 2003. Financial networks with electronic transactions: modeling, analysis, and computations. *Quantitative Finance* 3, 71-87.

Nagurney, A., Liu, Z., 2007. An evolutionary variational inequality formulation of supply chain networks with time-varying demands. In *Network Science, Nonlinear Science and Infrastructure Systems*. Friesz, T. L., Editor, Springer, Berlin, Germany, pp. 267-302.

Nagurney, A., Liu, Z., Cojocaru, M. G., Daniele, D., 2007. Dynamic electric power supply chains and transportation networks: An evolutionary variational inequality formulation. *Transportation Research E* 43, 624-642.

Nagurney, A., Liu, Z., Woolley, T., 2006. Optimal endogenous carbon taxes for electric power supply chains with power plants. *Mathematical and Computer Modelling* 44, 899-916.

Nagurney, A., Parkes, D., Daniele, P., 2007. The Internet, evolutionary variational inequality, and the time-dependent Braess Paradox. *Computational Management Science* 4, 355-375.

Nagurney, A., Qiang, Q., 2007a. A network efficiency measure for congested networks. *Europhysics Letters* 79, 38005, 1-5.

Nagurney, A., Qiang, Q., 2007b. A transportation network efficiency measure that captures flows, behavior, and costs with applications to network component importance identification and vulnerability. In *Proceedings of the 18th Annual POMS Conference*, Dallas, Texas.

Nagurney, A., Qiang, Q., 2007c. Robustness of transportation networks subject to degradable links. *Europhysics Letters* 80, 68001, 1-6.

Nagurney, A., Qiang, Q., 2008a. A network efficiency measure with application to critical infrastructure networks. *Journal of Global Optimization* 40, 261-275.

Nagurney, A., Qiang, Q., 2008b. Identification of critical nodes and links in financial networks with intermediation and electronic transactions. In *Computational Methods in Financial Engineering*. Kontoghiorghes, E. J., Rustem, B., Winker, P., Editors, Springer, Berlin, Germany, pp. 273-297.

Nagurney, A., Qiang, Q., 2008c. An efficiency measure for dynamic networks with application to the Internet and vulnerability analysis. *Netnomics* 9, 1-20.

Nagurney, A., Qiang, Q., 2009. A relative total cost index for the evaluation of transportation network robustness in the presence of degradable links and alternative travel behavior. *International Transactions in Operational Research* 16, 49-67.

Nagurney, A., Qiang, Q., Nagurney, L. S., 2008. Environmental impact assessment of transportation networks with degradable links in an era of climate change. To appear in the *International Journal of Sustainable Transportation.*

Nagurney, A., Siokos, S., 1997. *Financial Networks: Statics and Dynamics.* Springer-Verlag, Heidelberg, Germany.

Nagurney, A., Takayama, T., Zhang, D., 1995. Massively parallel computation of spatial price equilibrium problems as dynamical systems. *Journal of Economic Dynamics and Control* 18, 3-37.

Nagurney, A., Toyasaki, F., 2003. Supply chain supernetworks and environmental criteria. *Transportation Research D* 8, 185-213.

Nagurney, A., Toyasaki, F., 2005. Reverse supply chain management and electronic waste recycling: A multitiered network equilibrium framework for e-cycling. *Transportation Research E* 41, 1-28.

Nagurney, A., Wakolbinger, T., and Zhao, L., 2006. The evolution and emergence of integrated social and financial networks with electronic transactions: A dynamic supernetwork theory for the modeling, analysis, and computation of financial flows and relationship levels. *Computational Economics* 27, 353-393.

Nagurney, A., Woolley, T., 2008. Environmental and cost synergy in supply chain network integration in mergers and acquisitions. To appear in *Multiple Criteria Decision Making for Sustainable Energy and Transportation Systems, Proceedings of the 19th International Conference on Multiple Criteria Decision Making, Auckland, New Zealand, January 7-12, 2008.* Ehrgott, M., Naujoks, B., Stewart, T. J., Wallenius, J., Editors, Springer, Berlin, Germany.

Nagurney, A., Woolley, T., Qiang, Q., 2008. Multiproduct supply chain horizontal network integration: Models, theory, and computational results. To appear in the *International Journal of Operational Research.*

Nagurney, A., Zhang, D., 1996. *Projected Dynamical Systems and Variational Inequalities with Applications.* Kluwer Academic Publishers, Boston, Massachusetts.

Nagurney, A., Zhang, D., 1997. Projected dynamical systems in the formulation, stability analysis, and computation of fixed demand traffic network equilibria. *Transportation Science* 31, 147-158.

Nash, J. F., 1950. Equilibrium points in n-person games. *Proceedings of the National Academy of Sciences* 36, 48-49.

Nash, J. F., 1951. Noncooperative games. *Annals of Mathematics* 54, 286-298.

National Assessment Team, 2001. *Climate Change Impacts on the United States: The Potential Consequences of Climate Variability and Change. US Global Change Research Program.* Cambridge University Press, Cambridge, England.

Newman, M. E. J., 2004. Analysis of weighted networks. *Physical Review E* 70, Article no. 056131.

Newman, M. E. J., Barabási, A. L., Watts, D. J., Editors, 2006. *The Structure and Dynamics of Networks.* Princeton University Press, Princeton, New Jersey.

Nicholson, A., Du, Z. P., 1997. Degradable transportation systems: an integrated equilibrium model. *Transportation Research B* 31, 209-223.

Niehaus, G., 2002. The allocation of catastrophe risk. *Journal of Banking and Finance* 6, 585-596.

Nijkamp, P., Reggiani, A., 1998. *The Economics of Complex Spatial Systems.* Elsevier, Amsterdam, The Netherlands.

Odell, K., Phillips, R. J., 2001. Testing the ties that bind: Financial networks and the 1906 San Francisco earthquake. Department of Economics, Colorado State University, Fort Collins, Colorado.

Office of the Speaker of the U.S. House of Representatives, 2008. Stabilizing housing is key to America's economic recovery.
http://speaker.house.gov/newsroom/reports?id=0037 Accessed on November 2, 2008.

O'Kelly, M. E., Kim, H., Kim, C., 2006. Internet reliability with realistic peering. *Environment and Planning B* 33, 325-343.

Okuguchi, K., Szidarovski, F., 1990. *The Theory of Oligopoly with Multi-Product Firms. Lecture Notes in Economics and Mathematical Systems.* Springer-Verlag, Berlin, Germany.

Onnela, J. P., Kaski, K., and Kertész, J., 2004. Clustering and information in correlation based financial networks. *The European Physical Journal B* 38, 353-362.

Ozdaglar, A., Srikant, R., 2008. Incentives and pricing in communication networks. In *Algorithmic Game Theory.* Nisan, N., Roughgarden, T., Tardos, E., Vazirani, V., Editors, Cambridge University Press, Cambridge, England.

Papadimitrou, C. H., 2001. Algorithms, games, and the Internet. In *Proceedings of the 33rd Annual ACM Symposium on Theory of Computing (STOC)*, pp. 749-753.

Parlar, M., 1997. Continuous review inventory problem with random supply interruptions. *European Journal of Operations Research* 99, 366-385.

Parlar, M., Perry, D., 1996. Inventory models of future supply uncertainty with single and multiple suppliers. *Naval Research Logistics* 43, 191-210.

Pas, E. I., Principio, S. L., 1997. Braess paradox: Some new insights. *Transportation Research B* 31, 265-276.

Patriksson, M., 1994. *The Traffic Assignment Problem - Models and Methods.* VSP, Utrecht, The Netherlands.

Pepall, L., Richards, D., Norman, G., 1999. *Industrial Organization: Contemporary Theory and Practice.* South-Western College Publishing, Cleveland, Ohio.

Perakis, G., 2007. The price of anarchy when costs are non-separable and asymmetric. *Mathematics of Operations Research* 32, 614-628.

Perea-Lopez, E., Ydtsie, B. E., Grossmann, I. E., 2003. A model predictive control strategy for supply chain management. *Computers & Chemical Engineering* 27, 1201-1218.

Perelman, L. J., 2007. Shifting security paradigms: Towards resilience. In *Critical Thinking: Moving from Infrastructure Protection to Infrastructure Resilience*. Critical Infrastructure Discussion Paper Series, George Mason University School of Law, Fairfax, Virginia, pp. 23-48.

Perry, M., 2005. Logistics nightmare as aid arrives. *Reuters*, January 1.

Perry, M. K., Porter, R. H., 1985. Oligopoly and the incentive for horizontal merger. *American Economic Review* 75, 219-227.

Phoenix Business Journal, 2009. Wells Fargo, Wachovia complete merger. January 2.

Pigou, A. C., 1920. *The Economics of Welfare*. Macmillan, London, England.

Power, R., 2002. CSI/FBI computer crime and security survey. *Computer Security Issues and Trends* 8, 1-22.

Qiang, Q., Nagurney, A., 2008. A unified network performance measure with importance identification and the ranking of network components. *Optimization Letters* 2, 127-142.

Qiang, Q., Nagurney, A., Dong, G., 2009. Modeling of supply chain risk under disruptions with performance measurement and robustness analysis. In *Managing Supply Chain Risk and Vulnerability: Tools and Methods for Supply Chain Decision Makers*. Wu, T., Blackhurst, J., Editors, Springer, Berlin, Germany, in press.

Quesnay, F., 1758. *Tableau Economique*, reproduced in facsimile with an introduction by H. Higgs by the British Economic Society, 1895.

Ran, B., Boyce, D. E., 1996. *Modeling Dynamic Transportation Networks*. Springer-Verlag, Berlin, Germany.

Rao, P., 2002. Greening the supply chain: A new initiative in south east Asia. *International Journal of Operations & Production Management* 22, 632-655.

Reed, B., 2008. Internet cable cuts raise alarms over infrastructure vulnerabilities. *Network World*, January 31.

Reilly, J., Prinn, R., Harnisch, J., Fitzmaurice, J., Jacoby, H., Kicklighterë, D., Melillo, J., Stone, P., Sokolov, A. Wang, C., 1999. Multi-gas assessment of the Kyoto Protocol. *Nature* 401, 549-555.

Resende, M. G. C., Pardalos, P. M., Editors, 2006. *Handbook of Optimization in Telecommunications*. Springer Science and Business Media, New York.

Richardson, H. W., Gordon, P., Moore II, J. E., Editors, 2006. *The Economic Impacts of Terrorist Attacks*. Edward Elgar Publishers, Cheltenham, England.

Rilett, L. R., Benedek, C. M., 1994. Traffic assignment under environmental and equity objective. *Transportation Research Record* 1443, 92-99.

Robinson, C. P., Woodard, J. B., Varnado, S. G., 1998. Critical infrastructure: interlinked and vulnerable. *Issues in Science and Technology* 15, 61-67.

Roper Starch Worldwide Inc., 1997. Green gauge report. New York, New York.

Roughgarden, T., 2003. The price of anarchy is independent of the network topology. *Journal of Computer and System Sciences* 67, 341-364.

Roughgarden, T., 2005. *Selfish Routing and the Price of Anarchy.* MIT Press, Cambridge, Massachusetts.

Royal Dutch Shell plc, 2008. Form 20-F.
http://sec.gov/Archives/edgar/data/1306965/000115697308000312/u55159e20vf.htm Accessed on January 12, 2009.

Rustem, B., Howe, M., 2002. *Algorithms for Worst-Case Design and Risk Management.* Princeton University Press, Princeton, New Jersey.

Sakakibara, H., Kajitani, Y., Okada, N., 2004. Road network robustness for avoiding functional isolation in disasters. *Journal of Transportation Engineering* 130, 560-567.

Salant, S., Switzer, S., Reynolds, R., 1983. Losses due to merger: The effects of an exogenous change in industry structure on Cournot-Nash equilibrium. *Quarterly Journal of Economics* 48, 185-200.

Samuelson, P. A., 1952. Spatial price equilibrium and linear programming. *American Economic Review* 42, 283-303.

Samuelson, P. A., 1983. *Foundations of Economic Analysis.* Enlarged edition. Harvard University Press, Cambridge, Massachusetts.

Sandholm, W. H., Dokumaci, E., Lahkar, R., 2008. The projection dynamic and the replicator dynamic. *Games and Economic Behavior* 64, 666-683.

Sarkis, J., 2003. A strategic decision framework for green supply chain management. *Journal for Cleaner Production* 11, 397-409.

Scalingi, P. L., 2007. Moving beyond critical infrastructure protection to disaster resilience. In *Critical Thinking: Moving from Infrastructure Protection to Infrastructure Resilience.* Critical Infrastructure Discussion Paper Series, George Mason University School of Law, Fairfax, Virginia, pp. 49-71.

Schillo, M., Bürchert, H., Fischer, K., Klusch, M., 2001. Towards a definition of robustness for market-style open multi-agent systems. In *Proceedings of the 5th International Conference on Autonomous Agents*, pp. 75-76.

Schulz, C., 2007. Identification of critical transportation infrastructures. Forum DKKV/CEDM, Disaster Reduction in Climate Change, Karlsruhe University, Germany.

Scott, D. M., Novak, D., Aultman-Hall, L., Guo, F., 2006. Network robustness index: a new method for identifying critical links and evaluating the performance of transportation networks. *Journal of Transport Geography* 14, 215-227.

Shakkottai, S., Srikant, R., Ozdaglar, A., Acemoglu, D., 2008. The price of simplicity. *IEEE Journal on Selected Areas of Communications* 26.

Sheffi, Y., 1985. *Urban Transportation Networks - Equilibrium Analysis with Mathematical Programming Methods.* Prentice-Hall, Englewood Cliffs, New Jersey.

Sheffi, Y., 2005. *The Resilient Enterprise: Overcoming Vulnerability for Competitive Advantage.* MIT Press, Cambridge, Massachusetts.

Silberberg, E., Suen, W., 2000. *The Structure of Economics: A Mathematics Analysis.* McGraw-Hill, New York.

Smith, M. J., 1979. Existence, uniqueness, and stability of traffic equilibria. *Transportation Research B* 13, 259-304.

Smith, O. P., Levasseur, G., 2002. Impacts of climate change on transportation infrastructure in Alaska. http://www.climate.dot.gov/workshop1002/smith.doc Accessed on January 15, 2008.

Snyder, L. V., 2003. *Supply Chain Robustness and Reliability: Models and Algorithms.* Ph.D. Dissertation, Northwestern University, Department of Industrial Engineering and Management Sciences, Evanston, Illinois.

Snyder, L. V., Daskin, M. S., 2005. A reliability model for facility location: the expected failure cost case. *Transportation Science* 39, 400-416.

Snyder, L. V., Shen, Z. J. M., 2006. Supply chain management under the threat of disruptions. *The Bridge* 36, 39-45.

Soylu, A., Oruc, C., Turkay, M., Fujita, K., Asakura, T., 2006. Synergy analysis of collaborative supply chain management in energy systems using multi-period MILP. *European Journal of Operational Research* 174, 387-403.

Soyster, A. L., 1973. Convex programming with set-inclusive constraints and applications to inexact linear programming. *Operations Research* 21, 1154-1157.

Stanwick, P. A., Stanwick, S. D., 2002. Overcoming M&A environmental problems. *Journal of Corporate Accounting & Finance* 13, 33-37.

Sugawara, S., Niemeier, D. A., 2002. How much can vehicle emissions be reduced? *Transportation Research Record* 1815, 29-37.

Talluri, S., Baker, R. C., 2002. A multi-phase mathematical programming approach for effective supply chain design. *European Journal of Operational Research* 141, 544-558.

Tang, C. S., 2006a. Robust strategies for mitigating supply chain disruptions. *International Journal of Logistics* 9, 33-45.

Tang, C. S., 2006b. Perspectives in supply chain risk management. *International Journal of Production Economics* 103, 451-488.

Target Corporation, 2008. Form 10-K.
http://sec.gov/Archives/edgar/data/27419/000104746908002681/a2183529z10-k.htm
Accessed on January 11, 2009.

Taylor, M. A. P., Sekhar, V. C., D'Este, G. M., 2006. Application of accessibility based methods for vulnerability analysis of strategic road networks. *Networks and Spatial Economics* 6, 267-291.

Thomas, A. S., Kopczak, L. R., 2005. From logistics to supply chain management. The path forward in the humanitarian sector. Fritz Institute, San Francisco, California.

Thomas, D. J., Griffin, P. M., 1996. Coordinated supply chain management. *European Journal of Operational Research* 94, 1-15.

Thore, S., 1969. Credit networks. *Economica* 36, 42-57.

Thore, S., Kydland, F., 1972. Dynamic flow-of-funds networks. In *Applications of Management Science in Banking and Finance.* Eilon, S., Fowkes, T. R., Editors, Epping, England, pp. 259-276.

Thore, S., 1980. *Programming the Network of Financial Intermediation.* Universitetsforlaget, Oslo, Norway.

Tirole, J., 1988. *The Theory of Industrial Organization.* MIT Press, Cambridge, Massachusetts.

Tomlin, B. T., 2006. On the value of mitigation and contingency strategies for managing supply chain disruption risks. *Management Science* 52, 639-657.

TradingMarkets.com, 2008. Anheuser-Busch shareholders okays InBev merger. November 12.

UN News Center, 2004. As pledges for tsunami victims top $1.1 billion, Annan says logistics is biggest challenge. December 31.

U.S.-Canada Power System Outage Task Force, 2004. Final report on the August 14th, 2003 blackout in the United States and Canada: Causes and recommendations. April.

U.S. Customs and Border Protection, 2007. Import trade trends, FY 2007 mid-year report.

U.S. Department of Commerce, 2006. Hurricane Katrina service assessment report. http://www.weather.gov/om/assessments/pdfs/Katrina.pdf Accessed on November 12, 2008.

U.S. Department of Transportation, 2002. The potential impacts of climate change on transportation. Workshop summary and proceedings.
http://climate.dot.gov/workshop1002.index.html Accessed on October 28, 2007.

U.S. Department of Transportation Federal Highway Administration, 2004. Highway Statistics 2004. http://www.fhwa.dot.gov Accessed on January 15, 2008.

U.S. Department of Transportation Federal Highway Administration, 2006a. An initial assessment of freight bottlenecks on highways.
http://www.fhwa.dot.gov/policy/otps/bottlenecks/ Accessed on October 18, 2007.

U.S. Department of Transportation Federal Highway Administration, 2006b. The freight story: A national perspective on enhancing freight transportation.
http://ops.fhwa.dot.gov/freight/freight_analysis/freight_story/freight.pdf Accessed on October 19, 2007.

U.S. Department of Transportation Federal Highway Administration, 2006c. Air quality fact book. Publication No. FHWA-HEP-05-045 HEP/12-05(8M)E, Washington, DC.

U.S. Environmental Protection Agency. 2005. Emission facts: Metrics for expressing greenhouse gas emissions: Carbon equivalents and carbon dioxide equivalents.
http://www.epa.gov/OMS/climate/420f05002.htm Accessed on January 5, 2008.

U.S. Environmental Protection Agency, 2008. Inventory of U.S. greenhouse gas emissions and sinks: 1990-2006.
http://www.epa.gov/climatechange/emissions/downloads/08_CR.pdf Accessed on January 15, 2009.

Van Wassenhove, L. N., 2006. Humanitarian logistics: Supply chain management in high gear. *Journal Operational Research Society* 57, 475-489.

Vlasic, B., Wayne, L., 2008. Auto suppliers share anxiety over a bailout. *New York Times*, December 11.

Wade, R., 2000. National power, coercive liberalism and 'global' finance. In *International Politics: Enduring Concepts and Contemporary Issues.* Art, R. J., Jervis, R., Editors. Addison Wesley, New York, pp. 482-489.

Wagner, S. M., Bode, C. 2007. An empirical investigation into supply chain vulnerability. *Journal of Purchasing and Supply Management* 12, 301-312.

Wallace, C. E., Courage, K. G., Hadi, M. A., Gan, A. G., 1998. TRANSYT-7F user's guide. University of Florida, Gainesville, Florida

Wang, Z., Zhang, F., Wang, Z., 2007. Research of return supply chain supernetwork model based on variational inequalities. In *Proceedings of the IEEE International Conference on Automation and Logistics*, Jinan, China, pp. 25-30.

Wardrop, J. G., 1952. Some theoretical aspects of road traffic research. In *Proceedings of the Institution of Civil Engineers*, Part II 1, pp. 325-378.

Watts, D. J., Strogatz, S. H., 1998. Collective dynamics of 'small-world' networks. *Nature* 393, 440-442.

Wilson, M. C., 2007. The impact of transportation disruptions on supply chain performance. *Transportation Research E* 43, 295-320.

Winters, J. A., 1998. Asia December 1998: Asia and the 'magic' of the marketplace. *Current History* winter, 418-425.

Wolfe, A., 2007. Taiwan quake exposes Internet vulnerability. *Power and Interest News Report*, January 15.

Wong, G., 2007. After credit crisis, new forces drive deals. CNNMoney.com, October 10.

Woolley, T., Nagurney, A., Stranlund, J., 2009. Spatially differentiated trade of permits for multipollutant electric power supply chains. In *Optimization in the Energy Industry.* Kallrath, J., Pardalos, P. M., Rebennack, S., Scheidt, M., Editors, Springer, Berlin, Germany, pp. 277-296.

World Resources Institute, 1998. *World Resources 1998-99: Environmental Change and Human Health.* Oxford University Press, New York.

Wu, T., Blackhurst, J., Chidambaram, V., 2006. The model for inbound supply risk analysis. *Computers in Industry* 57, 350-365.

Wu, K., Nagurney, A., Liu, Z., Stranlund, J. K., 2006. Modeling generator power plant portfolios and pollution taxes in electric power supply chain networks: A transportation network equilibrium transformation. *Transportation Research D* 11, 171-190.

Xu, S., 2007. *Supply Chain Synergy in Mergers and Acquisitions: Strategies, Models and Key Factors.* Ph.D. Dissertation, University of Massachusetts, Amherst, Massachusetts.

Yerra, B. M., Levinson, D. M., 2005. The emergence of hierarchy in transportation networks, 2005. *The Annals of Regional Science* 39, 541-553.

Yin, Y., Lawphongpanich, L., 2006. Internalizing emission externality on road networks. *Transportation Research D* 11, 292-301.

Yook, S. H., Jeong, H., Barabási, A. L., Tu, Y., 2001. Weighted evolving networks. *Physical Review Letters* 86, 5835-5838.

Yu, P. L., 1985. *Multiple Criteria Decision Making - Concepts, Techniques, and Extensions*. Plenum Press, New York.

Zhang, D., 2006. A network economic model for supply chain vs. supply chain competition. *Omega* 34, 283-295.

Zhang, D., Dong, J., Nagurney, A., 2003. A supply chain network economy: Modeling and qualitative analysis. In: Nagurney, A., Editor, *Innovations in Financial and Economic Networks*. Edward Elgar Publishing, Cheltenham, England, pp. 197-213.

Zhang, D., Nagurney, A., 1997. Formulation, stability, and cmputation of traffic network equilibria as projected dynamical systems. *Journal of Optimization Theory and Applications* 93, 417-444.

Zhu, D., Hu, Y., Li, Y., Yu, B., 2006. A new measure for airline networks performance evaluation and critical cities identification. School of Management, Fudan University, Shanghai, China.

Zimmerman, R., 2003. Global climate change and transportation infrastructure: lessons from the New York area. In *The Potential Impacts of Climate Change on Transportation: Workshop Summary and Proceedings*. Washington, DC, U.S. DOT, pp. 91-101.

Zsidisin, G. A., Siferd, S. P., 2001. Environmental purchasing: A framework for theory development. *European Journal of Purchasing & Supply Management* 7, 61-73.

Glossary

This is a glossary of symbols used in this book. Others symbols are defined in the text. A vector is assumed to be a column vector unless noted otherwise.

$\in$	an element of
$\subset$	subset of
$\subseteq$	subset of or equal to
$\cup, \cap$	union, intersection
$\forall$	for all
$\exists$	there exists
R	the real line
R^N	Euclidean N-dimensional space
R^N_+	Euclidean N-dimensional space on the nonnegative orthant
$:$	such that; also $\mid$
$\equiv$	is equivalent to
$\mapsto$	maps to
$\rightarrow$	tends to
$\circ$	composition
$\|x\| = (\sum_{i=1}^{n} x_i^2)^{\frac{1}{2}}$	length of $x \in R^N$ with components $(x_1, x_2, \ldots, x_N)$
x^T	transpose of a vector x

$\langle x^T, x \rangle$	inner product of vector x in R^N where $\langle x^T, x \rangle = x_1^2 + \ldots + x_N^2$
$x^T \cdot x$	also denotes the inner product of x
$\vert y \vert$	absolute value of y
$[a, b]$; (a, b)	a closed interval; an open interval in R
∇f	gradient of $f : R^N \mapsto R$
∇F	the $N \times N$ Jacobian of a mapping $F : R^N \mapsto R^N$
$\frac{\partial f}{\partial x}$	partial derivative of f with respect to x
$\text{argmin}_{x \in \mathcal{K}} f(x)$	the set of $x \in \mathcal{K}$ attaining the minimum of $f(x)$
A^T	transpose of the matrix A
A^{-1}	the inverse of the matrix A
$\int$	integral
$\mathcal{H}$	A Hilbert space is a vector space $\mathcal{H}$ with an inner product $\langle f^T, g \rangle$ such that the norm defined by $\vert g \vert = \sqrt{\langle f^T, f \rangle}$ turns $\mathcal{H}$ into a complete metric space
L^2	the set of all functions $f : R \mapsto R$ such that the integral of f^2 over the whole real line is finite; L^2 is an example of an infinite-dimensional Hilbert space; R^N is an example of a finite-dimensional Hilbert space

INDEX